T0094070

Introduction to Recursive Programming

Introduction to

Recursive Programming

Manuel Rubio-Sánchez

CRC Press is an imprint of the
Taylor & Francis Group, an **informa** business
A CHAPMAN & HALL BOOK

CRC Press
Taylor & Francis Group
6000 Broken Sound Parkway NW, Suite 300
Boca Raton, FL 33487-2742

Printed on acid-free paper
Version Date: 20170817

International Standard Book Number-13: 978-1-4987-3528-5 (Paperback)
International Standard Book Number-13: 978-1-138-10521-8 (Hardback)

Visit the Taylor & Francis Web site at
http://www.taylorandfrancis.com

and the CRC Press Web site at
http://www.crcpress.com

To the future generations

Contents

Preface

Recursion is one of the most fundamental concepts in computer science and a key programming technique that, similarly to iteration, allows computations to be carried out repeatedly. It accomplishes this by employing methods that invoke themselves, where the central idea consists of designing a solution to a problem by relying on solutions to smaller instances of the same problem. Most importantly, recursion is a powerful problem-solving approach that enables programmers to develop concise, intuitive, and elegant algorithms.

Despite the importance of recursion for algorithm design, most programming books do not cover the topic in detail. They usually devote just a single chapter or a brief section, which is often insufficient for assimilating the concepts needed to master the topic. Exceptions include *Recursion via Pascal*, by J. S. Rohl (Cambridge University Press, 1984); *Thinking Recursively with Java*, by E. S. Roberts (Wiley, 2006); and *Practicing Recursion in Java*, by I. Pevac (CreateSpace Independent Publishing Platform, 2016), which focus exclusively on recursion. The current book provides a comprehensive treatment of the topic, but differs from the previous texts in several ways.

Numerous computer programming professors and researchers in the field of computer science education agree that recursion is difficult for novice students. With this in mind, the book incorporates several elements in order to foster its pedagogical effectiveness. Firstly, it contains a larger collection of simple problems in order to provide a solid foundation of the core concepts, before diving into more complex material. In addition, one of the book's main assets is the use of a step-by-step methodology, together with specially designed diagrams, for guiding and illustrating the process of developing recursive algorithms. The book also contains specific chapters on combinatorial problems and mutual recursion. These topics can broaden students' understanding of recursion by forcing them to apply the learned concepts differently, or in a more sophisticated manner. Lastly, introductory programming courses usually focus on the imperative programming paradigm, where students primarily learn iteration, understanding and controlling *how* programs work. In contrast, recursion requires adopting a completely different way of thinking, where the emphasis should be on *what* programs compute. In this regard, several studies encourage instructors to avoid, or postpone, covering how recursive programs work (i.e., control flow, recursion trees, the program stack, or the relationship between iteration and tail recursion) when introducing recursion, since the concepts and abilities learned for iteration may actually hamper the acquisition of skills related to recursion and declarative programming. Therefore, topics related to iteration and program execution are covered towards the end of the book, when the reader should have mastered the design of recursive algorithms from a purely declarative perspective.

The book also includes a richer chapter on the theoretical analysis of the computational cost of recursive programs. On the one hand, it contains a broad treatment of mathematical recurrence relations, which constitute the fundamental tools for analyzing the runtime or recursive algorithms. On the other hand, it includes a section on mathematical preliminaries that reviews concepts and properties that are not only needed for solving recurrence relations, but also for understanding the statements and solutions to the computational problems in the book. In this regard, the text also offers the possibility to learn some basic mathematics along the way. The reader is encouraged to embrace this material, since it is essential in many fields of computer science.

The code examples are written in Python 3, which is arguably today's most popular introductory programming language in top universities. In particular, they were tested on Spyder (Scientific PYthon Development EnviRonment). The reader should be aware that the purpose of the book is not to teach Python, but to transmit skills associated with recursive thinking for problem solving. Thus, aspects such as code simplicity and legibility have been prioritized over efficiency. In this regard, the code does not contain advanced Python features. Therefore, students with background in other programming languages such as C++ or Java should be able to understand the code without effort. Of course, the methods in the book can be implemented in several ways, and readers are encouraged to write more efficient versions, include more sophisticated Python constructs, or design alternative algorithms. Lastly, the book provides recursive variants of iterative algorithms that usually accompany other well-known recursive algorithms. For instance, it contains recursive versions of Hoare's partition method used in the quicksort algorithm, or of the merging method within the merge sort algorithm.

The book proposes numerous exercises at the end of the chapters, whose fully worked-out solutions are included in an instructor's manual available at the book's official website (see www.crcpress.com). Many of them are related to the problems analyzed in the main text, which make them appropriate candidates for exams and assignments.

The code in the text will also be available for download at the book's website. In addition, I will maintain a complementary website related to the book: https://sites.google.com/view/recursiveprogrammingintro/. Readers are more than welcome to send me comments, suggestions for improvements, alternative (clearer or more efficient) code, versions in other programming languages, or detected errata. Please send emails to: recursion.book@gmail.com.

INTENDED AUDIENCE

The main goal of the book is to teach students how to think and program recursively, by analyzing a wide variety of computational problems. It is intended mainly for undergraduate students in computer science or related technical disciplines that cover programming and algorithms (e.g., bioinformatics, engineering, mathematics, physics, etc.). The book could also be useful for amateur programmers, students of massive open online courses, or more experienced professionals who would like to refresh the material, or learn it in a different or clearer way.

Students should have some basic programming experience in order to understand the code in the book. The reader should be familiar with notions introduced in a first programming course such as expressions, variables, conditional and loop constructs, methods, parameters, or elementary data structures such as arrays or lists. These concepts are not explained in the book. Also, the code in the book is in accordance with the procedural programming paradigm, and does not use object oriented programming features. Regarding Python, a basic background can be helpful, but is not strictly necessary. Lastly, the student should be competent in high school mathematics.

Computer science professors can also benefit from the book, not just as a handbook with a large collection and variety of problems, but also by adopting the methodology and diagrams described to build recursive solutions. Furthermore, professors may employ its structure to organize their classes. The book could be used as a required textbook in introductory (CS1/2) programming courses, and in more advanced classes on the design and analysis of algorithms (for example, it covers topics such as divide and conquer, or backtracking). Additionally, since the book provides a solid foundation of recursion, it can be used as a complementary text in courses related to data structures, or as an introduction to functional programming. However, the reader should be aware that the book does not cover data structures or functional programming concepts.

BOOK CONTENT AND ORGANIZATION

The first chapter assumes that the reader does not have any previous background on recursion, and introduces fundamental concepts, notation, and the first coded examples.

The second chapter presents a methodology for developing recursive algorithms, as well as diagrams designed to help thinking recursively, which illustrate the original problem and its decomposition into smaller instances of the same problem. It is one of the most important chapters since the methodology and recursive diagrams will be used throughout the rest of the book. Readers are encouraged to read the chapter, regardless of their previous background on recursion.

Chapter 3 reviews essential mathematical fundamentals and notation. Moreover, it describes methods for solving recurrence relations, which are the main mathematical tools for theoretically analyzing the computational cost of recursive algorithms. The chapter can be skipped when covering recursion in an introductory course. However, it is included early in the book in order to provide a context for characterizing and comparing different algorithms regarding their efficiency, which would be essential in a more advanced course on design and analysis of algorithms.

The fourth chapter covers "linear recursion." This type of recursion leads to the simplest recursive algorithms, where the solutions to computational problems are obtained by considering the solution to a single smaller instance of the problem. Although the proposed problems can also be solved easily through iteration, they are ideal candidates for introducing fundamental recursive concepts, as well as examples of how to use the methodology and recursive diagrams.

The fifth chapter covers a particular type of linear recursion called "tail recursion," where the last action performed by a method is a recursive call, invoking itself. Tail recursion is special due to its relationship with iteration. This connection will nevertheless be postponed until Chapter 11. Instead, this chapter focuses on solutions from a purely declarative approach, relying exclusively on recursive concepts.

The advantages of recursion over iteration are mainly due to the use of "multiple recursion," where methods invoke themselves several times, and the algorithms are based on combining several solutions to smaller instances of the same problem. Chapter 6 introduces multiple recursion through methods based on the eminent "divide and conquer" algorithm design paradigm. While some examples can be used in an introductory programming course, the chapter is especially appropriate in a more advanced class on algorithms. Alternatively, Chapter 7 contains challenging problems, related to puzzles and fractal images, which can also be solved through multiple recursion, but are not considered to follow the divide and conquer approach.

Recursion is used extensively in combinatorics, which is a branch of mathematics related to counting that has applications in advanced analysis of algorithms. Chapter 8 proposes using recursion for solving combinatorial counting problems, which are usually not covered in programming texts. This unique chapter will force the reader to apply the acquired recursive thinking skills to a different family of problems. Lastly, although some examples are challenging, many of the solutions will have appeared in earlier chapters. Thus, some examples can be used in an introductory programming course.

Chapter 9 introduces "mutual recursion," where several methods invoke themselves indirectly. The solutions are more sophisticated since it is necessary to think about several problems simultaneously. Nevertheless, this type of recursion involves applying the same essential concepts covered in earlier chapters.

Chapter 10 covers how recursive programs work from a low-level point of view. It includes aspects such as tracing and debugging, the program stack, or recursion trees. In addition, it contains a brief introduction to memoization and dynamic programming, which is another important algorithm design paradigm.

Tail-recursive algorithms can not only be transformed to iterative versions; some are also designed by thinking iteratively. Chapter 11 examines the connection between iteration and tail recursion in detail. In addition, it provides a brief introduction to "nested recursion," and includes a strategy for designing simple tail-recursive functions that are usually defined by thinking iteratively, but through a purely declarative approach. These last two topics are curiosities regarding recursion, and should be skipped in introductory courses.

The last chapter presents backtracking, which is another major algorithm design technique that is used for searching for solutions to computational problems in large discrete state spaces. The strategy is usually applied for solving constraint satisfaction and discrete optimization problems. For example, the chapter will cover classical problems such as the N-queens puzzle, finding a path through a maze, solving sudokus, or the 0-1 knapsack problem.

POSSIBLE COURSE ROAD MAPS

It is possible to cover only a subset of the chapters. The road map for introductory programming courses could be Chapters 1, 2, 4, 5, and 10. The instructor should decide whether to include examples from Chapters 6–9, and whether to cover the first section of Chapter 11.

If students have previously acquired skills to develop linear-recursive methods, a more advanced course on algorithm analysis and design could cover Chapters 2, 3, 5, 6, 7, 9, 11, and 12. Thus, Chapters 1, 4, and 10 could be proposed as readings for refreshing the material. In both of these suggested road maps Chapter 8 is optional. Finally, it is important to cover Chapters 10 and 11 after the previous ones.

ACKNOWLEDGEMENTS

The content of this book has been used to teach computer programming courses at Universidad Rey Juan Carlos, in Madrid (Spain). I am grateful to the students for their feedback and suggestions. I would also like to thank Ángel Velázquez and the members of the LITE (Laboratory of Information Technologies in Education) research group for providing useful insights regarding the content of the book. I would also like to express my gratitude to Luís Fernández, computer science professor at Universidad Politécnica de Madrid, for his advice and experience related to teaching recursion. A special thanks to Gert Lanckriet and members of the Computer Audition Laboratory at University of California, San Diego.

Manuel Rubio-Sánchez
July, 2017

List of Figures

List of Tables

List of Listings

CHAPTER 1

Basic Concepts of Recursive Programming

To iterate is human, to recurse divine.

— Laurence Peter Deutsch

RECURSION is a broad concept that is used in diverse disciplines such as mathematics, bioinformatics, or linguistics, and is even present in art or in nature. In the context of computer programming, recursion should be understood as a powerful problem-solving strategy that allows us to design simple, succinct, and elegant algorithms for solving computational problems. This chapter presents key terms and notation, and introduces fundamental concepts related to recursive programming and thinking that will be further developed throughout the book.

1.1 RECOGNIZING RECURSION

An entity or concept is said to be recursive when simpler or smaller self-similar instances form part of its constituents. Nature provides numerous examples where we can observe this property (see Figure 1.1). For instance, a branch of a tree can be understood as a stem, plus a set of smaller branches that emanate from it, which in turn contain other smaller branches, and so on, until reaching a bud, leaf, or flower. Blood vessels or rivers exhibit similar branching patterns, where the larger structure appears to contain instances of itself at smaller scales. Another related recursive example is a romanesco broccoli, where it is

Tree branches

Branching rivers

Romanesco broccoli

Spiral Droste effect

Sierpiński's triangle

Matryoshka dolls

Figure 1.1 Examples of recursive entities.

apparent that the individual florets resemble the entire plant. Other examples include mountain ranges, clouds, or animal skin patterns.

Recursion also appears in art. A well-known example is the Droste effect, which consists of a picture appearing within itself. In theory the process could be repeated indefinitely, but naturally stops in practice when the smallest picture to be drawn is sufficiently small (for example, if it occupies a single pixel in a digital image). A computer-generated fractal is another type of recursive image. For instance, Sierpiński's triangle is composed of three smaller identical triangles that are subsequently decomposed into yet smaller ones. Assuming that the process is infinitely repeated, each small triangle will exhibit the same structure as the original's. Lastly, a classical example used to illustrate the concept of recursion is a collection of matryoshka dolls. In this craftwork each doll has a different size and can fit inside a larger one. Note that the recursive object is not a single hollow doll, but a full nested collection. Thus, when thinking recursively, a collection of dolls can be described as a single (largest) doll that contains a smaller collection of dolls.

While the recursive entities in the previous examples were clearly tangible, recursion also appears in a wide variety of abstract concepts. In this regard, recursion can be understood as the process of defining concepts by using the definition itself. Many mathematical formulas and definitions can be expressed this way. Clear explicit examples include sequences for which the n-th term is defined through some formula or procedure involving earlier terms. Consider the following recursive definition:

$$s_n = s_{n-1} + s_{n-2}. \tag{1.1}$$

The formula states that a term in a sequence (s_n) is simply the sum of the two previous terms (s_{n-1} and s_{n-2}). We can immediately observe that the formula is recursive, since the entity it defines, s, appears on both sides of the equation. Thus, the elements of the sequence are clearly defined in terms of themselves. Furthermore, note that the recursive formula in (1.1) does not describe a particular sequence, but an entire family of sequences in which a term is the sum of the two previous ones. In order to characterize a specific sequence we need to provide more information. In this case, it is enough to indicate any two terms in the sequence. Typically, the first two terms are used to define this type of sequence. For instance, if $s_1 = s_2 = 1$ the sequence is:

$$1, 1, 2, 3, 5, 8, 13, 21, 34, 55, \ldots$$

which is the well-known Fibonacci sequence. Lastly, sequences may also be defined starting at term s_0.

The sequence s can be understood as a function that receives a positive integer n as an argument, and returns the n-th term in the sequence. In this regard, the Fibonacci function, in this case simply denoted as F, can be defined as:

$$F(n) = \begin{cases} 1 & \text{if } n = 1, \\ 1 & \text{if } n = 2, \\ F(n-1) + F(n-2) & \text{if } n > 2. \end{cases} \tag{1.2}$$

Throughout the book we will use this notation in order to describe functions, where the definitions include two types of expressions or cases. The **base cases** correspond to scenarios where the function's output can be obtained trivially, without requiring values of the function on additional arguments. For Fibonacci numbers the base cases are, by definition, $F(1) = 1$, and $F(2) = 1$. The **recursive cases** include more complex recursive expressions that typically involve the defined function applied to smaller input arguments. The Fibonacci function has one recursive case: $F(n) = F(n-1) + F(n-2)$, for $n > 2$. The base cases are necessary in order to provide concrete values for the function's terms in the recursive cases. Lastly, a recursive definition may contain several base and recursive cases.

Another function that can be expressed recursively is the factorial of some nonnegative integer n:

$$n! = 1 \times 2 \times \cdots \times (n-1) \times n.$$

In this case, it is not immediately obvious whether the function can be expressed recursively, since there is not an explicit factorial on the right-hand side of the definition. However, since $(n-1)! = 1 \times 2 \times \cdots \times (n-1)$, we can rewrite the formula as the recursive expression $n! = (n-1)! \times n$. Lastly, by convention $0! = 1$, which follows from plugging in the value $n = 1$ in the recursive formula. Thus, the factorial function can be defined recursively as:

$$n! = \begin{cases} 1 & \text{if } n = 0, \\ (n-1)! \times n & \text{if } n > 0. \end{cases} \tag{1.3}$$

Similarly, consider the problem of calculating the sum of the first n positive integers. The associated function $S(n)$ can be obviously defined as:

$$S(n) = 1 + 2 + \cdots + (n-1) + n. \tag{1.4}$$

Again, we do not observe a term involving S on the right-hand side of the definition. However, we can group the $n-1$ smallest terms in order to form $S(n-1) = 1+2+ \cdots + (n-1)$, which leads to the following recursive definition:

$$S(n) = \begin{cases} 1 & \text{if } n = 1, \\ S(n-1) + n & \text{if } n > 1. \end{cases} \tag{1.5}$$

Note that $S(n-1)$ is a self-similar **subproblem** to $S(n)$, but is **simpler**, since it needs fewer operations in order to calculate its result. Thus, we say that the subproblem has a smaller **size**. In addition, we say we have **decomposed** the original problem ($S(n)$) into a **smaller** one, in order to form the recursive definition. Lastly, $S(n-1)$ is a smaller **instance** of the original problem.

Another mathematical entity for which how it can be expressed recursively may not seem immediately obvious is a nonnegative integer. These numbers can be decomposed and defined recursively in several ways, by considering smaller numbers. For instance, a nonnegative integer n can be expressed as its predecessor plus a unit:

$$n = \begin{cases} 0 & \text{if } n = 0, \\ predecessor(n) + 1 & \text{if } n > 0. \end{cases}$$

Note that n appears on both sides of the equals sign in the recursive case. In addition, if we consider that the *predecessor* function necessarily returns a nonnegative integer, then it cannot be applied to 0. Thus, the definition is completed with a trivial base case for $n = 0$.

Another way to think of (nonnegative) integers consists of considering them as ordered collections of digits. For example, the number 5342 can be the concatenation of the following pairs of smaller numbers:

$$(5, 342), \qquad (53, 42), \qquad (534, 2).$$

In practice, the simplest way to decompose these integers consists of considering the least significant digit individually, together with the rest of the number. Therefore, an integer can be defined as follows:

$$n = \begin{cases} n & \text{if } n < 10, \\ (n//10) \times 10 + (n\%10) & \text{if } n \geq 10, \end{cases}$$

where // and % represent the quotient and remainder of an integer division, respectively, which corresponds to Python notation. For example,

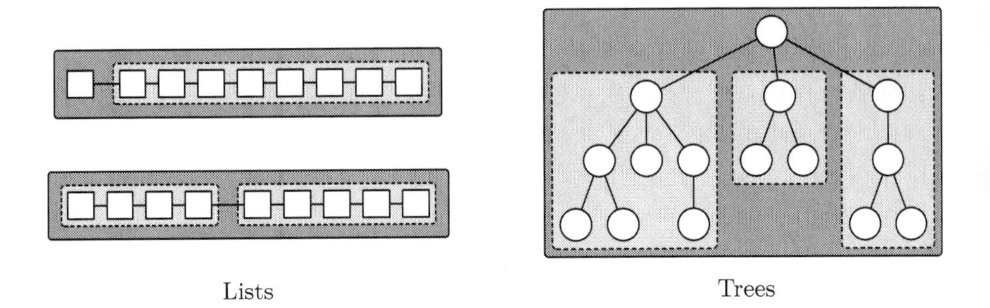

Figure 1.2 Recursive decomposition of lists and trees.

if $n = 5342$ then the quotient is $(n//10) = 534$, while the remainder is $(n\%10) = 2$, which represents the least significant digit of n. Clearly, the number n can be recovered by multiplying the quotient by 10 and adding the remainder. Finally, the base case considers numbers with only one digit that naturally cannot be decomposed any further.

Recursive expressions are abundant in mathematics. For instance, they are often used in order to describe properties of functions. The following recursive expression indicates that the derivative of a sum of functions is the sum of their derivatives:

$$[f(x) + g(x)]' = [f(x)]' + [g(x)]'.$$

In this case the recursive entity is the derivative function, denoted as $[\cdot]'$, but not the functions $f(x)$ and $g(x)$. Observe that the formula explicitly indicates the decomposition that takes place, where some initial function (which is the input argument to the derivative function) is broken up into the sum of the functions $f(x)$ and $g(x)$.

Data structures can also be understood as recursive entities. Figure 1.2 shows how lists and trees can be decomposed recursively. On the one hand, a list can consist of a single element plus another list (this is the usual definition of a list as an abstract data type), or it can be subdivided into several lists (in this broader context a list is any collection of data elements that are arranged linearly in an ordered sequence, as in lists, arrays, tuples, etc.). On the other hand, a tree consists of a parent node and a set (or list) of subtrees, whose root node is a child of the original parent node. The recursive definitions of data structures are completed by considering empty (base) cases. For instance, a list that contains only one element would consist of that element plus an empty list. Lastly, observe that in these diagrams the darker boxes represent a

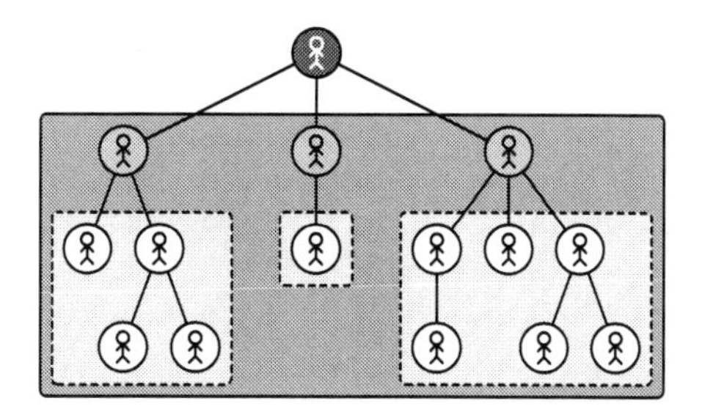

Figure 1.3 Family tree representing the descendants of a person, which are its children plus the descendants of its children.

full recursive entity, while the lighter ones indicate smaller self-similar instances.

Recursion can even be used to define words in dictionaries. This might seem impossible since we are told in school that the description of a word in a dictionary should not contain the word itself. However, many concepts can be defined correctly this way. Consider the term "descendant" of a specific ancestor. Notice that it can be defined perfectly as: someone who is either a child of the ancestor, or a *descendant* of any of the ancestor's children. In this case, we can identify a recursive structure where the set of descendants can be organized in order to form a (family) tree, as shown in Figure 1.3. The darker box contains all of the descendants of a common ancestor appearing at the root of the tree, while the lighter boxes encompass the descendants of the ancestor's children.

1.2 PROBLEM DECOMPOSITION

In general, when programming and thinking recursively, our main task will consist of providing our own recursive definitions of entities, concepts, functions, problems, etc. While the first step usually involves establishing the base cases, the main challenge consists of describing the recursive cases. In this regard, it is important to understand the concepts of: (a) problem decomposition, and (b) induction, which we will cover briefly in this chapter.

The book will focus on developing recursive algorithms for solving **computational problems**. These can be understood as questions that computers could possibly answer, and are defined through statements that describe relationships between a collection of known input values or parameters, and a set of output values, results, or solutions. For example,

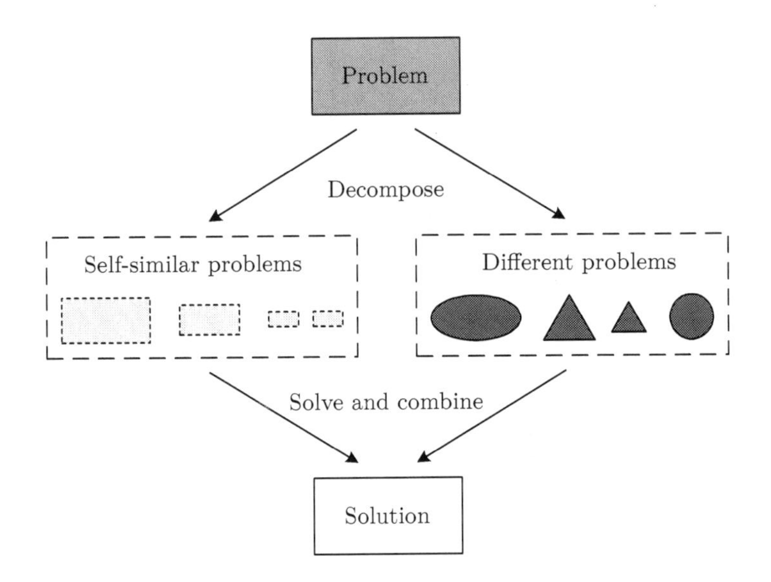

Figure 1.4 Recursive problem solving.

"given some positive integer n, calculate the sum of the first n positive integers" is the statement of a computational problem with one input parameter (n), and one output value defined as $1 + 2 + \cdots + (n-1) + n$. An instance of a problem is a specific collection of valid input values that will allow us to compute a solution to the problem. In contrast, an **algorithm** is a logical procedure that describes a step-by-step set of computations needed in order to obtain the outputs, given the initial inputs. Thus, an algorithm determines how to solve a problem. It is worth mentioning that computational problems can be solved by different algorithms. The goal of this book is to explain how to design and implement recursive algorithms and programs, where a key step involves decomposing a computational problem.

Decomposition is an important concept in computer science and plays a major role not only in recursive programming, but also in general problem solving. The idea consists of breaking up complex problems into smaller, simpler ones that are easier to express, compute, code, or solve. Subsequently, the solutions to the subproblems can be processed in order to obtain a solution to the original complex problem.

In the context of recursive problem solving and programming, decomposition involves breaking up a computational problem into several subproblems, some of which are self-similar to the original, as illustrated in Figure 1.4. Note that obtaining the solution to a problem may re-

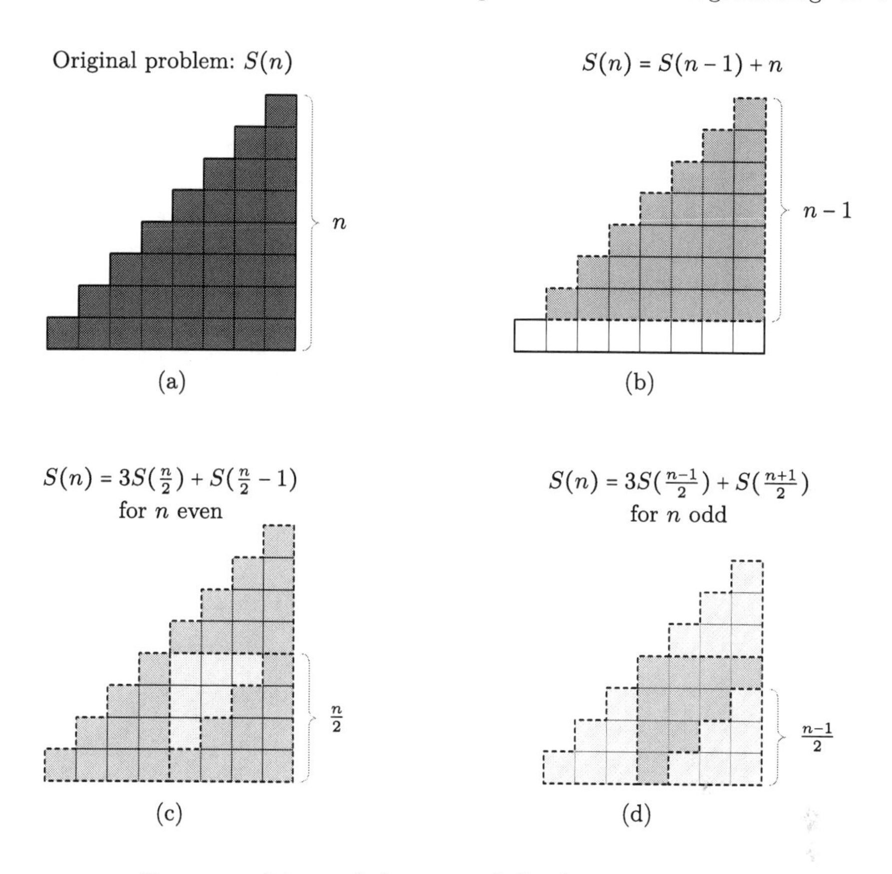

Figure 1.5 Decompositions of the sum of the first positive integers.

quire solving additional different problems that are not self-similar to the original one. We will tackle several of these problems throughout the book, but in this introductory chapter we will examine examples where the original problems will only be decomposed into self-similar ones.

For the first example we will reexamine the problem of computing the sum of the first n positive integers, denoted as $S(n)$, which can be formally expressed as in (1.4). There are several ways to break down the problem into smaller subproblems and form a recursive definition of $S(n)$. Firstly, it only depends on the input parameter n, which also specifies the size of the problem. In this example, the base case is associated with the smallest positive integer $n = 1$, where clearly $S(1) = 1$ is the smallest instance of the problem. Furthermore, we need to get closer to the problem's base case when considering subproblems. Therefore, we have to think of how we can reduce the input parameter n.

A first possibility consists of decreasing n by a single unit. In that case, the goal would be to define $S(n)$ somehow by using the subproblem $S(n-1)$. The corresponding recursive solution, derived in Section 1.1 algebraically, is given in (1.5). We can also obtain the recursive case by analyzing a graphical description of the problem. For instance, the goal could consist of counting the total number of blocks in a "triangular" structure that contains n blocks in its first level, $n-1$ in its second, and so on (the n-th level would therefore have a single block), as shown in Figure 1.5(a) for $n = 8$. In order to decompose the problem recursively we need to find self-similar problems. In this case, it is not hard to find smaller similar triangular shapes inside the original. For example, Figure 1.5(b) shows a triangular structure of height $n-1$ that contains all of the blocks of the original, except for the n blocks on the first level. Since this smaller triangular shape contains exactly $S(n-1)$ blocks, it follows that $S(n) = S(n-1) + n$.

Another option that involves using a problem that adds $n-1$ terms consists of considering the sum $2 + \cdots + (n-1) + n$. However, it is important to note that it is not a self-similar problem to $S(n)$. Clearly, it is not the sum of the first positive integers. Instead, it is a special case of a more general problem consisting of adding all of the integers from some initial value m up to another larger n: $m + (m+1) + \cdots + (n-1) + n$, for $m \leq n$. The difference between both problems can also be understood graphically. Regarding the illustrations in Figure 1.5, this general problem would define right trapezoid structures instead of triangular shapes. Finally, it is possible to compute the sum of the first n positive integers by using this more general problem, since we could simply set $m = 1$. Nevertheless, its recursive definition is slightly more complex since it requires two input parameters (m and n) instead of one.

Other possibilities concern decreasing n by larger quantities. For example, the input value n can be divided by two in order to obtain the decomposition illustrated in Figures 1.5(c) and (d). When n is even we can fit three triangular structures of height $n/2$ inside the larger one corresponding to $S(n)$. Since the remaining blocks also form a triangular shape of height $n/2-1$ the recursive formula can be expressed as $S(n) = 3S(n/2) + S(n/2-1)$. Alternatively, when n is odd it is possible to fit three triangular structures of height $(n-1)/2$, and one of size $(n+1)/2$. Thus, in that case the recursive formula is

$S(n) = 3S((n-1)/2) + S((n+1)/2)$. The final recursive function is:

$$S(n) = \begin{cases} 1 & \text{if } n = 1, \\ 3 & \text{if } n = 2, \\ 3S\left(\frac{n}{2}\right) + S\left(\frac{n}{2} - 1\right) & \text{if } n > 2 \text{ and } n \text{ is even}, \\ 3S\left(\frac{n-1}{2}\right) + S\left(\frac{n+1}{2}\right) & \text{if } n > 2 \text{ and } n \text{ is odd}. \end{cases} \tag{1.6}$$

It is important to note that the definition needs an additional base case for $n = 2$. Without it we would have $S(2) = 3S(1)+S(0)$ due to the recursive case when n is even. However, $S(0)$ is not defined since, according to the statement of the problem, the input to S must be a positive integer. The new base case is therefore necessary in order to avoid using the recursive formula for $n = 2$.

When dividing the size of a problem the resulting subproblems are considerably smaller than the original, and can therefore be solved much faster. Roughly speaking, if the number of subproblems to be solved is small and it is possible to combine their solutions efficiently, this strategy can lead to substantially faster algorithms for solving the original problem. However, in this particular example, code for (1.6) is not necessarily more efficient than the one implementing (1.5). Intuitively, this is because (1.6) requires solving two subproblems (with different arguments), while the decomposition in (1.5) only involves one subproblem. Chapter 3 covers how to analyze the runtime cost of these recursive algorithms.

The previous idea for adding the first n positive integers broke up the problem into two subproblems of smaller size, where the new input parameters approach the specified base cases. In general, we can decompose problems into any number of simpler subproblems, as long as the new parameters get closer to the values specified in the base cases. For instance, consider the following alternative recursive definition of the Fibonacci function (it is equivalent to (1.2)):

$$F(n) = \begin{cases} 1 & \text{if } n = 1, \\ 1 & \text{if } n = 2, \\ 1 + \sum_{i=1}^{n-2} F(i) & \text{if } n > 2, \end{cases} \tag{1.7}$$

where $\sum_{i=1}^{n-2} F(i)$ is the sum $F(1)+F(2)+\cdots+F(n-2)$ (see Section 3.1.4). In this example, for some value n that determines the size of the problem,

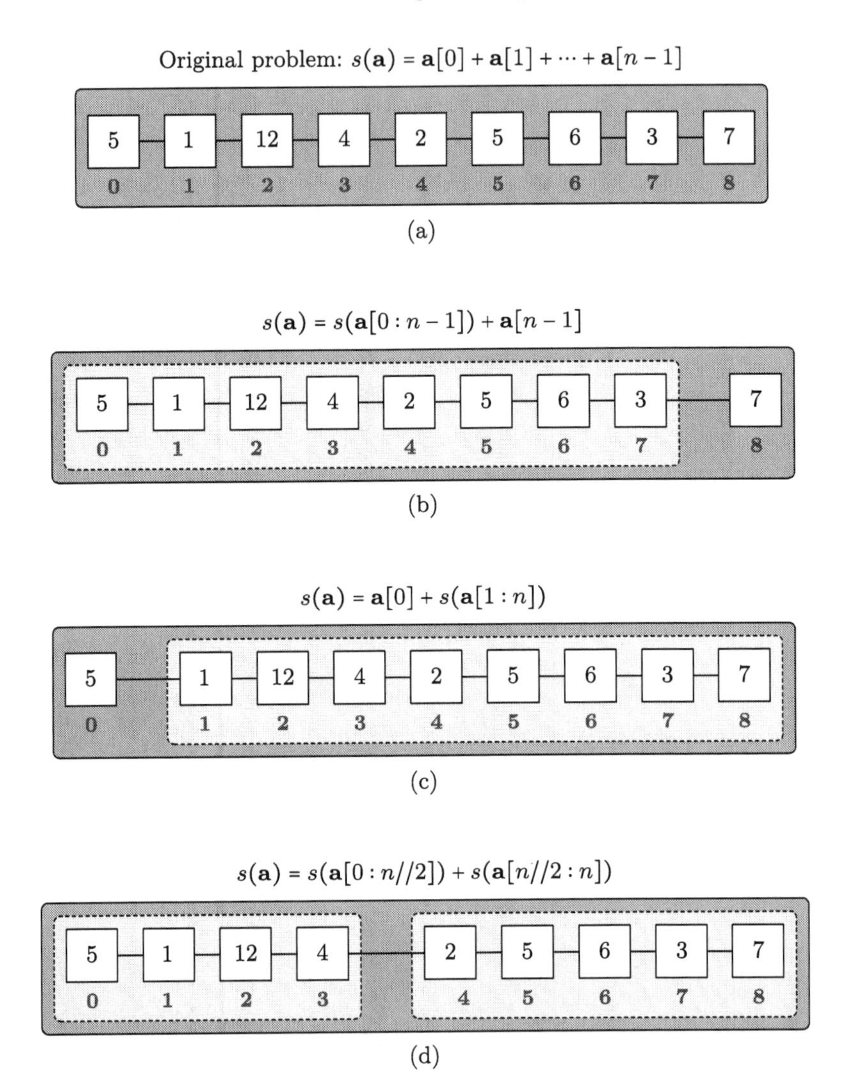

Figure 1.6 Decompositions of the sum of the elements in a list, denoted as $\mathbf{a}$, of $(n = 9)$ numbers.

the recursive case relies on decomposing the original problem into every smaller problem of size 1 up to $n-2$.

For the last example of problem decomposition we will use lists that allow us to access individual elements by using numerical indices (in many programming languages this data structure is called an "array," while in Python it is simply a "list"). The problem consists of adding the elements of a list, denoted as $\mathbf{a}$, containing n numbers. Formally, the problem can be expressed as:

$$s(\mathbf{a}) = \sum_{i=0}^{n-1} \mathbf{a}[i] = \mathbf{a}[0] + \mathbf{a}[1] + \cdots + \mathbf{a}[n-1], \tag{1.8}$$

where $\mathbf{a}[i]$ is the $(i+1)$-th element in the list, since the first one is not indexed by 1, but by 0. Figure 1.6(a) shows a diagram representing a particular instance of 9 elements.

Regarding notation, in this book we will assume that a **sublist** of some list $\mathbf{a}$ is a collection of *contiguous* elements of $\mathbf{a}$, unless explicitly stated otherwise. In contrast, in a **subsequence** of some initial sequence $\mathbf{s}$ its elements appear in the same order as in $\mathbf{s}$, but they are not required to be contiguous in $\mathbf{s}$. In other words, a subsequence can be obtained from an original sequence $\mathbf{s}$ by deleting some elements of $\mathbf{s}$, and without modifying the order of the remaining elements.

The problem can be decomposed by decreasing its size by a single unit. On the one hand, the list can be broken down into the sublist containing the first $n-1$ elements ($\mathbf{a}[0:n-1]$, where $\mathbf{a}[i:j]$ denotes the sublist from $\mathbf{a}[i]$ to $\mathbf{a}[j-1]$, following Python's notation) and a single value corresponding to the last number on the list ($\mathbf{a}[n-1]$), as shown in Figure 1.6(b). In that case, the problem can be defined recursively through:

$$s(\mathbf{a}) = \begin{cases} 0 & \text{if } n = 0, \\ s(\mathbf{a}[0:n-1]) + \mathbf{a}[n-1] & \text{if } n > 0. \end{cases} \tag{1.9}$$

In the recursive case the subproblem is naturally applied to the sublist of size $n-1$. The base case considers the trivial situation when the list is empty, which does not require any addition. Another possible base case can be $s(\mathbf{a}) = \mathbf{a}[0]$ when $n = 1$. However, it would be redundant in this decomposition, and therefore not necessary. Note that if $n = 1$ the function adds $\mathbf{a}[0]$ and the result of applying the function on an empty list, which is 0. Thus, it can be omitted for the sake of conciseness.

On the other hand, we can also interpret that the original list is its first element $\mathbf{a}[0]$, together with the smaller list $\mathbf{a}[1:n]$, as illustrated in Figure 1.6(c). In this case, the problem can be expressed recursively through:

$$s(\mathbf{a}) = \begin{cases} 0 & \text{if } n = 0, \\ \mathbf{a}[0] + s(\mathbf{a}[1:n]) & \text{if } n > 0. \end{cases} \tag{1.10}$$

Although both decompositions are very similar, the code for each one can be quite different depending on the programming language used.

Section 1.3 will show several ways of coding algorithms for solving the problem according to these decompositions.

Another way to break up the problem consists of considering each half of the list separately, as shown in Figure 1.6(d). This results in two subproblems of roughly half the size of the original's. The decomposition produces the following recursive definition:

$$s(\mathbf{a}) = \begin{cases} 0 & \text{if } n = 0, \\ \mathbf{a}[0] & \text{if } n = 1, \\ s(\mathbf{a}[0:n//2]) + s(\mathbf{a}[n//2:n]) & \text{if } n > 1. \end{cases} \tag{1.11}$$

Unlike the previous definitions, this decomposition requires a base case when $n = 1$. Without it the function would never return a concrete value for a nonempty list. Observe that the definition would not add or return any element of the list. For a nonempty list the recursive case would be applied repeatedly, but the process would never halt. This situation is denoted as **infinite recursion**. For instance, if the list contained a single element the recursive case would add the value associated with an empty list (0), to the result of the same initial problem. In other words, we would try to calculate $s(\mathbf{a}) = 0 + s(\mathbf{a}) = 0 + s(\mathbf{a}) = 0 + s(\mathbf{a}) \ldots$, which would be repeated indefinitely. The obvious issue in this scenario is that the original problem $s(\mathbf{a})$ is not decomposed into smaller and simpler ones when $n = 1$.

1.3 RECURSIVE CODE

In order to use recursion when designing algorithms it is crucial to learn how to decompose problems into smaller self-similar ones, and define recursive methods by relying on induction (see Section 1.4). Once these are specified it is fairly straightforward to convert the definitions into code, especially when working with basic data types such as integers, real numbers, characters, or Boolean values. Consider the function in (1.5) that adds the first n positive integers (i.e., natural numbers). In Python the related function can be coded as shown in Listing 1.1. The analogy between (1.5) and the Python function is evident. As in many of the examples covered throughout the book, a simple `if` statement is the only control flow structure needed in order to code the function. Additionally, the name of the function appears within its body, implementing a **recursive call**. Thus, we say that the function **calls** or **invokes** itself, and is therefore recursive (there exist recursive functions that do not call themselves directly within their body, as explained in Chapter 9).

C, Java:

```
1 int sum_first_naturals(int n)
2 {
3     if (n==1)
4         return 1;
5     else
6         return sum_first_naturals(n-1) + n;
7 }
```

Pascal:

```
1 function sum_first_naturals(n: integer): integer;
2 begin
3     if n=1 then
4         sum_first_naturals := 1
5     else
6         sum_first_naturals := sum_first_naturals(n-1) + n;
7 end;
```

MATLAB®:

```
1 function result = sum_first_naturals(n)
2     if n==1
3         result = 1;
4     else
5         result = sum_first_naturals(n-1) + n;
6     end
```

Scala:

```
1 def sum_first_naturals(n: Int): Int = {
2     if (n==1)
3         return 1
4     else
5         return sum_first_naturals(n-1) + n
6 }
```

Haskell:

```
1 sum_first_naturals 1 = 1
2 sum_first_naturals n = sum_first_naturals (n - 1) + n
```

Figure 1.7 Functions that compute the sum of the first n natural numbers in several programming languages.

Listing 1.1 Python code for adding the first n natural numbers.

```
1 def sum_first_naturals(n):
2     if n == 1:
3         return 1  # Base case
4     else:
5         return sum_first_naturals(n - 1) + n  # Recursive case
```

Coding the function in other programming languages is also straightforward. Figure 1.7 shows equivalent code in several programming languages. Again, the resemblance between the codes and the function definition is apparent. Although the code in the book will be in Python, translating it to other programming languages should be fairly straightforward.

An important detail about the function in (1.5) and the associated codes is that they do not check if $n > 0$. This type of condition on an input parameter, which is is specified in the statement of a problem or definition of a function, is known as a **precondition**. Programmers can assume that the preconditions always hold, and therefore do not have to develop code in order to detect or handle them.

Listing 1.2 Alternative Python code for adding the first n natural numbers.

```
1  def sum_first_naturals_2(n):
2      if n == 1:
3          return 1
4      elif n == 2:
5          return 3
6      elif n % 2:
7          return (3 * sum_first_naturals_2((n - 1) / 2)
8                  + sum_first_naturals_2((n + 1) / 2))
9      else:
10         return (3 * sum_first_naturals_2(n / 2)
11                 + sum_first_naturals_2(n / 2 - 1))
```

Listing 1.2 shows the recursive code associated with (1.6). The function uses a cascaded `if` statement in order to differentiate between the two base cases (lines 2–5) and the two recursive cases (lines 6–11), which each make two recursive calls to the defined Python function.

It is also straightforward to code a function that computes the n-th Fibonacci number by relying on the standard definition in (1.2). List-

Listing 1.3 Python code for computing the n-th Fibonacci number.

```
1 def fibonacci(n):
2     if n == 1 or n == 2:
3         return 1
4     else:
5         return fibonacci(n - 1) + fibonacci(n - 2)
```

ing 1.3 shows the corresponding code, where both base cases are considered in the Boolean expression of the `if` statement.

Listing 1.4 Alternative Python code for computing the n-th Fibonacci number.

```
1 def fibonacci_alt(n):
2     if n == 1 or n == 2:
3         return 1
4     else:
5         aux = 1
6         for i in range(1, n - 1):
7             aux += fibonacci_alt(i)
8         return aux
```

Implementing the Fibonacci function defined in (1.7) requires more work. While the base cases are identical, the summation in the recursive case entails using a loop or another function in order to compute and add the values $F(1)$, $F(2)$,..., $F(n-2)$. Listing 1.4 shows a possible solution that uses a `for` loop. The result of the summation can be stored in an auxiliary accumulator variable `aux` that can be initialized to 1 (line 5), taking into account the extra unit term in the recursive case. The `for` loop simply adds the terms $F(i)$, for $i = 1,\dots,n-2$, to the auxiliary variable (lines 6–7). Finally, the function returns the computed Fibonacci number stored in `aux`.

When data types are more complex there may be more variations among the codes of different programming languages due to low-level details. For instance, when working with a data structure similar to a list it is usually necessary to access its length, since recursive algorithms will use that value in order to define sublists. Figure 1.8 illustrates three combinations of lists (or similar data structures) together with parameters that are necessary in order to use sublists in recursive programs. In (a) the data structure (denoted as $\mathbf{a}$) allows us to recover its length

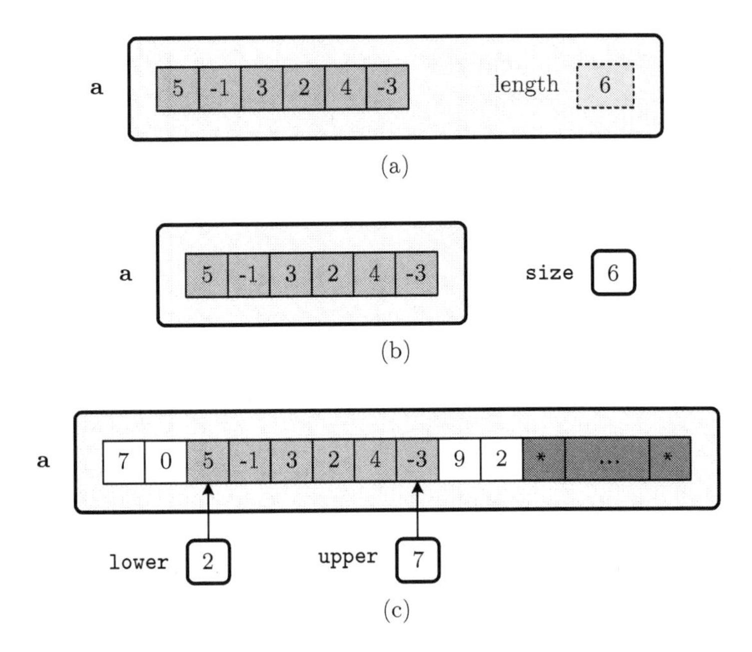

Figure 1.8 Data structures similar to lists, and parameters necessary for defining sublists.

without using any extra variables or parameters. For instance, it could be a property of the list, or it could be recovered through some method. In Python, the length of a list can be obtained through the function `len`. However, it is not possible to access the length of a standard array in other programming languages such as C or Pascal. If the length of the list cannot be recovered directly from the data structure an additional parameter, denoted as `size` (which contains the length of the list) is needed in order to store and access it, as shown in (b). This approach can be used when working with arrays of a fixed size or partially filled arrays. In these cases a more efficient alternative consists of using starting and finishing indices for determining the limits of a sublist, as shown in (c). Note that in this scenario the fixed size of the data structure may be a large constant (enough for the requirements of the application), but the true length of the lists and sublists may be much smaller. The graphic in (c) shows a list where only the first 10 elements are relevant (the rest should therefore be ignored). Furthermore, a sublist of six elements is defined through the `lower` and `upper` index variables, which delimit its boundaries. Note that the elements at these indices are included in the sublist.

Listing 1.5 Recursive functions for adding the elements in a list **a**, where the only input to the recursive function is the list.

```
1 # Decomposition: s(a) => s(a[0:n-1]), a[n-1]
2 def sum_list_length_1(a):
3     if len(a) == 0:
4         return 0
5     else:
6         return (sum_list_length_1(a[0:len(a) - 1])
7                 + a[len(a) - 1])
8
9
10 # Decomposition: s(a) => a[0], s(a[1:n])
11 def sum_list_length_2(a):
12     if len(a) == 0:
13         return 0
14     else:
15         return a[0] + sum_list_length_2(a[1:len(a)])
16
17
18 # Decomposition: s(a) => s(a[0:n//2]), s(a[n//2:n])
19 def sum_list_length_3(a):
20     if len(a) == 0:
21         return 0
22     elif len(a) == 1:
23         return a[0]
24     else:
25         middle = len(a) // 2
26         return (sum_list_length_3(a[0:middle])
27                 + sum_list_length_3(a[middle:len(a)]))
28
29
30 # Some list:
31 a = [5, -1, 3, 2, 4, -3]
32
33 # Function calls:
34 print(sum_list_length_1(a))
35 print(sum_list_length_2(a))
36 print(sum_list_length_3(a))
```

The constructs and syntax of Python allow us to focus on algorithms at a high level, avoiding the need to understand low-level mechanisms such as parameter passing. Nevertheless, its flexibility permits a wide variety of coding possibilities. Listing 1.5 shows three solutions to the problem of adding the elements in a list, corresponding to the three decomposition strategies in Figure 1.6, where the only input parameter is the list, which is the scenario in Figure 1.8(a). Functions `sum_list_length_1`, `sum_list_length_2`, and `sum_list_length_3` implement the recursive definitions in (1.9), (1.10), and (1.11), respectively. The last lines of the example declare a list `v` and print the sum of its elements using the three functions. Note that the number of elements in the list n is recovered through `len`. Finally, recall that `a[lower,upper]` is the sublist of `a` from index `lower` to `upper`-1, while `a[lower:]` is equivalent to `a[lower:len(a)]`. If the size of the list cannot be obtained from the list directly it can be passed to the functions through an additional parameter, as shown in Figure 1.8(b). The corresponding code is similar to Listing 1.5 and is proposed as an exercise at the end of the chapter.

Alternatively, Listing 1.6 shows the corresponding functions when using large lists and two parameters (`lower` and `upper`) in order to determine sublists within it, as illustrated in Figure 1.8(c). Observe the analogy between these functions and the ones in Listing 1.5. In this case, empty lists occur when `lower` is greater than `upper`. Also, the sublist contains a single element when both indices are equal (recall that both parameters indicate positions of elements that belong to the sublist).

1.4 INDUCTION

Induction is another concept that plays a fundamental role when designing recursive code. The term has different meanings depending on the field and topic where it is used. In the context of recursive programming it is related to mathematical proofs by induction. The key idea is that programmers must assume that the recursive code they are trying to implement already works for simpler and smaller problems, even if they have not yet written a line of code! This notion is also referred to as the recursive "leap of faith." This section reviews these crucial concepts.

1.4.1 Mathematical proofs by induction

In mathematics, proofs by induction constitute an important tool for showing that some statement is true. The simplest proofs involve formulas that depend on some positive (or nonnegative) integer n. In these cases, the proofs verify that the formulas are indeed correct for every possible value of n. The approach involves two steps:

a) Base case (basis). Verify that the formula is valid for the smallest value of n, say n_0.

Listing 1.6 Alternative recursive functions for adding the elements in a sublist of a list **a**. The boundaries of the sublist are specified by two input parameters that mark lower and upper indices in the list.

```
1  # Decomposition: s(a) => s(a[0:n-1]), a[n-1]
2  def sum_list_limits_1(a, lower, upper):
3      if lower > upper:
4          return 0
5      else:
6          return a[upper] + sum_list_limits_1(a, lower, upper - 1)
7
8
9  # Decomposition: s(a) => a[0], s(a[1:n])
10 def sum_list_limits_2(a, lower, upper):
11     if lower > upper:
12         return 0
13     else:
14         return a[lower] + sum_list_limits_2(a, lower + 1, upper)
15
16
17 # Decomposition: s(a) => s(a[0:n//2]), s(a[n//2:n])
18 def sum_list_limits_3(a, lower, upper):
19     if lower > upper:
20         return 0
21     elif lower == upper:
22         return a[lower]  # or a[upper]
23     else:
24         middle = (upper + lower) // 2
25         return (sum_list_limits_3(a, lower, middle)
26                 + sum_list_limits_3(a, middle + 1, upper))
27
28
29 # Some list:
30 a = [5, -1, 3, 2, 4, -3]
31
32 # Function calls:
33 print(sum_list_limits_1(a, 0, len(a) - 1))
34 print(sum_list_limits_2(a, 0, len(a) - 1))
35 print(sum_list_limits_3(a, 0, len(a) - 1))
```

b) Inductive step. Firstly, assume the formula is true for some general value of n. This assumption is referred to as the **induction hypothesis**. Subsequently, by relying on the assumption, show that if the formula holds for some value n, then it will also be true for $n+1$.

If it is possible to prove both steps, then, by induction, it follows that the statement holds for all $n \geq n_0$. The statement would be true for n_0, and then for $n_0 + 1$, $n_0 + 2$, and so on, by applying the inductive step repeatedly.

Again, consider the sum of the first n positive numbers (see (1.4)). We will try to show if the following identity (which is the induction hypothesis) involving a quadratic polynomial holds:

$$S(n) = \sum_{i=1}^{n} i = \frac{n(n+1)}{2}. \tag{1.12}$$

The base case is trivially true, since $S(1) = 1(2)/2 = 1$. For the induction step we need to show whether

$$S(n+1) = \sum_{i=1}^{n+1} i = \frac{(n+1)(n+2)}{2} \tag{1.13}$$

holds, by assuming that (1.12) is true. Firstly, $S(n+1)$ can be expressed as:

$$S(n+1) = \sum_{i=1}^{n} i + (n+1).$$

Furthermore, assuming (1.12) holds we can substitute the summation by the polynomial:

$$S(n+1) \overset{\substack{\text{induction}\\\text{hypothesis}\\\downarrow}}{=} \frac{n(n+1)}{2} + n + 1 = \frac{n^2 + n + 2n + 2}{2} = \frac{(n+1)(n+2)}{2},$$

showing that (1.13) is true, which completes the proof.

1.4.2 Recursive leap of faith

Recursive functions typically call themselves in order to solve smaller subproblems. It is reasonable for beginner programmers to doubt if a recursive function will really work as they code, and to question whether it is legitimate to call the function being written within its body, since it has not even been finished! However, not only can functions call themselves (in programming languages that support recursion), it is crucial to assume that they work correctly for subproblems of smaller size. This assumption, which plays a similar role as the induction hypothesis in proofs by induction, is referred to as the recursive "leap of faith." It is one of the cornerstones of recursive thinking, but also one of the hardest

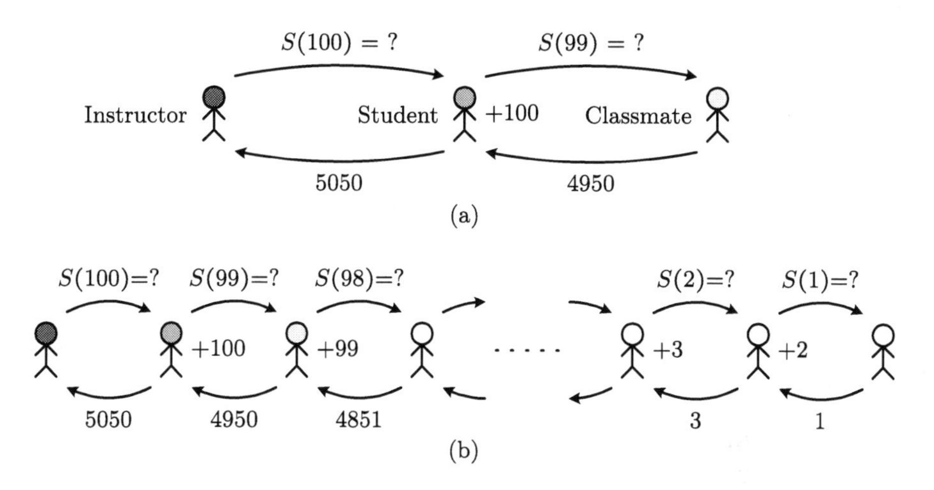

Figure 1.9 Thought experiment in a classroom, where an instructor asks a student to add the first 100 positive integers. $S(n)$ represents the sum of the first n positive integers.

concepts for novice programmers to wrap their heads around. Naturally, it is not a mystical leap of faith that should be accepted without understanding it. The following thought experiment illustrates its role in recursive thinking.

Consider you are in a large classroom and the instructor asks you to add up the integers from 1 to 100. Assuming you are unaware of the formula in (1.12), you would have to carry out the 99 involved additions. However, imagine you are allowed to talk to classmates and there is a "clever" one who claims he can add the first n positive integers, for $n \leq 99$. In that case you could take a much simpler approach. You could simply ask him to tell you the sum of the first 99 positive integers, which is 4950, and all you would need to do is add 100 to your classmate's response, obtaining 5050. Although it appears that you are cheating or applying some trick, this strategy (illustrated in Figure 1.9(a)) is a valid way to think (recursively) and solve the problem. However, it is necessary to assume that the clever classmate is indeed providing a correct answer to your question. This assumption corresponds to the recursive leap of faith.

There are two possibilities regarding your classmate's ability to respond to your question. Perhaps he is indeed clever, and can somehow provide the answer without any help (for this problem it is straightforward to apply (1.12)). However, this would be unlikely for complicated problems. Instead, a second possibility is that he is not as clever as he claims to be, and is also using your strategy (i.e., asking another "clever" classmate) to come up with his answer. In other words, you can interpret that your friend is simply a clone of yourself. In fact, in this last scenario, represented in Figure 1.9(b), all of the students in the class would implement your approach. Note that the process is clearly recursive, where each student asks another one to solve a smaller instance of the problem (decreasing its size by one at each step), until a final student can simply reply 1 in the trivial base case. From there on, each student is able to provide an answer to whoever asked him to compute a full sum of integers by calculating a single addition. This process is carried out sequentially until you receive the sum of the first 99 positive integers, and you will be able to respond correctly to the instructor's question by adding 100. It is worth mentioning that the approach does not halt when reaching the base case (this is one of students' major misconceptions). Note that there is a first stage where students successively ask questions to other classmates until one of them responds the trivial 1, and a second phase where they calculate an addition and supply the result to a classmate.

This approach can be formalized precisely by (1.5) and coded as in Listing 1.1, where using $S(n-1)$ or `sum_first_naturals(n-1)` is equivalent to asking your classmate to solve a subproblem. The recursive leap of faith consists of assuming that $S(n-1)$ in the definition, or `sum_first_naturals(n-1)` in the code, will indeed return the correct answer.

Similarly to proofs by induction, we can reason that the entire procedure has to be correct. Firstly, your response will be correct as long as your classmate returns the right answer. But this also applies to your classmate, and to the student he is asking to solve a subproblem. This reasoning can be applied repeatedly to every student until finally some student asks another one to add up the integers from 1 to 1. Since the last student returns 1, which is trivially correct, all of the responses from the students must therefore be correct.

All of this implies that programmers can – and *should* – construct valid recursive cases by assuming that recursive function calls, involving self-similar subproblems of smaller size, work correctly. Informally, we must think that we can obtain their solutions "for free." Therefore, when breaking up a problem as illustrated in Figure 1.4, while we will have to solve the different problems resulting from the decomposition, we will not need to solve the self-similar subproblems, since we can assume that these solutions are already available. Naturally, we will have to process (modify, extend, combine, etc.) them somehow in order to construct the recursive cases. Lastly, recursive algorithms are completed by incorporating base cases that are not only correct, but that also allow the algorithms to terminate.

1.4.3 Imperative vs. declarative programming

Programming paradigms are general strategies for developing software. The imperative programming paradigm focuses on **how** programs work, where the code explicitly describes the control flow and the instructions that modify variable values (i.e., the program state). Iterative code follows this paradigm. In contrast, recursive code relies on the declarative programming paradigm, which focuses on **what** programs should perform without describing the control flow explicitly, and can be considered as the opposite of imperative programming. Functional programming follows this paradigm, which prevents side effects, and where the computation is carried out by evaluating mathematical functions.

Therefore, when designing recursive code, programmers are strongly encouraged to think of **what** functions accomplish, instead of **how** they carry out a task. Note that this is related to functional abstraction and the use of software libraries, where programmers do not need to worry about implementation details, and can focus exclusively on functionality by considering methods as black boxes that work properly. In this regard, programmers can – and *should* – consider calling recursive functions from a high-level point of view as invoking black boxes that return correct results (as in Figure 1.9(a)), instead of thinking about all of the recursive steps taken until reaching a base case (as in Figure 1.9(b)). More often than not, focusing on lower-level details involving the sequence of function calls will be confusing.

1.5 RECURSION VS. ITERATION

The computational power of computers is mainly due to their ability to carry out tasks repeatedly, either through iteration or recursion. The former uses constructs such as `while` or `for` loops to implement repetitions, while recursive functions invoke themselves successively, carrying out tasks repeatedly in each call, until reaching a base case. Iteration and recursion are equivalent in the sense that they can solve the same kinds of problems. Every iterative program can be converted into a recursive one, and vice versa. Choosing which one to use may depend on several factors, such as the computational problem to solve, efficiency, the language, or the programming paradigm. For instance, iteration is preferred in imperative languages, while recursion is used extensively in functional programming, which follows the declarative paradigm.

The examples shown so far in the book can be coded easily through loops. Thus, the benefit of using recursion may not be clear yet. In practice, the main advantage of using recursive algorithms over iterative ones is that for many computational problems they are much simpler to design and comprehend. A recursive algorithm could resemble more closely the logical approach we would take to solve a problem. Thus, it would be more intuitive, elegant, concise, and easier to understand.

In addition, recursive algorithms use the program stack implicitly to store information, where the operations carried out on it (e.g., push and pop) are transparent to the programmer. Therefore, they constitute clear alternatives to iterative algorithms where it is the programmer's responsibility to explicitly manage a stack (or similar) data structure. For instance, when the structure of the problem or the data resembles a tree, recursive algorithms may be easier to code and comprehend than iterative versions, since the latter may need to implement breadth- or depth-first searches, which use queues and stacks, respectively.

In contrast, recursive algorithms are generally not as efficient as iterative versions, and use more memory. These drawbacks are related to the use of the program stack. In general, every call to a function, whether it is recursive or not, allocates memory on the program stack and stores information on it, which entails a higher computational overhead. Thus, a recursive program cannot only be slower than an iterative version, a large number of calls could cause a stack overflow runtime error. Furthermore, some recursive definitions may be considerably slow. For instance, the Fibonacci codes in Listings 1.3 and 1.4 run in exponential time, while Fibonacci numbers can be computed in (much faster) logarithmic time. Lastly, recursive algorithms are harder to debug (i.e., analyze a program step by step in order to detect errors and fix them), especially if the functions invoke themselves several times, as in Listings 1.3 and 1.4.

Finally, while in some functional programming languages loops are not allowed, many other languages support both iteration and recursion. Thus, it is possible to combine both programming styles in order to build algorithms that are not only powerful, but also clear to understand (e.g., backtracking). Listing 1.4 shows a simple example of a recursive function that contains a loop.

1.6 TYPES OF RECURSION

Recursive algorithms can be categorized according to several criteria. This last section briefly describes the types of recursive functions and procedures that we will use and analyze throughout the book. Each type will be illustrated through recursive functions that, when invoked with certain specific arguments, can be used for computing Fibonacci numbers ($F(n)$). Lastly, recursive algorithms may belong to several categories.

1.6.1 Linear recursion

Linear recursion occurs when methods call themselves only once. There are two types of linear-recursive methods, but we will use the term "linear recursion" when referring to methods that process the result of the recursive call somehow before producing or returning its own output. For example, the factorial function in (1.3) belongs to this category since it only carries out a single recursive call, and the output of the subproblem is multiplied by n in order to generate the function's result. The functions in (1.5), (1.9), and (1.10) are also clear examples of linear recursion. The following function provides another example:

$$f(n) = \begin{cases} 1 & \text{if } n = 1 \text{ or } n = 2, \\ \left\lfloor \Phi \cdot f(n-1) + \frac{1}{2} \right\rfloor & \text{if } n > 2, \end{cases} \tag{1.14}$$

where $\Phi = (1 + \sqrt{5})/2 \approx 1.618$ is a constant also known as the "golden ratio," and where $\lfloor \cdot \rfloor$ denotes the floor function. In this case, $F(n) = f(n)$ is the n-th Fibonacci number. Chapter 4 covers this type of linear-recursive algorithms in depth.

1.6.2 Tail recursion

A second type of linear recursion is called "tail recursion." Methods that fall into this category also call themselves once, but the recursive call is the last operation carried out in the recursive case. Therefore, they do not manipulate the result of the recursive call. For example, consider the following tail-recursive function:

$$f(n, a, b) = \begin{cases} b & \text{if } n = 1, \\ f(n-1, a+b, a) & \text{if } n > 1. \end{cases} \tag{1.15}$$

Observe that in the recursive case the method simply returns the result of a recursive call, which does not appear within a mathematical or logical expression. Therefore, the recursive case is simply specifying relationships between sets of arguments for which the function returns the same value. As the algorithm carries out recursive calls it modifies the arguments until it is possible to compute the solution easily in a base case. In this example Fibonacci numbers can be recovered through $F(n) = f(n,1,1)$. This type of recursive algorithm will be covered in Chapter 5.

1.6.3 Multiple recursion

Multiple recursion occurs when a method calls itself several times in some recursive case (see Chapter 7). If the method invokes itself twice it is also called "binary recursion." Examples seen so far include the functions in (1.2), (1.6), (1.7), and (1.11). The following function uses multiple recursion in order to provide an alternative definition of Fibonacci numbers $(F(n) = f(n))$:

$$f(n) = \begin{cases} 1 & \text{if } n = 1 \text{ or } n = 2, \\ \left[f\left(\frac{n}{2}+1\right)\right]^2 - \left[f\left(\frac{n}{2}-1\right)\right]^2 & \text{if } n > 2 \text{ and } n \text{ even}, \\ \left[f\left(\frac{n+1}{2}\right)\right]^2 + \left[f\left(\frac{n-1}{2}\right)\right]^2 & \text{if } n > 2 \text{ and } n \text{ odd}. \end{cases} \tag{1.16}$$

This type of recursion appears in algorithms based on the "divide and conquer" design strategy, covered in Chapter 6.

1.6.4 Mutual recursion

A set of methods are said to be mutually recursive when they can call each other in a cyclical order. For example, a function f may call a second function g, which in turn can call a third function h, which can end up calling the initial function f. This type of recursion is also called "indirect" since a method might not invoke itself directly within its body, but through a cyclical chain of calls. For example, consider the two following functions:

$$A(n) = \begin{cases} 0 & \text{if } n = 1, \\ A(n-1) + B(n-1) & \text{if } n > 1, \end{cases} \tag{1.17}$$

$$B(n) = \begin{cases} 1 & \text{if } n = 1, \\ A(n-1) & \text{if } n > 1. \end{cases} \tag{1.18}$$

It is clear that A is recursive since it calls itself, but the recursion in B is achieved indirectly by calling A, which in turn calls B. Thus, recursion arises since invoking B can produce other calls to B (on simpler instances of the problem). These functions can also be used to generate Fibonacci numbers. In particular, $F(n) = B(n) + A(n)$. Mutual recursion will be covered in Chapter 9.

1.6.5 Nested recursion

Nested recursion occurs when an argument of a recursive function is defined through another recursive call. Consider the following function:

$$f(n,s) = \begin{cases} 1+s & \text{if } n = 1 \text{ or } n = 2, \\ f(n-1, s+f(n-2,0)) & \text{if } n > 2. \end{cases} \tag{1.19}$$

The second parameter of the recursive call is an expression that involves yet another recursive function call. In this case, Fibonacci numbers can be recovered through $F(n) = f(n,0)$. This type of recursion is rare, but appears in some problems and contexts related to tail recursion. Finally, an overview of nested recursion is covered in Chapter 11.

1.7 EXERCISES

Exercise 1.1 — What does the following function calculate?

$$f(n) = \begin{cases} 1 & \text{if } n = 0, \\ f(n-1) \times n & \text{if } n > 0. \end{cases}$$

Exercise 1.2 — Consider a sequence defined by the following recursive definition: $s_n = s_{n-1} + 3$. Calculate its first four terms, by considering that: (a) $s_0 = 0$, and (b) $s_0 = 4$. Provide a nonrecursive definition of s_n in both cases.

Exercise 1.3 — Consider a sequence defined by the following recursive definition: $s_n = s_{n-1} + s_{n-2}$. If we are given the initial values of s_1 and s_2 we have enough information to build the entire sequence. Show that it is also possible to construct the sequence when given two arbitrary values s_i and s_j, where $i < j$. Finally, find the elements of the sequence between $s_1 = 1$ and $s_5 = 17$.

Exercise 1.4 — The set "descendants of a person" can be defined recursively as the children of that person, together with the descendants of those children. Provide a mathematical description of the concept using set notation. In particular, define a function $D(p)$, where D denotes descendants, and the argument p refers to a particular person. Also, consider that you can use the function $C(p)$, which returns the set containing the children of a person p.

Exercise 1.5 — Let $F(n) = F(n-1) + F(n-2)$ represent a recursive function, where n is a positive integer, and with arbitrary initial values for $F(1)$ and $F(2)$. Show that it can be written as $F(n) = F(2) + \sum_{i=1}^{n-2} F(i)$, for $n \geq 2$.

Exercise 1.6 — Implement the factorial function, defined in (1.3).

Exercise 1.7 — When viewed as an abstract data type, a list can be either empty, or it can consist of a single element, denoted as the "head" of the list, plus another list (which may be empty), referred to as the "tail" of the list. Let `a` represent a list in Python. There are several ways to check whether it is empty. For example, the condition `a==[]` returns `True` if the list is empty, and `False` otherwise. In addition, the head of the list is simply `a[0]`, while the tail can be specified as `a[1:]`. Write an alternative version of the function `sum_list_length_2` in Listing 1.5 that avoids using `len` by using the previous elements.

Exercise 1.8 — Implement three functions that calculate the sum of elements of a list of numbers using the three decompositions in Figure 1.6. The functions will receive two input parameters: a list, and its size (i.e., length), according to the scenario in Figure1.8(b). In addition, indicate an example calling the functions and printing their results.

Exercise 1.9 — Show by mathematical induction the following identity related to a geometric series, where n is a positive integer, and $x \neq 1$ is some real number:

$$\sum_{i=0}^{n-1} x^i = \frac{x^n - 1}{x - 1}.$$

Exercise 1.10 — Code the five recursive functions defined in Section 1.6 that compute Fibonacci numbers $F(n)$. Since some require several parameters besides n, or do not compute a Fibonacci number directly, implement additional "wrapper" functions that only receive the parameter n, and call the coded functions in order to produce Fibonacci numbers. Test that they produce correct outputs for $n = 1, \ldots, 10$.

CHAPTER 2

Methodology for Recursive Thinking

Science is much more than a body of knowledge. It is a way of thinking.

— Carl Sagan

WHEN thinking declaratively it is possible to design recursive algorithms for solving a wide variety of computational problems by following a systematic approach. This chapter describes a general template for deriving recursive solutions from a declarative perspective that unravels the process of recursive thinking into a sequence of steps. In addition, the chapter introduces several useful diagrams that programmers can utilize when designing recursive cases. These are beneficial since they force us to think about problem decomposition and induction, the two most important concepts underlying recursive (declarative) thinking.

Finally, for some tasks (e.g., debugging) or certain complex problems it may be useful to think about the lower-level details regarding the sequence of operations carried out by recursive procedures. We will nevertheless postpone covering these lower-level aspects until Chapter 10. Instead, this chapter focuses exclusively on higher-level declarative thinking.

2.1 TEMPLATE FOR DESIGNING RECURSIVE ALGORITHMS

The recursive functions introduced in Chapter 1, as well as many of the algorithms that will appear in the rest of the book, can be designed by following the methodology described in Figure 2.1. This template em-

1. Determine the **size** of the problem.

2. Define **bases cases**.

3. **Decompose** the computational problem into self-similar subproblems of smaller size, and possibly additional different problems.

4. Define **recursive cases** by relying on **induction** and **diagrams**.

5. **Test** the code.

Figure 2.1 General template for designing recursive algorithms.

phasizes declarative thinking due to its fourth step, since using induction implies focusing on what an algorithm performs, rather than on how it solves the problem. The following sections explain each step and mention common pitfalls and misunderstandings.

2.2 SIZE OF THE PROBLEM

Naturally, the first step when tackling any computational problem is to understand it. Given the problem statement, the inputs, outputs, and relationships between them should be clearly identified in order to proceed. Regardless of the type of algorithm we intend to design, it is crucial to determine the size of the problem (also denoted as input size). It can be thought of as a mathematical expression involving quantities related to the input parameters that determine the problem's complexity in terms of the number of operations that algorithms need in order to solve it. When using recursion it is especially relevant since the base and recursive cases clearly depend on it.

In many problems the size only depends on one factor, which also happens to be one of the input parameters. For instance, the size of calculating the n-th Fibonacci number, the factorial of n, or the sum of the first n positive integers (see Chapter 1) is precisely the input parameter n. For these problems it is apparent that solving smaller instances (e.g., for $n-1$ or $n/2$) requires less operations. In other cases the size of the problem might not be specified explicitly by the inputs, but the programming language will allow us to retrieve the particular size through its syntax and constructs. For example, when working with lists, their length often determines the size of the problem, which can be recovered through the function `len`.

In other problems the size may be expressed as a function of the input parameters. For instance, consider the problem of computing the sum of the digits of some positive integer n. Although the input parameter is n, the size of the problem is not n, but rather the number of digits of n, which determines the number of operations to carry out. Formally this quantity can be expressed as $\lfloor \log_{10}(n) \rfloor + 1$.

The size of a problem may also depend on several input parameters. Consider the problem of adding the elements of a sublist within a list (see Listing 1.6), determined by "lower" and "upper" indices. Solving the problem requires adding $n-m+1$ elements, where n and m are the upper and lower indices, respectively. Thus, it needs to compute $n-m$ additions. The small (unit) difference between both expressions is irrelevant. It is worth mentioning that it is not necessary to know exactly how many operations will be carried out in order to solve a problem. Instead, it is enough to understand when a base case is reached, and how to reduce the size of the problem in order to decompose it into smaller subproblems. For this last problem, the size is decreased by reducing the difference between m and n.

Moreover, the size of a problem does not necessarily need to specify the number of operations that algorithms must perform. Consider the problem of adding the elements of a square $n \times n$-dimensional matrix. Since it contains n^2 elements, the sum requires $n^2 - 1$ additions, which is a function of n. However, in this case the size of the problem is simply n, and not n^2. Note that smaller subproblems arise by decreasing n, not n^2. If the matrix were $n \times m$-dimensional, the problem size would depend on both parameters n and m. In particular, the size of the problem would be nm, where we could obtain subproblems of smaller size by decreasing n, m, or both.

The size is generally a property of a problem, not of a particular algorithm that solves it. Thus, it is not the actual number of computations carried out by a specific algorithm, since problems can be solved by different algorithms whose runtime may vary considerably. Consider the searching problem that consists of determining whether or not some number appears in a sorted list of n numbers. In the worst case (when the list does not contain the number and the result is `False`) it can be solved naively in n steps, or more efficiently in $\lfloor \log_2(n) \rfloor + 1$ operations by the "binary search" algorithm. Regardless of the algorithm used, the function that describes the runtime only depends on n. Thus, the size of the problem is n.

Nevertheless, for some problems the size can be defined in several ways. Moreover, the choice of the size affects the rest of the decisions in the recursive design process, and ultimately leads to different algorithms for solving the problem. For example, consider the problem of adding two nonnegative integers a and b by only using unit increments or decrements (see Section 4.1.2). The size of the problem can be $a+b$, a, b, or $\min(a,b)$, and the final algorithm will be different depending on which problem size we choose to work with.

2.3 BASE CASES

Base cases are instances of problems that can be solved without using recursion; or more precisely, without carrying out recursive calls. The most common type of base case is associated with the smallest instances of a problem, for which the result can be determined trivially, and sometimes may not even require carrying out computations. For instance, the value of the first and second Fibonacci numbers is simply 1, which follows directly from the definition. The "sum" of the first positive integer is evidently 1. Similarly, the "sum" of the digits of a nonnegative number n that contains only one digit (i.e., $n < 10$) is obviously that digit (n). Observe that these functions simply return a value in the base cases, and do not perform any addition or other mathematical operation.

Some methods may require several base cases. For example, (1.2) defines a base case for $n = 1$ and another for $n = 2$. Since the associated value (1) is the same, the Python function in Listing 1.3 can use a single logical expression (`(n==1) or (n==2)`) in order to take into account both cases. Other methods need to use a cascaded `if` structure in order to separate each base case. For instance, the function in (1.6) defines a base case for $n = 1$ and another for $n = 2$ since it returns different values for both inputs. Logically, the associated code in Listing 1.2 separates both cases.

One of the advantages of recursive programming is that the resulting code is often concise, which helps in understanding it. In this regard, consider the factorial function defined in (1.3). One could argue that if base cases correspond to trivial instances of a problem, then the definition of the function could include the case $1! = 1$ as well. However, $1!$ can be

Listing 2.1 Misconceptions regarding base cases through the factorial function.

```
def factorial(n):
    if n == 0:
        return 1
    else:
        return factorial(n - 1) * n


def factorial_redundant(n):
    if n == 0 or n == 1:
        return 1
    else:
        return factorial_redundant(n - 1) * n


def factorial_missing_base_case(n):
    if n == 1:
        return 1
    else:
        return factorial_missing_base_case(n - 1) * n


def factorial_no_base_case(n):
    return factorial_no_base_case(n - 1) * n
```

obtained through the recursive case correctly since $1! = 0! \times 1 = 1 \times 1 = 1$. Therefore, although 1! is a valid base case (it is obviously correct), including it would be redundant, and therefore unnecessary. Moreover, these dispensable cases could be misleading, since programmers trying to understand the recursive code could interpret that the functions require them in order to be correct. Thus, for clarity and conciseness, it is often recommended to include the least number of base cases that lead to correct recursive definitions. Lastly, redundant base cases usually do not have a relevant impact on the efficiency of a recursive program. In general, although they can reduce the number of recursive function calls for a given input, the saved computational time is often negligible. For instance, if 1! is included in the factorial function as a base case the recursive process will only perform one recursive call less, which barely has any effect on computing time (moreover, the code would need two conditions, and could even be more inefficient if both are evaluated before reaching a recursive case).

Listing 2.1 uses the factorial function to illustrate this and other basic pitfalls regarding bases cases. Firstly, the `factorial` method provides a perfect implementation of the mathematical function. The `factorial_redundant` method is correct since it produces a valid output for any nonnegative integer input argument. However, it contains an extra base case that is unnecessary.

In addition, we should also strive for generality, which is an important desirable software feature. More general functions are those that can operate correctly on a wider range of inputs, and are therefore more applicable and useful. The factorial function in (1.3) receives a nonnegative integer as its input argument, and is defined adequately since it can be applied correctly not only to every positive integer, but also to 0. Replacing the base case $0! = 1$ by $1! = 1$ would lead to a less general function, since it would not be defined for $n = 0$. The method `factorial_missing_base_case` implements this function. If it were called with `n` = 0 as its input argument it would fail to produce a valid result since it would run into an infinite recursion. In particular, `factorial_missing_base_case(0)` would call `factorial_missing_base_case(-1)`, which would call `factorial_missing_base_case(-2)`, and so on. In practice the process textcolorredwould halt producing a runtime error message (e.g., "stack overflow," or "maximum recursion depth exceeded") after performing a large number of function calls (see Chapter 10). Finally, `factorial_no_base_case` always generates an infinite recursion since it does not contain any base case, and can therefore never halt.

For many problems, there are instances for which we can provide a result without using recursion, even if their size is not small. For example, consider the problem of determining whether some element appears in a list. The size of the problem is the number of elements in the list. A first base case occurs when the list is empty (this is the smallest instance of the problem, of zero size), where the result is obviously `False`. In addition, the algorithm can check if the element appears in some position (e.g., the first, middle, or last) of the list, and return a true value immediately if it does, even if the list is very large (Chapter 5 covers this type of searching problems). In both cases there is no need to carry out a recursive call.

2.4 PROBLEM DECOMPOSITION

The next step in the recursive design methodology consists of identifying self-similar problems smaller than the original (and possibly other different problems), as illustrated in Figure 1.4. These smaller subproblems will be used in the fourth step of the template in order to establish the recursive cases. The process involves decreasing the size of the problem in order to consider smaller instances that are closer to the base cases. Therefore, it is essential to correctly establish the size of the problem and its base cases before proceeding with this step.

Many recursive solutions reduce the size of a problem by decreasing it a single unit, or by diving it by two, as shown graphically in Figures 1.5 and 1.6. Representing problems through diagrams is useful and highly recommended since it may help us to recognize self-similar subproblems (i.e., recursion) visually, and can also facilitate designing recursive cases. Figure 2.2 shows additional diagrams that illustrate the decomposition of the sum of the first positive integers when the problem size is decreased by a unit. Firstly, since the recursive method is a mathematical function that can be written as a formula, we can simply expand the expression in order to identify the original problem and the corresponding subproblem, as shown in (a). In (b) the diagram also contains a formula, but uses boxes that enclose the formulas instead of using curly braces. In this case, the darker outer box represents a full problem, while the lighter box contained in it symbolizes a subproblem. In (c) the diagram uses a similar scheme, where a problem consists of adding the small circles arranged in a triangular pattern. This illustration is analogous to that of Figure 1.5(b). Finally, in (d) the numbers to be added are depicted inside rectangles, which are concatenated in order to represent the problem and subproblem.

These apparently simple diagrams are appropriate since they allow us to identify the subproblem within the original one. However, it is crucial to observe that the subproblem is not decomposed into yet smaller ones. For instance, the sum of the positive integers up to $n-2$ does not appear explicitly as a subproblem in any of the diagrams. In other words, the diagrams avoid illustrating subproblems in a nested manner, or in a tree structure. This is one of the main pitfalls and sources of misunderstandings when thinking recursively. Figures 2.3(a) and (c) illustrate a proper decomposition of the sum of the first positive integers ($S(n)$). Instead, while the nested diagram in (b), or the tree representing the sequence of recursive calls in (d), represent the problem correctly, they

$$\underbrace{\overbrace{1+2+3+\cdots+(n-2)+(n-1)}^{\text{Subproblem}} \quad + \quad n}_{\text{Original problem}}$$

(a)

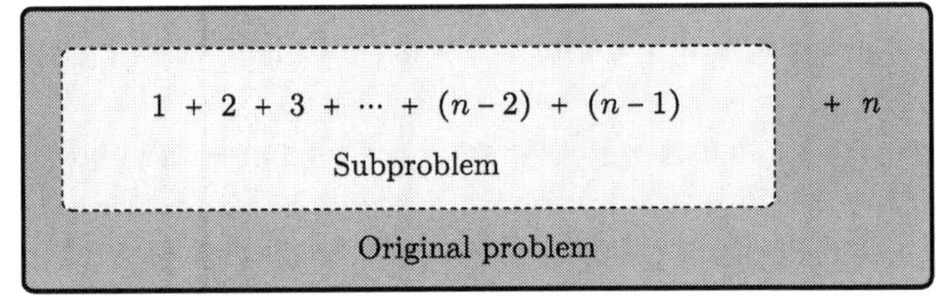

(b)

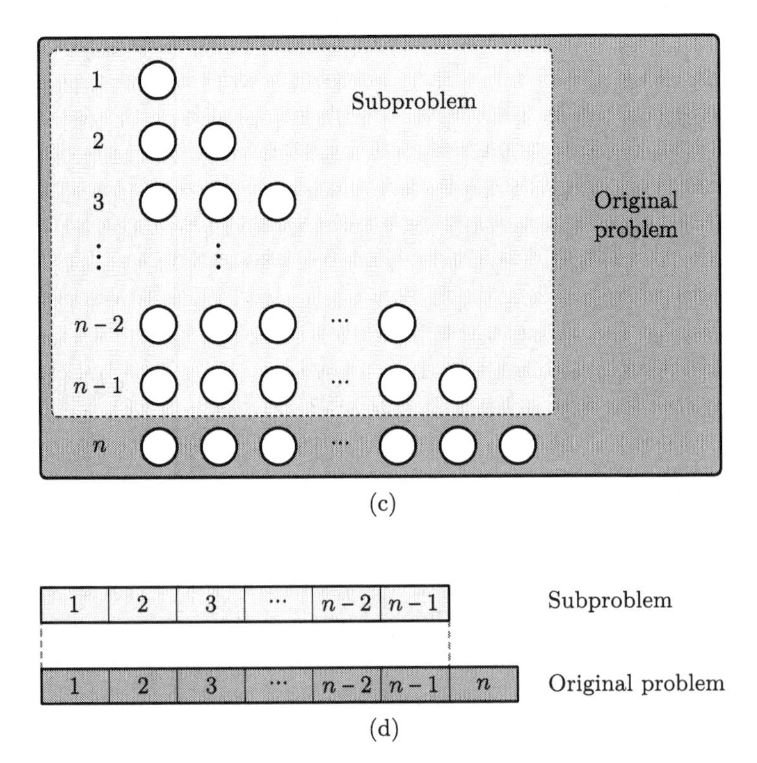

(c)

(d)

Figure 2.2 Additional diagrams that illustrate the decomposition of the sum of the first positive integers when the problem size is decreased by a unit.

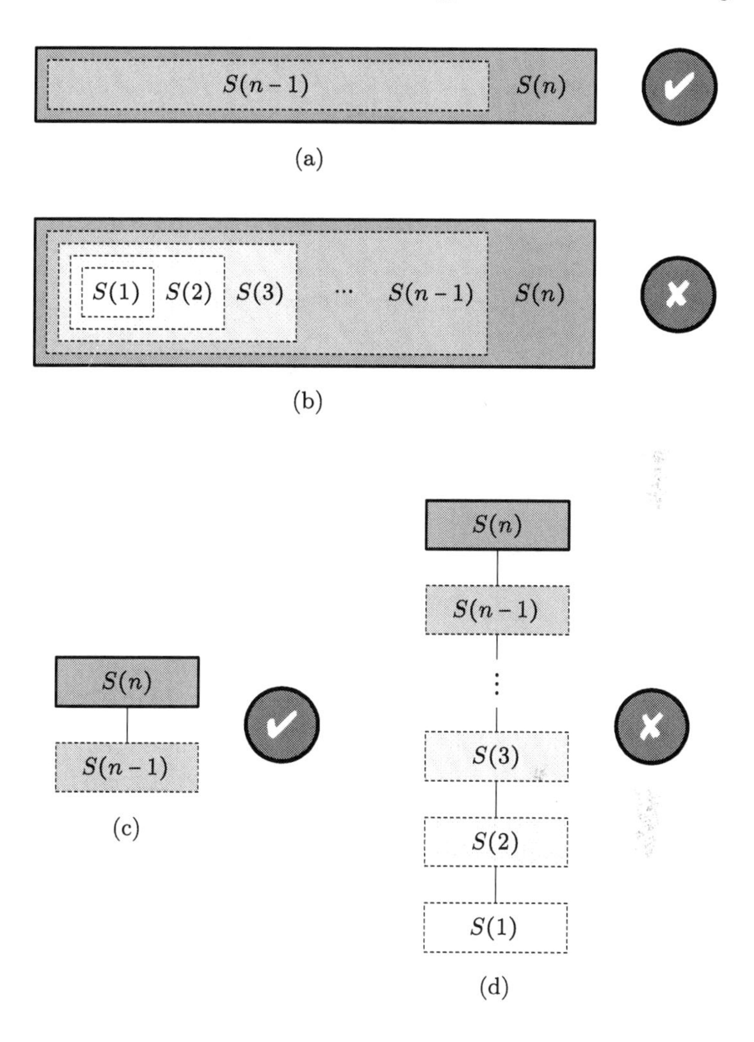

Figure 2.3 When thinking recursively we generally do not need to decompose a problem into every instance of smaller size.

incorporate additional subproblems ($S(1)$, $S(2)$,..., $S(n-2)$) that do not appear in the recursive definition in (1.5). Therefore, these subproblems should be omitted from the diagrams, and therefore also from the recursive thinking process, in order to avoid confusion. The nested and tree-structured diagrams will nevertheless be useful for other tasks (see Chapter 10).

Avoiding nested diagrams does not mean that we should only consider one subproblem. Many recursive definitions require decomposing

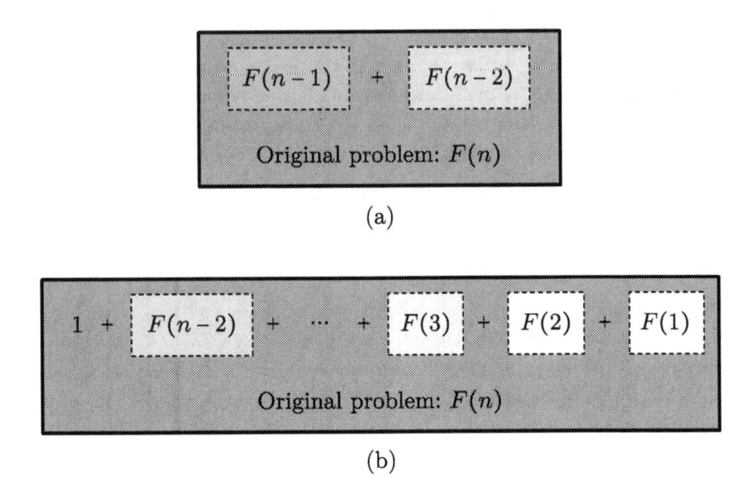

Figure 2.4 Decompositions of the Fibonacci function.

the original problem into numerous subproblems. For instance, Figure 2.4 shows two diagrams for decomposing a Fibonacci number. In particular, (a) is associated with the definition in (1.2), while (b) is related to (1.7). Although the diagram in (b) shows every subproblem from $F(1)$ until $F(n-2)$, it is a valid representation since the boxes that represent subproblems are not nested.

For most problems in this book the size will depend on only one parameter or factor. However, in some cases it is determined by several. For instance, problems on matrices usually depend on their height (number of rows) and width (number of columns). Problems of smaller size can therefore arise from decreasing one or both of these parameters. In particular, many recursive algorithms partition matrices in blocks that correspond to submatrices. These "block matrices" can be interpreted as the original matrix with vertical and horizontal subdivisions, as illustrated in the following example with a 4×7 matrix:

$$\mathbf{A} = \left[\begin{array}{c|c} \mathbf{A}_{1,1} & \mathbf{A}_{1,2} \\ \hline \mathbf{A}_{2,1} & \mathbf{A}_{2,2} \end{array}\right] = \left[\begin{array}{ccc|cccc} 3 & 1 & 2 & 0 & 3 & 4 & 9 \\ 7 & 2 & 3 & 3 & 1 & 7 & 5 \\ \hline 0 & 5 & 3 & 2 & 1 & 4 & 8 \\ 6 & 3 & 1 & 5 & 4 & 9 & 0 \end{array}\right],$$

$$\mathbf{A}_{1,1} = \begin{bmatrix} 3 & 1 & 2 \\ 7 & 2 & 3 \end{bmatrix}, \quad \mathbf{A}_{1,2} = \begin{bmatrix} 0 & 3 & 4 & 9 \\ 3 & 1 & 7 & 5 \end{bmatrix},$$

$$\mathbf{A}_{2,1} = \begin{bmatrix} 0 & 5 & 3 \\ 6 & 3 & 1 \end{bmatrix}, \quad \mathbf{A}_{2,2} = \begin{bmatrix} 2 & 1 & 4 & 8 \\ 5 & 4 & 9 & 0 \end{bmatrix}.$$

In this case, the number of rows and columns of each submatrix results from performing the integer division of the original height and width of the matrix by two.

Regarding efficiency, dividing the size of the problem by two, instead of decreasing it by a single unit, may lead to faster algorithms. Therefore, this strategy should be considered in general, and especially if a theoretical analysis shows that it is possible to improve the computational cost of an algorithm based on reducing the size of the problem by a unit. However, dividing the size of a problem by two may not lead to a reasonable recursive algorithm. For instance, it is difficult to obtain a simple recursive formula for the factorial function $n!$ by considering the subproblem $(n/2)!$.

2.5 RECURSIVE CASES, INDUCTION, AND DIAGRAMS

The next step in the template for designing recursive algorithms consists of defining the recursive cases, which involves figuring out how to build the full solution to an original problem by using the solutions to the self-similar subproblems provided by the decomposition stage. As mentioned in Section 1.4, we can assume that these simpler solutions are readily available by relying on induction. Thus, the real challenge when thinking recursively consists of determining how to modify, extend, or combine the solutions to subproblems in order to obtain the complete solution to an original problem, which corresponds to the last step in the scheme presented in Figure 1.4.

2.5.1 Thinking recursively through diagrams

It is often useful to represent the thought process or "mental model" of deriving recursive cases by means of diagrams. The general illustration in Figure 2.5 shows a recursive thinking procedure when a problem is decomposed into a single self-similar subproblem. The process starts by considering a general instance of the problem, of a certain size, determined by the input parameters, which are depicted in the top-left box. Applying the recursive method to those parameters would produce the results represented in the top-right box. Note that this top row is simply another way to specify the statement of the problem. In order to define recursive cases we first choose a particular decomposition that leads to a smaller instance of the problem. The new, simpler, parame-

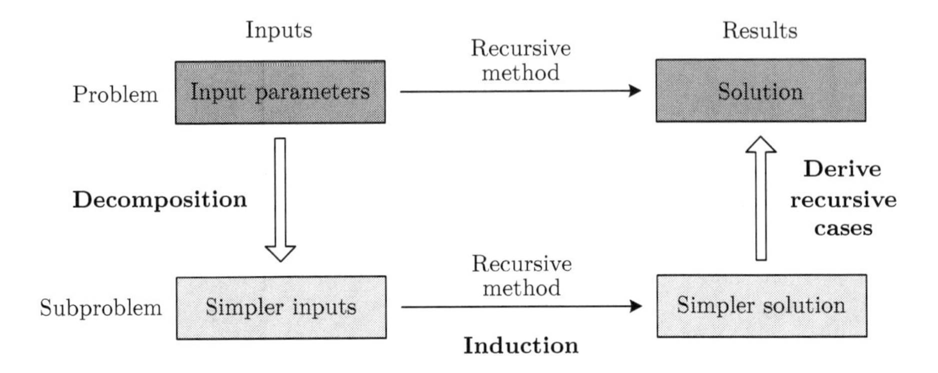

Figure 2.5 General diagram for thinking about recursive cases (when a problem is decomposed into a single self-similar subproblem).

ters for the self-similar subproblem are shown in the bottom-left box. A recursive call to the method with those parameters would therefore produce the results in the bottom-right box, which are obtained by applying the statement of the problem to the simpler inputs. Finally, since we can assume that these results are available by relying on induction, we derive recursive cases by determining how we can modify or extend them in order to obtain or arrive at the solution to the original problem (top-right box).

For the sum of the first n positive integers ($S(n)$) the diagram would be:

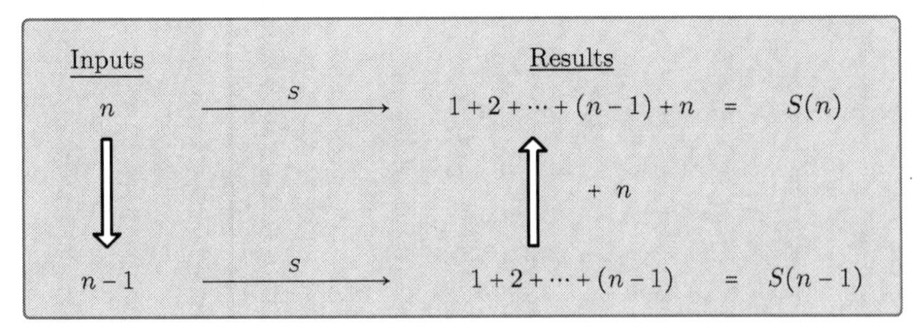

Firstly, we can choose to decompose the problem by reducing its size by a unit. For this particular decomposition our goal is to figure out how we can obtain $S(n)$ by modifying or extending the solution to the subproblem $S(n-1)$. In this case, it is easy to see that adding n to $S(n-1)$ yields $S(n)$. Therefore, the recursive case is defined as $S(n) = S(n-1) + n$.

It is important to understand that we are only required to define the decomposition, and establish how to obtain the results to the original problem given the results of the subproblem. Therefore, we only have to think about the processes associated with the thicker arrows. Alternatively, the thinner arrows simply symbolize the effect of solving particular instances of the problem (on different inputs), which is completely defined by the statement of the problem. Observe that we do not have to make any algorithmic design decisions regarding the relationships between inputs and results, since they are specified in the statement of the problem.

The advantage of the diagram in Figure 2.5 is its generality, since it can be applied to numerous problems for deriving recursive solutions. Therefore, the explanations to many of the examples covered throughout the book will rely on it. Naturally, other graphical representations, such as the diagrams in Figures 1.5, 1.6, and 2.2 can also be used instead. These alternative visualizations are especially useful since they not only show subproblems, but also their relationship with the original problem, which provides clues about how we can combine, increment, or modify the solutions to the subproblems in order to solve the original one. In other words, they visually suggest how we can define the recursive cases. For instance, it is apparent from the diagrams in Figure 2.2 that the solution to the problem consists of adding n to the result of the subproblem of size $n - 1$.

Indeed, more elaborate diagrams, especially tailored for the particular problem, may help us to find different decompositions that could eventually lead to more efficient algorithms. For instance, representing the sum of the first n positive integers through triangular pyramids of tiled blocks, as shown in Figure 1.5, allows us to derive recursive cases easily when the size of the subproblems is roughly half of the original one. Firstly, recall that the problem can be interpreted as counting the number of blocks in a triangular structure. Furthermore, since the representation allows us to superimpose several subproblems (i.e., smaller triangular structures) simultaneously on top of the original, the visualizations indicate how it is possible to "fill" a larger pyramid with smaller ones. Therefore, the diagrams illustrate how to derive recursive cases by adding the results (in this case, the number of blocks) of smaller subproblems. For example, in Figure 1.5(c) the output of three subproblems of size $n/2$ can be added to the result of the subproblem of size $n/2 - 1$ in order to obtain the total value of the original problem.

Nonetheless, the general diagram in Figure 2.5 can also be used in order to provide recursive solutions for $S(n)$ when the decomposition divides the size of the problem by two. In that case we have:

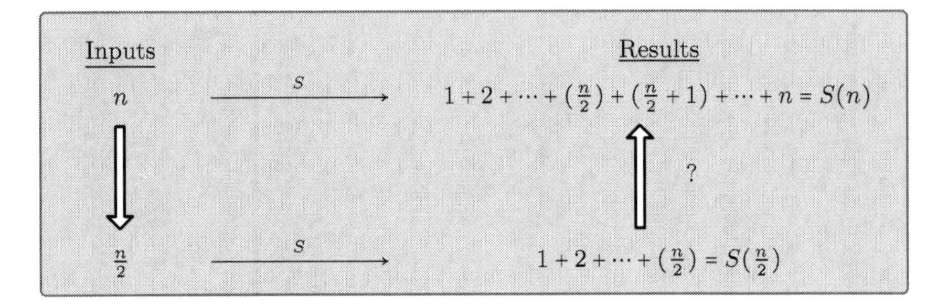

The question is to figure out how (and if) we can obtain $S(n)$ by modifying or extending $S(n/2)$. A first obvious idea leads to $S(n) = S(n/2) + (n/2+1) + \cdots + n$. However, although it is correct, its implementation requires either a loop, or using a different recursive function, in order to calculate the sum of the last $n/2$ terms. Observe that it is not possible to obtain $(n/2+1) + \cdots + n$ by using a recursive call to the function under construction (S), since that sum is not an instance of the problem. Nevertheless, it can be broken up, for example, in the following way:

$$\begin{array}{cccccccccc}
 & (n/2+1) & + & (n/2+2) & + & \cdots & + & n & & \\
\hline
 & n/2 & + & n/2 & + & \cdots & + & n/2 & = & (n/2)^2 \\
+ & 1 & + & 2 & + & \cdots & + & n/2 & = & S(n/2)
\end{array}$$

This not only simplifies the expression, but it also contains $S(n/2)$, which we can use in order to obtain a much simpler recursive case:

$$\begin{aligned}
S(n) &= S(n/2) + (n/2+1) + \cdots + n \\
&= S(n/2) + S(n/2) + (n/2)^2 \\
&= 2S(n/2) + (n/2)^2. \qquad (2.1)
\end{aligned}$$

The general diagram could be completed in the following way:

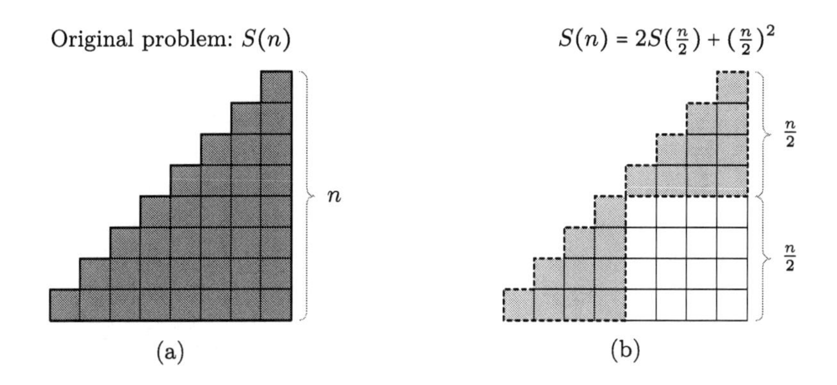

Figure 2.6 Diagram showing a decomposition of the sum of the first n positive integers $S(n)$ that uses two subproblems of half the size as the original.

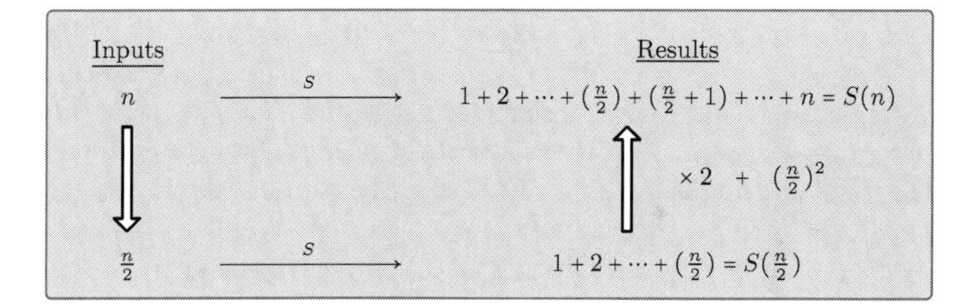

indicating that $S(n)$ can be obtained by multiplying the result of $S(n/2)$ times 2, and adding $(n/2)^2$. Although the decomposition of the problem is unique ($S(n/2)$), this example shows that we can use the result of the subproblem several times in order to arrive at the solution of the original problem. In this case, we can interpret that the recursive case uses $S(n/2)$ *twice*.

This recursive case can also be obtained through the diagrams that depict $S(n)$ as a triangular pyramid of blocks. In particular, Figure 2.6 illustrates the same reasoning that leads to $S(n) = 2S(n/2) + (n/2)^2$ through these representations. The nature of this particular diagram, where the solution to the problem consists of adding up blocks, facilitates obtaining recursive cases.

2.5.2 Concrete instances

In the previous example involving the sum of the first positive integers it is straightforward to express the result of the recursive method as a general function of the input parameter n. In other words, it is easy to obtain the expressions on the right-hand side of the general diagram. However, in other problems it can be more complicated to describe the results of the recursive methods (e.g., through some formula), and therefore to figure out how to arrive at the solution given the results to the subproblems. In these cases a useful initial approach consists of analyzing the problem through concrete instances.

Consider the problem of adding the digits of some nonnegative integer n, and assume that we choose the decomposition that decreases the number of digits by a unit, where the least significant digit is dropped from the original integer. This information provides the elements on the left-hand side of the general diagram. However, the method's output on the right-hand side can be complicated as a function of n. For example, one possible way to describe it is through the following summation:

$$\sum_{i=0}^{\lfloor \log_{10} n \rfloor} \left[(n//10^i)\%10 \right]. \tag{2.2}$$

Although we could use the formula to derive the recursive cases, we will proceed by analyzing concrete instances of the problem. For example (we can discard using the method's name in the results column, and when labeling the arrows, for the sake of simplicity):

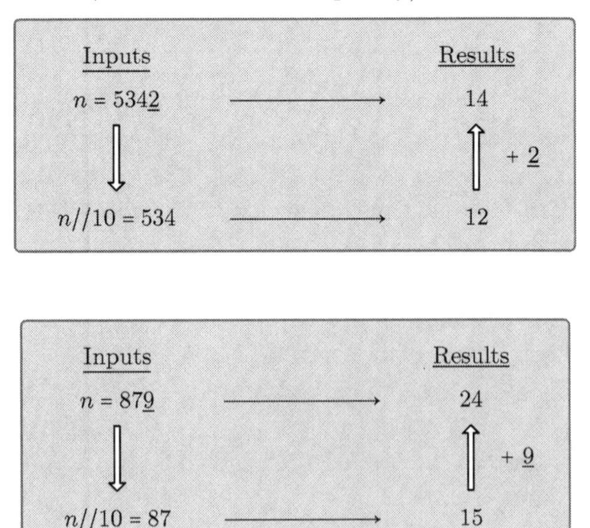

It is easy to see in these diagrams that we can obtain the result of the method by adding the last (underlined) digit of the original number to the output of the subproblem.

Listing 2.2 Code for computing the sum of the digits of a nonnegative integer.

```
def add_digits(n):
    if n < 10:
        return n
    else:
        return add_digits(n // 10) + (n % 10)
```

2.5.3 Alternative notations

Another approach consists of using some alternative general notation that facilitates constructing a diagram. Let $d_{m-1} \cdots d_1 d_0$ represent the sequence of digits that define some nonnegative integer n in base 10 (in other words, $n = d_{m-1} \cdot 10^{m-1} + \cdots + d_1 \cdot 10 + d_0$, with $0 \leq d_i \leq 9$ for all i, and $d_{m-1} \neq 0$ if $m > 1$). In that case the general diagram would be:

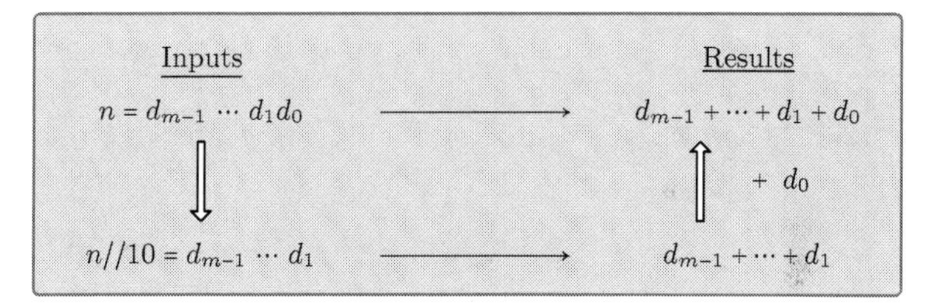

and the function can be coded as shown in Listing 2.2, where (n%10) represents d_0. Note that the base case for this problem occurs when n contains a single digit (i.e., $n < 10$), for which the result is obviously n.

Lastly, in practice more complicated problems may require analyzing multiple examples and scenarios (e.g., distinguishing when the input parameter is odd or even), that lead to different recursive cases.

2.5.4 Procedures

The methods seen so far correspond to functions that return values, where the results can be defined through formulas. Therefore, they can be used within expressions in other methods, where they would return specific values given a set of input arguments. However, there exist methods, called "procedures" in certain programming languages (e.g., Pascal), which do not return values. Instead, they may alter data structures passed as inputs to the method, or they can simply print information on a console. In these cases it is also possible to use the general diagram.

Consider the problem of printing the digits of some nonnegative integer n on the console, in reversed order, and vertically. In other words, the least significant digit will appear on the first line, the second least significant digit on the second line, and so on. For instance, if $n = 2743$ the program will print the following lines on the console:

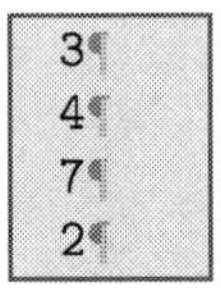

Firstly, the size of this problem is the number of digits of n. The base case occurs when n contains a single digit ($n < 10$), where the algorithm would simply print n. As in the previous example, the simplest decomposition considers $n//10$, where the least significant digit is removed from the original number. Figure 2.7 shows a possible decomposition diagram of the problem.

In addition, the general diagram can also be used for this type of procedure:

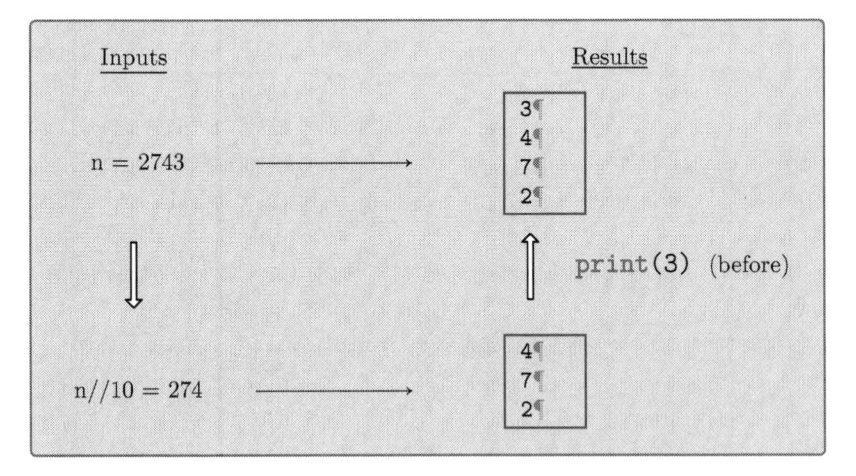

In this example the results are no longer numerical values, but a sequence of instructions, which correspond to printed lines on a console. For this problem it is possible to arrive at the solution by printing the least significant digit of the input number, and by calling the recursive method on the remaining digits. However, the order in which these operations are performed is crucial. In particular, the least significant digit

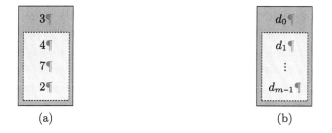

Figure 2.7 Diagram showing a decomposition of the problem consisting of printing the digits of a nonnegative integer on the console, in reversed order, and vertically. A particular ($n = 2743$) and a general m-digit ($n = d_{m-1} \cdots d_1 d_0$) case are shown in (a) and (b), respectively.

Listing 2.3 Code for printing the digits of a nonnegative integer vertically, and in reversed order.

```
1 def print_digits_reversed_vertically(n):
2     if n < 10:
3         print(n)
4     else:
5         print(n % 10)
6         print_digits_reversed_vertically(n // 10)
```

must be printed *before* the rest of the digits. The associated code is shown in Listing 2.3.

2.5.5 Several subproblems

Some algorithms need to decompose a problem into several self-similar subproblems, where the thought process of deriving recursive cases is analogous to the one illustrated in Figure 2.5. The diagrams simply need to include various subproblems and their corresponding solutions, according to the chosen decomposition of the problem, as shown in Figure 2.8. Besides extending and modifying the individual solutions to the subproblems, the recursive cases usually need to combine them as well.

Given a nonempty list of n integers ($n \geq 1$), consider the problem of finding the value of the largest one. For this example, we will decompose the problem by splitting it in two, in order to work with the first and second halves of the list, as shown in Figure 1.6(d). The following diagram illustrates the thought process with a concrete example:

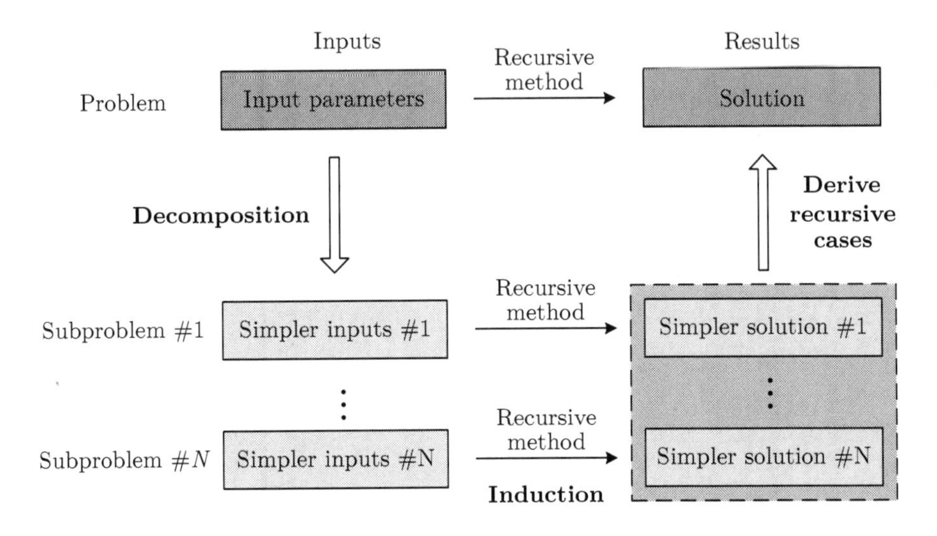

Figure 2.8 General diagram for thinking about recursive cases, when a problem is decomposed into several (N) self-similar subproblems.

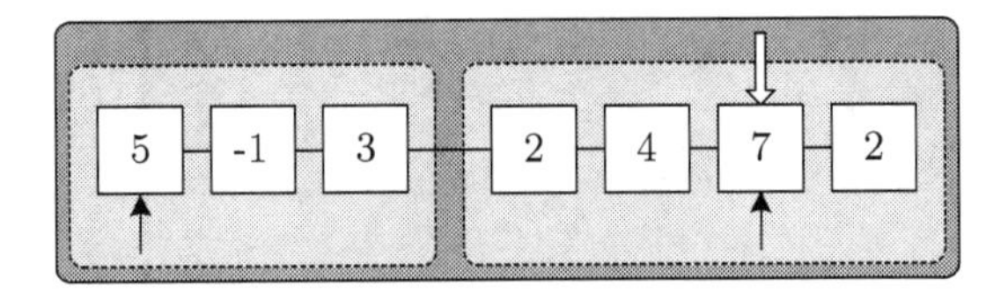

Figure 2.9 Alternative diagram showing a divide and conquer decomposition, and the recursive thought process, for the problem of finding the largest value in a list. The thick and thin arrows point to the solutions of the problem and subproblems, respectively.

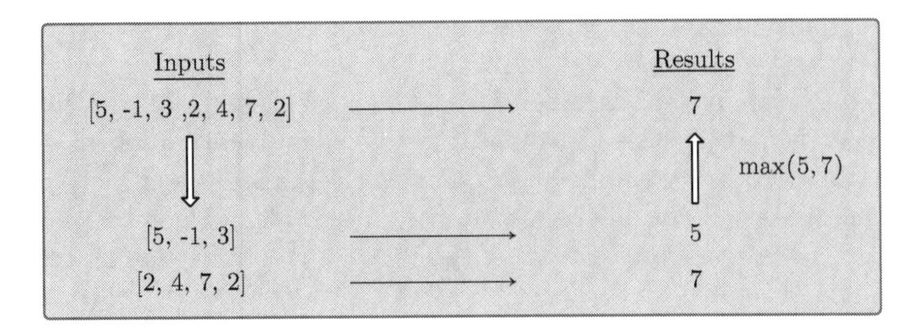

Figure 2.9 shows an alternative diagram that indicates the decomposition and recursive thought process for the problem. Both cases illustrate that each recursive call to one of the halves returns the greatest value in

Listing 2.4 Code for computing the maximum value in a list, through a divide and conquer approach.

```
def max_list_length_DaC(a):
    if len(a) == 1:
        return a[0]
    else:
        middle = len(a) // 2
        m1 = max_list_length_DaC(a[0:middle])
        m2 = max_list_length_DaC(a[middle:len(a)])
        return max(m1, m2)


def max_list_limits_DaC(a, lower, upper):
    if lower == upper:
        return a[lower]  # or a[upper]
    else:
        middle = (upper + lower) // 2
        m1 = max_list_limits_DaC(a, lower, middle)
        m2 = max_list_limits_DaC(a, middle + 1, upper)
        return max(m1, m2)


# Some list:
v = [5, -1, 3, 2, 4, 7, 2]

# Function calls:
print(max_list_length_DaC(v))
print(max_list_limits_DaC(v, 0, len(v) - 1)
```

the particular half. Therefore, the recursive case can simply return the maximum value in either half. The recursive function (f) is defined as follows:

$$f(\mathbf{a}) = \begin{cases} \mathbf{a}[0] & \text{if } n = 1, \\ \max\big(f(\mathbf{a}[0:n//2]), f(\mathbf{a}[n//2:n])\big) & \text{if } n > 1. \end{cases} \quad (2.3)$$

Listing 2.4 shows two ways of coding the function. The version that uses the lower and upper limits is usually faster. Naturally, this problem also allows recursive solutions based on a single subproblem whose size is decreased by a single unit. This approach is as efficient as the divide and conquer strategy. However, in practice it may produce runtime errors for large lists (see Chapter 10).

Listing 2.5 Erroneous Python code for determining if a nonnegative integer n is even.

```
1 def is_even_incorrect(n):
2     if n == 0:
3         return True
4     else:
5         return is_even_incorrect(n - 2)
```

Listing 2.6 Correct Python code for determining if a nonnegative integer n is even.

```
1 def is_even_correct(n):
2     if n == 0:
3         return True
4     elif n == 1:
5         return False
6     else:
7         return is_even_correct(n - 2)
```

2.6 TESTING

Testing is a fundamental stage in any software development process. In the context of this book its main purpose consists of discovering errors in the code. Testing therefore consists of running the developed software on different instances (i.e., inputs) of a problem in order to detect failures. Novice programmers are strongly encouraged to test their code since the ability to detect and correct errors (e.g., with a debugger) is a basic programming skill. In addition, it teaches valuable lessons in order to avoid pitfalls and code more efficiently and reliably.

Besides checking the basic correctness of the base and recursive cases, when testing recursive code, programmers should pay special attention to possible scenarios that lead to infinite recursions. These usually appear due to missing base cases or by erroneous recursive cases. For example, consider the function in Listing 2.5 whose goal consists of determining whether some nonnegative integer n is even. Both base and recursive cases are correct. Naturally, if a number n is even then so is $n-2$, and the function must return the same Boolean value for both integers. Nevertheless, `is_even_incorrect` only works for even numbers. Let $f(n)$ represent `is_even_incorrect(n)`. A call to $f(7)$ produces the following

Listing 2.7 Erroneous Python code for adding the first n positive numbers, which produces infinite recursions for most values of n.

```
1 def sum_first_naturals_3(n):
2     if n == 1:
3         return 1
4     else:
5         return 2 * sum_first_naturals_3(n / 2) + (n / 2)**2
```

recursive calls:

$$f(7) \to f(5) \to f(3) \to f(1) \to f(-1) \to f(-3) \to \cdots$$

which is an infinite recursion since the process does not halt at a base case. The fact that the function does not contain a base case that returns `False` provides a warning regarding its correction (not all Boolean functions need two base cases in order to return `True` or `False`). Indeed, the method can be fixed by adding that base case. Listing 2.6 shows a function that works for any valid argument ($n \geq 0$).

Another example is the function in Listing 2.7 that uses the recursive case described in (2.1) in order to compute the sum of the first n positive integers ($S(n)$). It is incomplete, and generates infinite recursions for values of n that are not a power of two. Firstly, since Python considers n to be a real number, $n/2$ is also a real number in general. Therefore, if the argument n is an odd integer in any recursive function call then $n/2$ will have a fractional part, and so will the arguments of the following recursive calls. Thus, the algorithm would not halt at the base case $n = 1$ (which is an integer with no fractional part), and would continue to make function calls with smaller and smaller arguments. For example, let $f(n)$ represent `sum_first_naturals_3(n)`. A call to $f(6)$ produces the following recursive calls:

$$f(6) \to f(3) \to f(1.5) \to f(0.75) \to f(0.375) \to f(0.1875) \to \cdots$$

never stopping at the base case. The only situation in which the algorithm works is when the first argument n is a power of two, since each of the divisions by two produces an even number, eventually reaching $n = 2$ and afterwards $n = 1$, where the function can finally return a concrete value instead of producing another function call.

The method `sum_first_naturals_3` does not work properly due to real-valued arguments. Thus, we could try replacing the real divi-

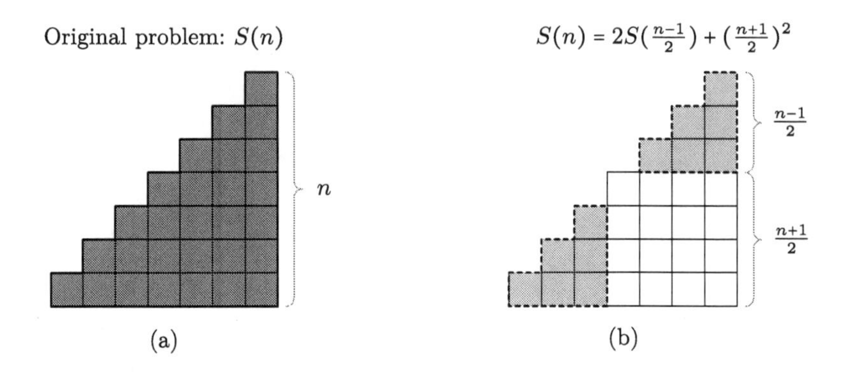

Figure 2.10 Diagram showing a decomposition of the sum of the first n positive integers $S(n)$ that uses two subproblems of (roughly) half the size as the original, when n is odd.

Listing 2.8 Incomplete Python code for adding the first n positive numbers.

```
1 def sum_first_naturals_4(n):
2     if n == 1:
3         return 1
4     else:
5         return 2 * sum_first_naturals_4(n // 2) + (n // 2)**2
```

sions by integer divisions, as shown in Listing 2.8. This forces the arguments to be integers, which prevents infinite recursions. Nevertheless, the function still does not work properly for arguments that are not powers of two. The issue with `sum_first_naturals_4` is that it is incomplete. In particular, although the recursive case is correct, it only applies when n is even. Figure 2.10 shows how to derive the recursive case ($S(n) = 2S((n-1)/2) + ((n+1)/2)^2$) for the problem and decomposition when n is odd. The complete function is therefore:

$$S(n) = \begin{cases} 1 & \text{if } n = 1, \\ 2S\left(\frac{n}{2}\right) + \left(\frac{n}{2}\right)^2 & \text{if } n > 1 \text{ and } n \text{ is even}, \\ 2S\left(\frac{n-1}{2}\right) + \left(\frac{n+1}{2}\right)^2 & \text{if } n > 1 \text{ and } n \text{ is odd}, \end{cases}$$

and the corresponding code is shown in Listing 2.9. With the new recursive case, every argument to function `sum_first_naturals_5` will also be an integer (given an initial integer input). Finally, replacing the

Listing 2.9 Python code for adding the first n positive numbers based on using two subproblems of (roughly) half the size as the original.

```
def sum_first_naturals_5(n):
    if n == 1:
        return 1
    elif n % 2 == 0:
        return 2 * sum_first_naturals_5(n / 2) + (n / 2)**2
    else:
        return (2 * sum_first_naturals_5((n - 1) / 2)
                + ((n + 1) / 2)**2)
```

real divisions (/) by integer divisions (//) would also lead to a correct algorithm.

2.7 EXERCISES

Exercise 2.1 — Let n be some positive integer. Consider the problem of determining the number of bits set to 1 in the binary representation of n (i.e., n expressed in base 2). For example, for $n = 25_{10} = 11001_2$ (the subscript indicates the base in which a number is expressed), the result is three bits set to 1. Indicate the size of the problem and provide a mathematical expression for such size.

Exercise 2.2 — Consider the function that adds the first n positive integers in (1.5). Define a more general function that it is also applicable to all nonnegative integers. In other words, modify the function considering that it can also receive $n = 0$ as an input argument. Finally, code the function.

Exercise 2.3 — Use similar diagrams as in Figures 2.6 and 2.10 in order to derive recursive definitions of the sum of the first n positive integers ($S(n)$), where the recursive case will add the result of four subproblems of (roughly) half the size as the original. Finally, define and code the full recursive function.

Exercise 2.4 — Consider the problem of printing the digits of some nonnegative integer n on the console vertically, in "normal" order, where the most significant digit must appear on the first line, the second most significant digit on the second line, and so on. For instance, if $n = 2743$ the program must print the following lines on the console:

2¶
7¶
4¶
3¶

Indicate the size of the problem and its base case. Draw a diagram for a general nonnegative input integer $n = d_{m-1} \cdots d_1 d_0$, where m is the number of digits of n, in order to illustrate the decomposition of the problem, and how to recover the solution to the problem given the result of a subproblem. Finally, derive the recursive case and code the method.

Exercise 2.5 — Define a general diagram when using a divide and conquer approach for the problem of calculating the largest value in a list **a** of n elements. Use appropriate general notation instead of concrete examples.

Exercise 2.6 — Define a recursive function that calculates the largest value in a list **a** of n elements, where the decomposition simply reduces the size of the problem by a unit.

CHAPTER 3

Runtime Analysis of Recursive Algorithms

The faster you go, the shorter you are.

— Albert Einstein

ALGORITHM analysis is the field that studies how to theoretically estimate the resources that algorithms need in order to solve computational problems. This chapter focuses on analyzing the runtime, also denoted as "computational time complexity," of recursive algorithms that solve problems whose size depends on a single factor (which occurs in the majority of the problems covered in the book). This will provide a context that will enable us to characterize and compare different algorithms regarding their efficiency. In particular, the chapter describes two methods for solving recurrence relations, which are recursive mathematical functions that describe the computational cost of recursive algorithms. These tools are used in order to transform a recurrence relation into an equivalent nonrecursive function that is easier to understand and compare. In addition, the chapter provides an overview of essential mathematical fundamentals and notation used in the analysis of algorithms. Lastly, the analysis of memory (storage) or space complexity of recursive algorithms will be covered in Chapter 10.

3.1 MATHEMATICAL PRELIMINARIES

This section presents a brief introduction to basic mathematical definitions and properties that appear in the analysis of the computational complexity of algorithms, and in several problems covered throughout the book.

3.1.1 Powers and logarithms

The following list reviews essential properties of powers and logarithms:

- $b^1 = a$
- $b^0 = 1$
- $b^x b^y = b^{x+y}$
- $(b^x)^y = b^{xy} = (b^y)^x$
- $b^{-x} = 1/b^x$
- $(ab)^x = a^x b^x$
- $\log_b b = 1$
- $\log_b 1 = 0$
- $\log_b(xy) = \log_b(x) + \log_b(y)$
- $\log_b(x/y) = \log_b(x) - \log_b(y)$
- $\log_b(x^y) = y \log_b x$
- $\log_b x = \log_a x / \log_a b$
- $\log_b a = 1/\log_a b$
- $x^{\log_b y} = y^{\log_b x}$
- $\log_b(b^x) = x$
- $b^{\log_b a} = a$

where a, b, x, and y are arbitrary real numbers, with the exceptions that: (1) the base of a logarithm must be positive and different than 1, (2) a logarithm is only defined for positive numbers, and (3) the denominator in a fraction cannot be 0. For example, $\log_b x = \log_a x / \log_a b$ is only valid for $a > 0$ with $a \neq 1$, $b > 0$ with $b \neq 1$, and $x > 0$.

Logarithms and powers of positive numbers are monotonically increasing functions. Therefore, if $x \leq y$ then $\log_b x \leq \log_b x$, and $b^x \leq b^y$ (for valid values of x, y, and b).

3.1.2 Binomial coefficients

A binomial coefficient, denoted as $\binom{n}{m}$, is an integer that appears in the polynomial expansion of the binomial power $(1+x)^n$. It can be defined as:

$$\binom{n}{m} = \begin{cases} 1 & \text{if } m = 0 \text{ or } n = m, \\ \frac{n!}{m!(n-m)!} & \text{otherwise,} \end{cases} \tag{3.1}$$

where n and m are integers that satisfy $n \geq m \geq 0$. In addition, a binomial coefficient can be defined recursively through:

$$\binom{n}{m} = \begin{cases} 1 & \text{if } m = 0 \text{ or } n = m \\ \binom{n-1}{m-1} + \binom{n-1}{m} & \text{otherwise.} \end{cases} \tag{3.2}$$

Binomial coefficients play an important role in combinatorics. In particular, $\binom{n}{m}$ determines the number of ways it is possible to choose m elements from a set of n distinct ones, where the order of selection does not matter.

3.1.3 Limits and L'Hopital's rule

The computational cost of different algorithms can be compared through limits involving a quotient of functions. Firstly,

$$\frac{k}{\infty} = 0, \qquad \text{and} \qquad \frac{\infty}{k} = \infty,$$

where k is a constant, and where ∞ should be understood as the result of a limit. In addition, since functions that measure computational cost are generally increasing (with the exception of constants), the limit when their input approaches infinity will also be infinity. For example:

$$\lim_{n\to\infty} \log_b n = \infty, \qquad \text{or} \qquad \lim_{n\to\infty} n^a = \infty,$$

for a valid base b for the logarithm, and for $a > 0$. Therefore, the following indeterminate form appears frequently:

$$\lim_{n\to\infty} \frac{f(n)}{g(n)} = \frac{\infty}{\infty}.$$

It is usually solved by simplifying the fraction $f(n)/g(n)$ until it is possible to obtain a result that is not an indeterminate form. A well-known approach for simplifying the fraction is to use L'Hopital's rule:

$$\lim_{n\to\infty} \frac{f(n)}{g(n)} = \lim_{n\to\infty} \frac{f'(n)}{g'(n)}, \tag{3.3}$$

where $f'(n)$ and $g'(n)$ are the derivatives of $f(n)$ and $g(n)$, respectively. Formally, L'Hopital's rule is only valid when the limit on the right-hand side of (3.3) exists, which is usually the case.

3.1.4 Sums and products

It is important to be familiarized with sums since they not only appear in function definitions and formulas that we may need to code, but also when analyzing the efficiency of both iterative and recursive algorithms. A sum, or summation, is simply the addition of a collection of mathematical entities (integers, real numbers, vectors, matrices, functions, etc.). When a sum involves adding some function $f(i)$, evaluated at consecutive integer values from an initial integer index m up to a final index n, it can be expressed compactly by using "sigma notation":

$$\sum_{i=m}^{n} f(i) = f(m) + f(m+1) + \cdots + f(n-1) + f(n). \tag{3.4}$$

Thus, the result of the sum is simply the addition of terms that arise from substituting every occurrence of i in function $f(i)$ with integers from m up to n. For instance:

$$\sum_{i=0}^{4} ki^2 = k \cdot 0^2 + k \cdot 1^2 + k \cdot 2^2 + k \cdot 3^2 + k \cdot 4^2,$$

where $f(i) = ki^2$.

It is important to understand that the result of the sum does not depend on the index variable. For example, the sum in (3.4) logically depends on the function f, and on the particular limits m and n, but not on any specific value of i. For instance, notice that i does not appear on the right-hand side of (3.4). Therefore, the integer index is simply a "dummy" variable that allows us to indicate all of the terms that need to be added with a single expression. In this regard, we can choose any name for the index variable, say j or k:

$$\sum_{i=m}^{n} f(i) = \sum_{j=m}^{n} f(j) = \sum_{k=m}^{n} f(k).$$

Furthermore, we can express the same sum in different ways by performing a change of variable:

$$\sum_{i=m}^{n} f(i) = \sum_{i=m-1}^{n-1} f(i+1) = \sum_{i=m}^{n} f(n-i+m).$$

In the second sum the limits (and parameter of f) appear shifted. The third sum simply adds the terms in "reverse" order (when $i = m$ it adds $f(n)$, while when $i = n$ the term added is $f(m)$). Finally, if the lower limit m is greater than the upper limit n, then the sum evaluates to 0, by convention.

3.1.4.1 Basic properties of sums

The following basic properties are useful for simplifying and working with sums, and can be derived easily from the addition and multiplication properties of the type of elements to be added. Firstly,

$$\sum_{i=1}^{n} 1 = \underbrace{1 + 1 + \cdots + 1 + 1}_{n \text{ times}} = n.$$

Notice that $f(i)$ is a constant (1) that does not depend on the index variable i. Similarly,

$$\sum_{i=1}^{n} k = \underbrace{k + k + \cdots + k + k}_{n \text{ times}} = kn.$$

The previous example also illustrates that when a multiplicative term in f does not depend on the index variable it can be "pulled out" of the sum:

$$\sum_{i=1}^{n} k = \sum_{i=1}^{n} (k \cdot 1) = \sum_{i=1}^{n} k \cdot 1 = k \cdot \left(\sum_{i=1}^{n} 1 \right) = k \sum_{i=1}^{n} 1 = kn. \quad (3.5)$$

In this case the constant term k, which we can consider is multiplied by 1, does not depend on i, and can be pulled out of the sum, where it appears multiplying it (regarding notation, (3.5) shows that it is not necessary to use parenthesis inside a sum if f does not contain additive terms, nor if some term multiplies a sum). In general, the property can be expressed as:

$$\sum_{i=m}^{n} kf(i) = k \sum_{i=m}^{n} f(i),$$

which follows from the distributive law of multiplication over addition, where the expression has been simplified by extracting the common factor k from all of the terms being added. Naturally, the factor k can be the product of several terms that do not depend on the index, and may contain the upper and lower limits, as shown in the following example:

$$\sum_{i=m}^{n} amn^2 i^3 = amn^2 \sum_{i=m}^{n} i^3,$$

where a is some constant.

Finally, sums where the function f contains several additive terms can be broken up into simpler individual sums. Formally,

$$\sum_{i=m}^{n} \left(f_1(i) + f_2(i) \right) = \sum_{i=m}^{n} f_1(i) + \sum_{i=m}^{n} f_2(i).$$

3.1.4.2 Arithmetic series

An arithmetic series is the sum of the terms in a sequence s_i, for $i = 0, 1, 2, \ldots$, in which each term is equal to the previous one plus a particular constant d (which could be negative). In other words, $s_i = s_{i-1} + d$, for $i > 0$, which is a recursive definition. These terms can also be described nonrecursively as follows:

$$s_i = s_{i-1} + d = s_{i-2} + 2d = \cdots = s_0 + id. \tag{3.6}$$

While technically an arithmetic series is an infinite sum:

$$\sum_{i=0}^{\infty} s_i,$$

when analyzing the runtime of algorithms the quantities of interest only add the terms of finite sequences:

$$\sum_{i=m}^{n} s_i,$$

also referred to as partial sums.

The sum of the first n positive integers ($S(n)$) is used frequently when analyzing the efficiency of recursive and iterative algorithms, and can be understood as a partial sum (of n elements) of the arithmetic series with $s_0 = 1$, and $d = 1$:

$$\begin{aligned} S(n) &= \sum_{i=0}^{n-1} s_i = \sum_{i=0}^{n-1} (s_0 + id) = \sum_{i=0}^{n-1} (1 + i) \\ &= \sum_{i=0}^{n-1} 1 + \sum_{i=0}^{n-1} i = n + \sum_{i=0}^{n-1} i = \sum_{i=1}^{n} i. \end{aligned}$$

As shown in Section 1.4.1, its result can be expressed as a quadratic polynomial (see (1.12)). A simple way to derive the formula consists of adding two sums $S(n)$, one expressed with terms in increasing order, and the other in decreasing order:

$$\begin{array}{rrcccccccc} & S(n) & = & 1 & + & 2 & + \cdots + & (n-1) & + & n \\ + & S(n) & = & n & + & (n-1) & + \cdots + & 2 & + & 1 \\ \hline & 2S(n) & = & (n+1) & + & (n+1) & + \cdots + & (n+1) & + & (n+1) \end{array}$$

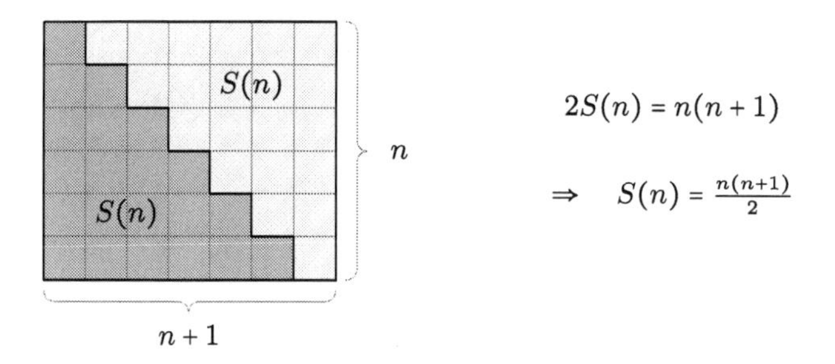

Figure 3.1 Graphical mnemonic for determining the quadratic formula for the sum of the first n positive integers ($S(n)$).

It follows from the result that $2S(n) = n(n+1)$, since there are n terms (each equal to $n+1$) on the right-hand side of the identity. Finally, dividing by 2 yields:

$$S(n) = \sum_{i=1}^{n} i = \frac{n(n+1)}{2}.$$

The idea can also be understood through the graphical mnemonic in Figure 3.1, where $S(n)$ is the area of a triangular pyramid of square (1×1) blocks. Two of these triangular structures can be joined together in order to form a rectangle of area $n \times (n+1)$. Therefore, it follows that the area ($S(n)$) of each triangular structure is $n(n+1)/2$.

A similar formula can be obtained for the general partial sum of n terms of an arithmetic sequence ($s_i = s_{i-1} + d$, for some initial s_0):

$$\sum_{i=0}^{n-1} s_i = \frac{n}{2}\left(s_0 + s_{n-1}\right),$$

which is the average between the first and last elements of the sequence, multiplied by the number of elements in the sequence.

3.1.4.3 Geometric series

A geometric series is the sum of the terms in a sequence s_i, for $i = 0, 1, 2, \ldots$, in which each term is equal to the previous one times a particular constant r. In other words, $s_i = r \cdot s_{i-1}$, for $i > 0$. These terms can also be described nonrecursively as follows:

$$s_i = r \cdot s_{i-1} = r^2 \cdot s_{i-2} = \cdots = r^i \cdot s_0. \tag{3.7}$$

The partial sum of a geometric series can be obtained through the following formula:

$$\sum_{i=m}^{n} s_i = \sum_{i=m}^{n} r^i \cdot s_0 = s_0 \sum_{i=m}^{n} r^i = s_0 \frac{r^m - r^{n+1}}{1-r}, \tag{3.8}$$

for $r \neq 1$. In practice the last equality can be derived easily through the following approach. First, let S represent the value of the sum. Subsequently, create another sum by multiplying S by r. Finally, subtract both sums (where most terms cancel out), and solve for S. The process (ignoring s_0) can be illustrated as follows:

$$\begin{array}{rclccccccccc} & S & = & r^m & + r^{m+1} & + & r^{m+2} & + \ldots + & r^{n-1} & + & r^n & \\ - & rS & = & & + r^{m+1} & + & r^{m+2} & + \ldots + & r^{n-1} & + & r^n & + r^{n+1} \\ \hline & S - rS & = & r^m & & & & & & & & - r^{n+1} \end{array}$$

Therefore, solving for S yields:

$$S = \sum_{i=m}^{n} r^i = \frac{r^m - r^{n+1}}{1-r} = \frac{r^{n+1} - r^m}{r-1}. \tag{3.9}$$

Finally, a geometric series converges to a constant value if $|r| < 1$:

$$\sum_{i=0}^{\infty} ar^i = a\frac{1}{1-r}.$$

3.1.4.4 Differentiation

Another useful sum is:

$$S = \sum_{i=1}^{n} ir^i = 1r^1 + 2r^2 + 3r^3 + \cdots + nr^n, \tag{3.10}$$

which can be interpreted as a hybrid between arithmetic and geometric series. It is possible to derive a formula for the sum as follows. Firstly, consider a general partial sum of a geometric series and its corresponding simplified formula:

$$T = 1 + r + r^2 + \ldots + r^n = \frac{r^{n+1} - 1}{r-1}.$$

We then differentiate with respect to r:

$$\frac{dT}{dr} = 1 + 2r + 3r^2 + \ldots + nr^{n-1} = \frac{nr^{n+1} - (n+1)r^n + 1}{(r-1)^2}, \tag{3.11}$$

and multiply by r in order to obtain a formula for (3.10):

$$S = \sum_{i=1}^{n} ir^i = r + 2r^2 + \ldots + nr^n = r\frac{nr^{n+1} - (n+1)r^n + 1}{(r-1)^2}. \tag{3.12}$$

We can use these formulas in order to simplify similar sums. For example, let

$$S = \sum_{i=1}^{N+1} i2^{i-1} = 1 \cdot 1 + 2 \cdot 2 + 3 \cdot 4 + 4 \cdot 8 + \cdots + (N+1) \cdot 2^N.$$

It is a special case of (3.11) for $r = 2$ and $n = N + 1$. Substituting, we obtain:

$$\begin{aligned} S = \frac{(N+1)2^{N+2} - (N+2)2^{N+1} + 1}{1} &= 2^{N+1}(2N + 2 - N - 2) + 1 \\ &= N2^{N+1} + 1. \end{aligned}$$

3.1.4.5 Products

Similarly to the notation used for sums, a product of several terms of some function $f(i)$, evaluated at consecutive integer values from an initial index m up to a final index n, can be written as follows:

$$\prod_{i=m}^{n} f(i) = f(m) \cdot f(m+1) \cdot \cdots \cdot f(n-1) \cdot f(n), \tag{3.13}$$

where by convention the product is 1 if $m > n$. For example, the factorial function can be expressed as:

$$n! = \prod_{i=1}^{n} i = 1 \cdot 2 \cdot \cdots \cdot (n-1) \cdot n,$$

where for mathematical convenience $0! = 1$.

Similarly to sums, in products multiplicative terms that do not depend on the index variable can also be pulled out of the product. However, if the product involves n terms, they have to be raised to the power of n:

$$\prod_{i=1}^{n} kf(i) = k^n \prod_{i=1}^{n} f(i).$$

In addition, the product of a sum is not the sum of the products in general:

$$\prod_{i=m}^{n} \left(f_1(i) + f_2(i)\right) \neq \prod_{i=m}^{n} f_1(i) + \prod_{i=m}^{n} f_2(i).$$

Lastly, the logarithm of a product is a sum of logarithms:

$$\log\left(\prod_{i=m}^{n} f(i)\right) = \sum_{i=m}^{n} \log f(i).$$

3.1.5 Floors and ceilings

The floor of a real number x, denoted as $\lfloor x \rfloor$, is the largest integer that is less than or equal to x. Similarly, the ceiling of x, denoted as $\lceil x \rceil$, is the smallest integer that is greater than or equal to x. Formally, they can be described as:

$$\lfloor x \rfloor = \max\{m \in \mathbb{Z} \mid m \leq x\},$$

$$\lceil x \rceil = \min\{m \in \mathbb{Z} \mid m \geq x\},$$

where $\mathbb{Z}$ represents the set of all integers, and $\mid$ can be read as "such that." The following list includes several basic properties of floors and ceilings:

- $\lfloor x \rfloor \leq x$
- $\lceil x \rceil \geq x$
- $\lfloor x + n \rfloor = \lfloor x \rfloor + n$
- $\lceil x + n \rceil = \lceil x \rceil + n$
- $\lfloor x \rfloor + \lfloor y \rfloor \leq \lfloor x + y \rfloor$
- $\lceil x \rceil + \lceil y \rceil \geq \lceil x + y \rceil$
- $n = \lfloor n/2 \rfloor + \lceil n/2 \rceil$
- $n - 2\lfloor n/2 \rfloor = 0 \Leftrightarrow n$ is even
- n // $2 = \lfloor n/2 \rfloor$
- n >> $m = \lfloor n/2^m \rfloor$
- $\lfloor \log_{10} p \rfloor + 1 =$ the number of decimal digits of p
- $\lfloor \log_2 p \rfloor + 1 =$ the number of binary digits (bits) of p

where x is a real number, n is an integer, m is a nonnegative integer, and p is a positive integer. In addition, // and >> represent operators in Python (and also in other programming languages) that compute the quotient of an integer division, and perform a right bit-shift, respectively. Finally, $\Leftrightarrow$ denotes "if and only if."

3.1.6 Trigonometry

Consider the right triangle in Figure 3.2. The following list reviews basic trigonometric definitions and properties:

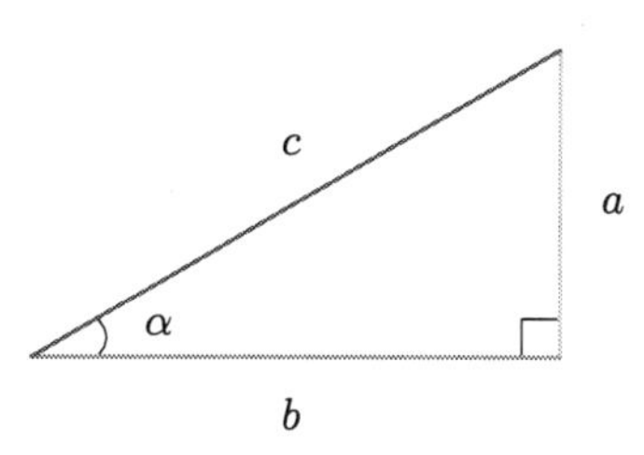

Figure 3.2 Right triangle used for showing trigonometric definitions.

- $\sin(\alpha) = a/c$
- $\cos(\alpha) = b/c$
- $\tan(\alpha) = \sin(\alpha)/\cos(\alpha) = a/b$
- $\cot(\alpha) = \cos(\alpha)/\sin(\alpha) = b/a$
- $\sin(0) = 0$
- $\cos(0) = 1$
- $\sin(30°) = \sin(\pi/6) = 1/2$
- $\cos(30°) = \cos(\pi/6) = \sqrt{3}/2$
- $\sin(45°) = \sin(\pi/4) = \sqrt{2}/2$
- $\cos(45°) = \cos(\pi/4) = \sqrt{2}/2$
- $\sin(60°) = \sin(\pi/3) = \sqrt{3}/2$
- $\cos(60°) = \cos(\pi/3) = 1/2$
- $\sin(90°) = \sin(\pi/2) = 1$
- $\cos(90°) = \cos(\pi/2) = 0$
- $\sin(\alpha) = -\sin(-\alpha)$
- $\cos(\alpha) = \cos(-\alpha)$

where sin, cos, tan, and cot denote sine, cosine, tangent, and cotangent, respectively. In most programming languages, the arguments of the trigonometric functions are in radians (one radian is equal to $180/\pi$ degrees).

3.1.7 Vectors and matrices

A **matrix** is a collection of numbers (when programming they can be other data types such as characters, Boolean values, etc.) that are arranged in rows and columns forming a rectangular structure. Formally, an $n \times m$-dimensional matrix $\mathbf{A}$ contains n rows and m columns of numbers, for $n \geq 1$ and $m \geq 1$. Typically, matrices are written within brackets or parentheses:

$$\mathbf{A} = \begin{bmatrix} a_{1,1} & a_{1,2} & \cdots & a_{1,m} \\ a_{2,1} & a_{2,2} & \cdots & a_{2,m} \\ \vdots & \vdots & \ddots & \vdots \\ a_{n,1} & a_{n,2} & \cdots & a_{n,m} \end{bmatrix}$$

where $a_{i,j}$ denotes the particular element or entry of matrix $\mathbf{A}$ in its i-th row and j-th column.

If one of the dimensions is equal to one the mathematical entity is called a **vector**, while if both dimensions are equal to one the object is a **scalar** (i.e., a number). In this book we will use a standard notation where matrices are represented by capital boldface letters ($\mathbf{A}$), vectors through boldface lower case letters ($\mathbf{a}$), and scalars by italic lower case letters or symbols (a).

The transpose of an $n \times m$-dimensional matrix $\mathbf{A}$ is the $m \times n$-dimensional matrix $\mathbf{A}^\mathsf{T}$ whose rows are the columns of $\mathbf{A}$ (and therefore its columns are the rows of $\mathbf{A}$). For example, if:

$$\mathbf{A} = \begin{bmatrix} 3 & 4 & 2 \\ 1 & 8 & 5 \end{bmatrix}, \qquad \text{then} \qquad \mathbf{A}^\mathsf{T} = \begin{bmatrix} 3 & 1 \\ 4 & 8 \\ 2 & 5 \end{bmatrix}.$$

It is possible to add and multiply matrices (and vectors). The sum $\mathbf{A} + \mathbf{B}$ is a matrix whose entries are $a_{i,j} + b_{i,j}$. In other words, matrices (and vectors) are added entrywise. Thus, $\mathbf{A}$ and $\mathbf{B}$ must share the same dimensions. For example:

$$\begin{bmatrix} 4 \\ -1 \\ 2 \end{bmatrix} + \begin{bmatrix} 2 \\ 3 \\ -7 \end{bmatrix} = \begin{bmatrix} 6 \\ 2 \\ -5 \end{bmatrix},$$

is a basic vector sum.

Instead, multiplication is more complex. Let $\mathbf{a}$ and $\mathbf{b}$ be two n-dimensional (column) vectors. Their **dot product**, expressed as $\mathbf{a}^\mathsf{T}\mathbf{b}$ (other notations include $\vec{\mathbf{a}} \cdot \vec{\mathbf{b}}$, or $\langle \mathbf{a}, \mathbf{b} \rangle$), is defined as the sum of their entrywise products:

$$\mathbf{a}^\mathsf{T}\mathbf{b} = \sum_{i=1}^{n} a_i b_i. \tag{3.14}$$

Vectors of n components (i.e., defined in $\mathbb{R}^n$) also have a geometrical interpretation. It can be shown that:

$$\mathbf{a}^\mathsf{T}\mathbf{b} = |a| \cdot |b| \cdot \cos(\alpha), \tag{3.15}$$

where α is the angle between the vectors $\mathbf{a}$ and $\mathbf{b}$, and $|\cdot|$ denotes the Euclidean norm:

$$|a| = \sqrt{a_1^2 + \cdots + a_n^2}. \tag{3.16}$$

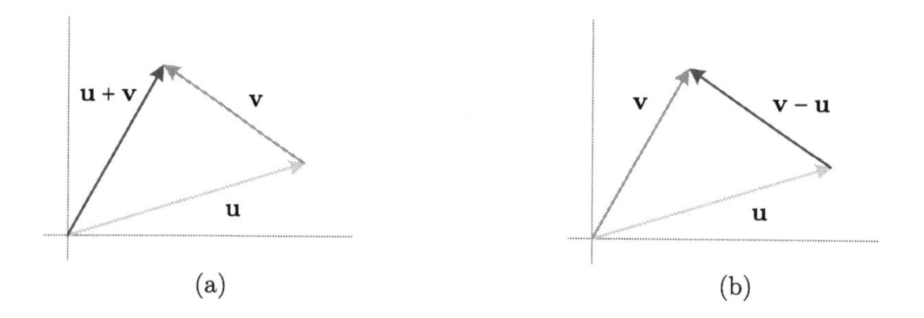

Figure 3.3 Geometric interpretation of vector addition and subtraction.

Alternatively, let **A** and **B** be $n \times m$ and $m \times p$-dimensional matrices, respectively. The product $\mathbf{C} = \mathbf{A} \cdot \mathbf{B}$ is an $n \times p$-dimensional matrix whose entries are defined as:

$$c_{i,j} = a_{i,1} \cdot b_{1,j} + a_{i,2} \cdot b_{2,j} + \cdots + a_{i,m} \cdot b_{m,j} = \sum_{k=1}^{m} a_{i,k} \cdot b_{k,j}.$$

In this case, the number of columns of **A** (i.e., m) must be the same as the number of rows of **B**. The entry $c_{i,j}$ corresponds to the dot product between the i-th row of **A**, and the j-th column of **B** (which are obviously both vectors). For example:

$$\mathbf{A}^\mathsf{T}\mathbf{A} = \begin{bmatrix} 3 & 1 \\ 4 & 8 \\ 2 & 5 \end{bmatrix} \cdot \begin{bmatrix} 3 & 4 & 2 \\ 1 & 8 & 5 \end{bmatrix} = \begin{bmatrix} 10 & 20 & 11 \\ 20 & 80 & 48 \\ 11 & 48 & 29 \end{bmatrix},$$

which is a **symmetric** matrix, since it is identical to its transpose.

Vectors can also be regarded as points. Geometrically, adding two vectors **u** and **v** can be understood as creating a new vector whose endpoint is the result of "concatenating" **u** and **v**, as shown in Figure 3.3(a). It therefore follows that the vector that begins at the endpoint of **u** and ends at the endpoint of **v** is the vector $\mathbf{v} - \mathbf{u}$, as illustrated in (b).

Lastly, a 2-dimensional vector can be rotated counterclockwise α degrees (or radians) by multiplying it times the following "rotation" matrix:

$$\mathbf{R} = \begin{bmatrix} \cos(\alpha) & -\sin(\alpha) \\ \sin(\alpha) & \cos(\alpha) \end{bmatrix}, \tag{3.17}$$

as shown in Figure 3.4, where **u** is a column vector (in most mathematical texts vectors are expressed as column vectors).

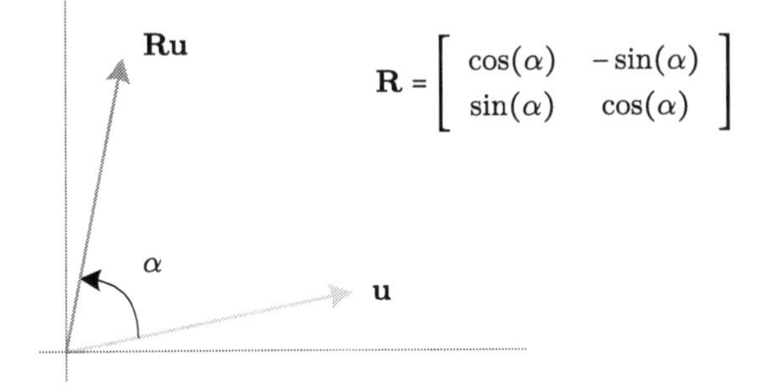

Figure 3.4 Rotation matrix (counterclockwise).

Listing 3.1 Measuring execution times through Python's `time` module.

```
import time

t = time.time()
# execute some code here
elapsed_time = time.time() - t
print(elapsed_time)
```

3.2 COMPUTATIONAL TIME COMPLEXITY

The computational time complexity of an algorithm is a theoretical measure of how much time, or how many operations, it needs to solve a problem (Listing 3.1 shows a simple way to measure execution times in practice, through Python's `time` module). It is determined by analyzing a function, say T, of the input size that quantifies this number of operations for a particular instance. In computer science the efficiency of algorithms is generally studied by contemplating how T behaves when the size of the problem is very large. Moreover, the key factor is the rate at which T grows as the input size tends to infinity. The following subsections explain these ideas and the mathematical notation typically used to characterize computational time complexity. Lastly, while the size of a problem may depend on several factors, in this introductory book we will analyze the computational time complexity of algorithms that solve problems whose size is determined by only one factor. Thus, the runtime cost of the covered algorithms will be determined by a function $T(n)$ of one parameter, where n usually represents the size of the problem.

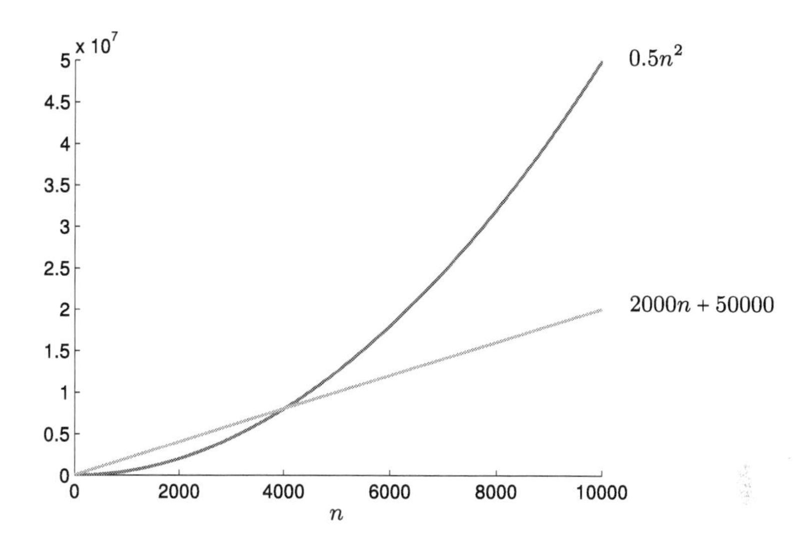

Figure 3.5 The highest-order term determines the order of growth of a function. For $T(n) = 0.5n^2 + 2000n + 50000$ the order is quadratic, since the term $0.5n^2$ clearly dominates the lower-order terms (even added up) for large values of n.

3.2.1 Order of growth of functions

The function $T(n)$ that quantifies the runtime of an algorithm may contain several additive terms that contribute differently to its value for a given input. For example, let $T(n) = 0.5n^2 + 2000n + 50000$. For moderate values of n all of the terms are relevant regarding the magnitude of $T(n)$. However, as n increases the leading term ($0.5n^2$) affects the growth of the function significantly more than the other two (even combined). Therefore, the **order of growth** of the function is characterized by its highest-order term. Figure 3.5 shows a plot of $0.5n^2$ together with $2000n + 50000$. For large inputs it is apparent that the quadratic term dominates the other two, and therefore characterizes the rate of growth of $T(n)$. The coefficients of the polynomial have been chosen in order to illustrate that they do not play a significant role in determining the function's rate of growth.

Figure 3.6 plots several orders of growth that appear frequently in computational complexity. They can be sorted as follows when considering large values of n:

$$1 \quad < \quad \log n \quad < \quad n \quad < \quad n \log n \quad < \quad n^2 \quad < \quad n^3 \quad < \quad 2^n \quad < \quad n!,$$

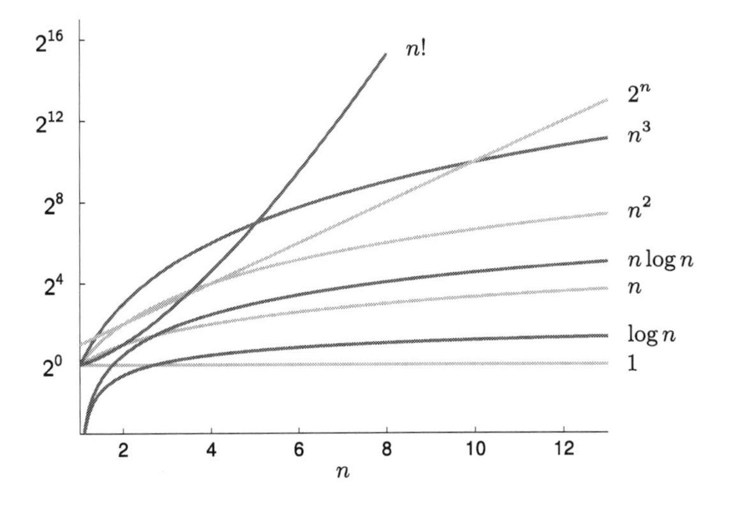

Figure 3.6 Orders of growth typically used in computational complexity.

where informally $f(n) < g(n)$ if:

$$\lim_{n \to \infty} \frac{f(n)}{g(n)} = 0.$$

The previous orders of growth are called (from left to right) "constant," "logarithmic," "linear," "n-log-n," "quadratic," "cubic," "exponential," and "factorial."

Since the scale of the Y axis in Figure 3.6 is logarithmic, the differences between the orders of growth may appear to be much smaller than

Table 3.1 Concrete values of common functions used in computational complexity.

1	$\log_2 n$	n	$n \log_2 n$	n^2	n^3	2^n	$n!$
1	0	1	0	1	1	2	1
1	1	2	2	4	8	4	2
1	2	4	8	16	64	16	24
1	3	8	24	64	512	256	40320
1	4	16	64	256	4096	65.536	$2.09 \cdot 10^{13}$
1	5	32	160	1024	32.768	4.295.967.296	$2.63 \cdot 10^{35}$

they actually are (the difference between consecutive tick marks means that an algorithm is 16 times slower/faster). Table 3.1 shows concrete values for the functions, where the fast growth rates of the exponential or factorial functions clearly stand out. Problems that cannot be solved by algorithms in polynomial time are typically considered to be intractable, since it would take too long for the methods to terminate even for problems of moderate size. In contrast, problems that can be solved in polynomial time are regarded as tractable. However, the line between tractable and intractable problems can be subtle. If the runtime of an algorithm is characterized by a polynomial order with a large degree, in practice it could take too long to obtain a solution, or for its intermediate results to be useful.

3.2.2 Asymptotic notation

An important detail about the orders mentioned so far is the lack of constant multiplicative terms. Similarly to lower-order terms, these are omitted since they are less relevant than the actual order of growth when determining the computational efficiency for large inputs. Moreover, it is not worth the effort to specify the efficiency of an algorithm with exact precision, since its runtime depends on numerous factors that include the computer's hardware, the programming language, the compiler or interpreter, and many others. Thus, it is sufficient to assume in practice that it simply takes a constant amount of time to execute a basic instruction, where its value is irrelevant.

The theoretical analysis of algorithms and problems relies on a type of notation called "asymptotic notation," which allows us to discard the lower-order terms and multiplicative constants when dealing with arbitrarily large inputs. In particular, asymptotic notation provides definitions of sets that we can use in order to specify "asymptotic bounds."

Big-O notation defines the following set:

$$\mathcal{O}(g(n)) = \left\{ f(n) : \exists\ c > 0 \text{ and } n_0 > 0 \ / \ 0 \le f(n) \le c \cdot g(n), \forall n \ge n_0 \right\}.$$

If a function $f(n)$ belongs to this set, then $g(n)$ will be an **asymptotic upper bound** for $f(n)$. This means that $g(n)$ will be greater than $f(n)$, but the definition only requires this to be true on an interval from some point $n_0 > 0$ until infinity, where in addition we can multiply $g(n)$ times a sufficiently large positive constant c. Figure 3.7(a) illustrates the idea graphically. If $g(n)$ is an asymptotic upper bound for $f(n)$ then it must be possible to find a positive constant c such that $cg(n) \ge f(n)$, but

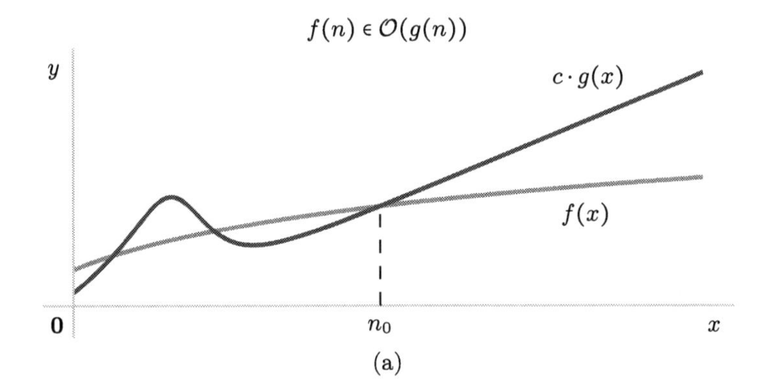

(a)

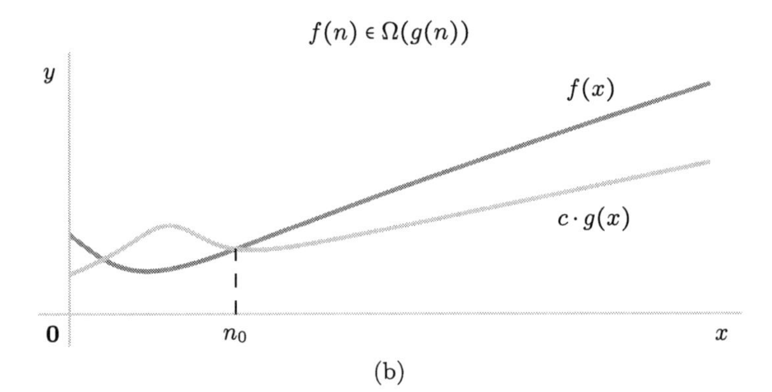

(b)

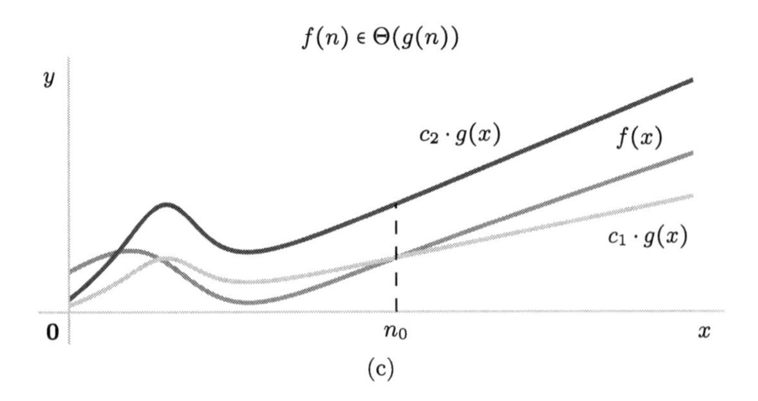

(c)

Figure 3.7 Graphical illustrations of asymptotic notation definitions for computational complexity.

from some positive value n_0 until infinity (whatever happens for $n < n_0$ is irrelevant). In order to prove that a function belongs to $\mathcal{O}(g(n))$ it is sufficient to show the existence of a pair of constants c and n_0 that will satisfy the definition, since they are not unique. For instance, if the definition is true for some particular c and n_0, then it will also be true for larger values of c and n_0. In this regard, it is not necessary to provide the lowest values of c and n_0 that satisfy the definition $\mathcal{O}$ (this also applies to the notations mentioned below).

Algorithms are often compared according to their efficiency in the **worst case**, which corresponds to an instance of a problem, amongst all that share the same size, for which the algorithm will require more resources (time, storage, etc.). Since Big-O notation specifies asymptotic upper bounds, it can be used in order to provide a guarantee that a particular algorithm will need at most a certain amount of resources, even in a worst-case scenario, for large inputs. For example, the running time for the quicksort algorithm that sorts a list or array of n elements (see Section 6.2.2) belongs to $\mathcal{O}(n^2)$ in general, since it requires carrying out on the order of n^2 comparisons in the worst case. However, quicksort can run faster (in $n \log n$ time) in the best and average cases.

In contrast, Big-Omega notation defines **asymptotic lower bounds**:

$$\Omega(g(n)) = \Big\{ f(n) : \exists\ c > 0 \text{ and } n_0 > 0 \ / \ 0 \le c \cdot g(n) \le f(n), \forall n \ge n_0 \Big\}.$$

Figure 3.7(b) illustrates the idea graphically. Big-Omega notation is useful for specifying a lower bound on the resources needed to solve a problem, no matter which algorithm is applied. For instance, it is possible to prove theoretically that any algorithm capable of sorting a list of n real numbers will require $\Omega(n \log n)$ comparisons in the worst case.

Finally, Big-Theta notation defines **asymptotic tight bounds**:

$$\Theta(g(n)) = \Big\{ f(n) : \exists\ c_1 > 0,\ c_2 > 0, \text{ and } n_0 > 0 \ / \\ 0 \le c_1 g(n) \le f(n) \le c_2 g(n), \forall n \ge n_0 \Big\}.$$

If $f(n) \in \Theta(g(n))$ then $f(n)$ and $g(n)$ will share the same order of growth. Thus, by choosing two appropriate constants c_1 and c_2, $g(n)$ will be both an upper and lower asymptotic bound of $f(n)$, as shown in Figure 3.7(c). In other words, $f(n) \in \mathcal{O}(g(n))$ and $f(n) \in \Omega(g(n))$. For example, the merge sort algorithm for sorting a list or array of n elements always requires (in the best and worst case) on the order of $n \log n$ comparisons. Therefore, we say its running time belongs to $\Theta(n \log n)$.

When specifying an asymptotic bound the constants and lower-order terms are dropped from the function $g(n)$. For example, although it is true that $3n^2 + 10n \in O(5n^2 + 20n)$, it is sufficient to indicate $3n^2 + 10n \in O(n^2)$, since these notations indicate orders of growth. In this regard, the reader may have noticed the lack of the base of a logarithm when describing the order of a function. The reason for this is that the difference between two logarithms of different bases is a constant multiplicative term. Let ρ represent some asymptotic bound, and a and b two bases for logarithms. The following identities indicate that all logarithms share the same order, regardless of their base:

$$\rho(\log_a g(n)) = \rho\left(\frac{\log_b g(n)}{\log_b a}\right) = \rho\left(\underbrace{\frac{1}{\log_b a}}_{\text{constant}} \log_b g(n)\right) = \rho(\log_b g(n)).$$

Thus, the base of a logarithm is not specified when indicating its order of growth.

Finally, we can also use limits to determine the order of functions, due to the following equivalent statements that relate them to the definitions of asymptotic bounds:

$$\begin{aligned}
f(n) \in O(g(n)) &\Leftrightarrow \lim_{n\to\infty} \frac{f(n)}{g(n)} < \infty \quad \text{(constant or zero)},\\
f(n) \in \Omega(g(n)) &\Leftrightarrow \lim_{n\to\infty} \frac{f(n)}{g(n)} > 0 \quad \text{(constant} > 0, \text{ or infinity)},\\
f(n) \in \Theta(g(n)) &\Leftrightarrow \lim_{n\to\infty} \frac{f(n)}{g(n)} = \text{constant} > 0.
\end{aligned}$$

3.3 RECURRENCE RELATIONS

The running time or number of operations carried out by a recursive algorithm is specified through a recurrence relation (or simply, "recurrence"), which is a recursive mathematical function, say T, that describes its computational cost. Consider the code in Listing 1.1 for adding the n first positive integers. Firstly, the number of operations that it needs to perform clearly depends on the input parameter n. Thus, T will be a function of n.

In the base case (when $n = 1$) the method performs the basic operations shown in Figure 3.8. Before carrying out instructions the program needs to store low-level information (e.g., regarding the parameters, or

```
def sum_first_naturals(n):                        a0
    if n==1:
        return 1
    else:
        return sum_first_naturals(n-1) + n

def sum_first_naturals(n):
    if n==1:                                      a1
        return 1
    else:
        return sum_first_naturals(n-1) + n

def sum_first_naturals(n):
    if n==1:                                      a2
        return 1
    else:
        return sum_first_naturals(n-1) + n

def sum_first_naturals(n):
    if n==1:
        return 1                                  a3
    else:
        return sum_first_naturals(n-1) + n
```

Figure 3.8 Sequence of operations carried by the function in Listing 1.1 in the base case.

the return address). Say this requires a_0 units of computing time, where a_0 is a simple constant. The next basic operation evaluates the condition, taking a_1 units of time. Since the result is `True`, the next operation is a "jump" to the third line of the method, which requires a_2 units of time. Finally, the method can return the value 1 in the last step, requiring a_3 units of time. In total, the method requires $a = a_0 + a_1 + a_2 + a_3$ units of time for $n = 1$. Thus, we can define $T(1) = a$. The exact value of a is irrelevant regarding the asymptotic computational complexity of the method. What is important is that a is a constant quantity that does not depend on n.

Alternatively, Figure 3.9 shows the operations carried out in the recursive case (when $n > 1$). Let $b = \sum_{i=0}^{5} b_i$ be the total computing time needed to carry out the basic operations (store low-level information, evaluate the condition, jump to the recursive case, subtract a unit from n, add n to the output of the recursive call, and return the result), which

```
def sum_first_naturals(n):                          b_0
    if n==1:
        return 1
    else:
        return sum_first_naturals(n-1) + n
```

```
def sum_first_naturals(n):
    if n==1:                                        b_1
        return 1
    else:
        return sum_first_naturals(n-1) + n
```

```
def sum_first_naturals(n):
    if n==1:                                        b_2
        return 1
    else:
        return sum_first_naturals(n-1) + n
```

```
def sum_first_naturals(n):
    if n==1:
        return 1
    else:
        return sum_first_naturals(n-1) + n          b_3
```

```
def sum_first_naturals(n):
    if n==1:
        return 1
    else:
        return sum_first_naturals(n-1) + n          T(n-1)
```

```
def sum_first_naturals(n):
    if n==1:
        return 1
    else:
        return sum_first_naturals(n-1) + n          b_4
```

```
def sum_first_naturals(n):
    if n==1:
        return 1
    else:
        return sum_first_naturals(n-1) + n          b_5
```

Figure 3.9 Sequence of operations carried by the function in Listing 1.1 in the recursive case.

is also a constant whose exact value is irrelevant regarding asymptotic computational complexity. In addition to b, we need to consider the time required by the recursive call. Since it solves a full problem of size $n-1$, we can define it as $T(n-1)$. Altogether, the recurrence relation $T(n)$ can be specified as follows:

$$T(n) = \begin{cases} a & \text{if } n = 1, \\ T(n-1) + b & \text{if } n > 1, \end{cases} \tag{3.18}$$

where, for example, $T(3) = T(2) + b = T(1) + b + b = a + 2b$.

Although T describes the algorithm's computational cost correctly, it is not trivial to figure out its order of growth since it is recursive. Therefore, the next step of the analysis consists of transforming the function into an equivalent nonrecursive form. This process is referred to as "solving" the recurrence relation. In this example, it is not hard to see that $T(n)$ is a linear function of n:

$$T(n) = b(n-1) + a = bn - b + a \in \Theta(n).$$

The next sections will cover methods for solving common recurrence relations.

In addition, in this introductory text we will simplify the recurrence relations in order to ease the analysis. Consider the code in Listing 2.9. Its associated runtime cost function can be defined as:

$$T(n) = \begin{cases} a & \text{if } n = 1, \\ T\left(\frac{n}{2}\right) + b & \text{if } n > 1 \text{ and } n \text{ is even}, \\ T\left(\left\lfloor\frac{n}{2}\right\rfloor\right) + c & \text{if } n > 1 \text{ and } n \text{ is odd}. \end{cases} \tag{3.19}$$

This recurrence relation is hard to analyze for two reasons. On the one hand, it contains more than one recursive case. On the other hand, although it is possible to work with the floor function, it is subject to technicalities and requires more complex mathematics (e.g., inequalities). Moreover, the extra complexity that stems from separating the odd and even cases, and dealing with a recurrence of finer detail, is unnecessary regarding the function's order of growth. Since the algorithm is based on solving subproblems of (approximately, in the odd case) half the size, we can work with the following recurrence relation instead:

$$T(n) = \begin{cases} a & \text{if } n = 1, \\ T\left(\frac{n}{2}\right) + b & \text{if } n > 1. \end{cases} \tag{3.20}$$

Using this function is analogous to utilizing (3.19) and assuming that the size of the problem (n) will be a power of two, which can be arbitrarily large.

Therefore, in this book we will cover recurrence relations with only one recursive case, which will not involve the floor or ceiling functions. This will allow us to determine exact nonrecursive definitions of recurrence relations whose order of growth can be characterized by the tight Θ asymptotic bound.

3.3.1 Expansion method

The expansion method, also known as the "iteration" or "backward substitution" method, can be used primarily to solve recurrence relations whose recursive case contains one reference to the recursive function (in some cases it can be applied to recurrences where the recursive function appears several times in the recursive definitions). The idea consists of simplifying the recurrence relation progressively step by step, until noticing a general pattern at some i-th stage. Subsequently, the function can take concrete values by considering that the base case is reached at that i-th step. The following examples illustrate the procedure. Finally, Section 10.2.1 illustrates a related visual approach called the "tree method."

3.3.1.1 Common recurrence relations

Consider the function defined in (3.18). Its recursive case:

$$T(n) = T(n-1) + b \tag{3.21}$$

can be applied repeatedly (to arguments of smaller size) in order to "expand" the T term of the right-hand side. For example, $T(n-1) = T(n-2)+b$, where all we have done is substitute n with $(n-1)$ in (3.21). Thus, we arrive at:

$$T(n) = \underbrace{[T(n-2)+b]}_{T(n-1)} + b = T(n-2) + 2b,$$

where the expression inside the square brackets is the expansion of $T(n-1)$. The idea can be applied again expanding $T(n-2)$, which is $T(n-3)+b$. Thus, at a third step we obtain:

$$T(n) = \underbrace{[T(n-3)+b]}_{T(n-2)} + 2b = T(n-3) + 3b.$$

1. Write down the recursive case of the recurrence relation.
2. Expand the recursive terms (T) on the right-hand side several times until detecting a general pattern for the i-th step.
3. Determine the value of i that allows us to reach a base case.
4. Substitute for i in the general pattern.

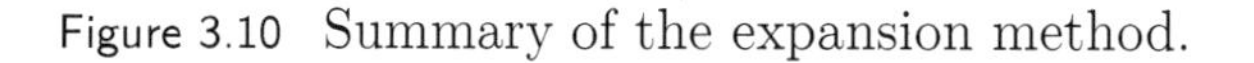

Figure 3.10 Summary of the expansion method.

After several of these expansions we should be able to detect a general pattern corresponding to the i-th step. For this function it is:

$$T(n) = T(n-i) + ib. \tag{3.22}$$

Finally, for some value of i the process will reach a base case. For function (3.18) it is defined for $T(1)$. Therefore, the term $T(n-i)$ will correspond to the base case when $n-i = 1$, or equivalently, when $i = n-1$. Substituting this result in (3.22) allows us to eliminate the variable i from the formula, and provide a full nonrecursive definition of $T(n)$:

$$T(n) = T(1) + (n-1)b = a + (n-1)b = bn - b + a \in \Theta(n).$$

Thus, Listing 1.1 runs in linear time with respect to n.

Figure 3.10 presents a summary of the expansion method that we will now apply to other common recurrence relations. Consider the function defined in (3.20). The expansion process is:

$$\begin{aligned}
T(n) &= T(n/2) + b && \text{(step 1)}\\
&= [T(n/4) + b] + b = T(n/4) + 2b && \text{(step 2)}\\
&= [T(n/8) + b] + 2b = T(n/8) + 3b && \text{(step 3)}\\
&= [T(n/16) + b] + 3b = T(n/16) + 4b && \text{(step 4)}\\
&\vdots
\end{aligned}$$

where the general pattern for the recurrence relation at step i is:

$$T(n) = T(n/2^i) + ib. \tag{3.23}$$

The base case $T(1)$ is reached when $n/2^i = 1$. Thus, it occurs when $n = 2^i$, or equivalently, when $i = \log_2 n$. By substituting in (3.23) we obtain:

$$T(n) = T(1) + b\log_2 n = a + b\log_2 n \in \Theta(\log n).$$

Since the order of growth is logarithmic, Listing 2.9 is faster than Listing 1.1, whose order is linear. This makes intuitive sense, since the former decomposes the original problem by dividing its size by two, while the latter is based on decrementing the size of the problem by a unit. Thus, Listing 2.9 needs fewer recursive function calls in order to reach the base case.

A subtle detail about recurrence relations that stem from dividing the size of a problem by an integer constant $k \geq 2$ is that they should not contain a single base case for $n = 0$. From a mathematical point of view, it would never be reached, and we would not be able to find a value for i in the method's third step. The pitfall is that the argument of $T(n)$ must be an integer. Thus, the fraction in $T(n/k)$ actually corresponds to an integer division. Notice that after reaching $T(1)$ the next expansion would correspond to $T(0)$ instead of $T(1/k)$. Therefore, for these recurrence relations we should include an additional base case, usually for $n = 1$, in order to apply the method correctly.

In the previous examples it was fairly straightforward to detect the general recursive pattern for the i-th stage of the expansion process. For the next recurrence relations the step is slightly more complex since it will involve computing sums like the ones presented in Section 3.1.4.

Consider the following recurrence relation:

$$T(n) = \begin{cases} a & \text{if } n = 0, \\ T(n-1) + bn + c & \text{if } n > 0. \end{cases} \tag{3.24}$$

The expansion process is:

$$\begin{aligned} T(n) &= T(n-1) + bn + c \\ &= [T(n-2) + b(n-1) + c] + bn + c \\ &= T(n-2) + 2bn - b + 2c \\ &= [T(n-3) + b(n-2) + c] + 2bn - b + 2c \\ &= T(n-3) + 3bn - b(1+2) + 3c \end{aligned}$$

$$\begin{aligned}
&= [T(n-4)+b(n-3)+c]+3bn-b(1+2)+3c\\
&= T(n-4)+4bn-b(1+2+3)+4c\\
&\vdots\\
&= T(n-i)+ibn-b(1+2+3+\cdots+(i-1))+ic,
\end{aligned}$$

where the square brackets indicate particular expansions of terms involving T. In addition, the last step contains the general recursive pattern, which can be written as:

$$T(n) = T(n-i)+ibn-b\sum_{j=1}^{i-1} j + ic = T(n-i)+ibn-b\frac{(i-1)i}{2}+ic. \quad (3.25)$$

Two common misconceptions when using sums in the pattern at the i-th stage are: (1) using i as the index variable of the sum, and (2) choosing n as its upper limit. It is important to note that $i-1$ is the upper limit of the sum, which implies that the index of the sum cannot be i.

Finally, the base case $T(0)$ is reached when $i=n$. Therefore, substituting in (3.25) yields:

$$T(n) = bn^2 - \frac{b}{2}n(n-1)+cn+a = \frac{b}{2}n^2+\left(c+\frac{b}{2}\right)n+a \in \Theta(n^2),$$

which is a quadratic polynomial.

The next recurrence relation appears in divide and conquer algorithms such as merge sort (see Chapter 6):

$$T(n) = \begin{cases} a & \text{if } n=1,\\ 2T(n/2)+bn+c & \text{if } n>1. \end{cases} \quad (3.26)$$

The expansion process is:

$$\begin{aligned}
T(n) &= 2T(n/2)+bn+c\\
&= 2[2T(n/4)+bn/2+c]+bn+c\\
&= 4T(n/4)+2bn+2c+c\\
&= 4[2T(n/8)+bn/4+c]+2bn+2c+c\\
&= 8T(n/8)+3bn+4c+2c+c
\end{aligned}$$

$$
\begin{aligned}
&= 8[2T(n/16) + bn/8 + c] + 3bn + 4c + 2c + c \\
&= 16T(n/16) + 4bn + 8c + 4c + 2c + c \\
&\vdots \\
&= 2^iT(n/2^i) + ibn + c(1 + 2 + 4 + \cdots + 2^{i-1}).
\end{aligned}
$$

Again, square brackets indicate expansions of terms involving T. In this case they are especially useful since each of the expanded terms needs to be multiplied by two. Students are advised to use them in general, since omitting them is a common source of mistakes.

In this case, the recursive pattern contains a partial sum of a geometric series.

$$
T(n) = 2^iT(n/2^i) + ibn + c\sum_{j=0}^{i-1} 2^j = 2^iT(n/2^i) + ibn + c(2^i - 1). \quad (3.27)
$$

Finally, we reach base case $T(1)$ when $n/2^i = 1$, which implies that $i = \log_2 n$. Therefore, substituting in (3.27) yields:

$$
T(n) = nT(1) + bn\log_2 n + c(n-1) = bn\log_2 n + n(a+c) - c \in \Theta(n\log n),
$$

whose highest-order term is $bn\log_2 n$.

3.3.1.2 Master theorem

The master theorem is a result that can be used as a quick recipe for determining the computational time complexity of algorithms based on the divide and conquer design strategy. In particular, it can be applied to recurrence relations of the following form:

$$
T(n) = \begin{cases} c & \text{if } n = 1, \\ aT(n/b) + f(n) & \text{if } n > 1, \end{cases}
$$

where $a \geq 1$, $b > 1$, $c \geq 0$, and f is an asymptotically positive function. Thus, these algorithms solve a subproblems whose size is equal to the original's divided by b. In addition, the further processing or combination of the subsolutions requires $f(n)$ operations. Depending on the relative contribution of the terms $aT(n/b)$ and $f(n)$ to the algorithm's runtime cost, the master theorem states, in its most general definition, that it is possible to determine an asymptotic tight bound for T in the following three cases:

1. If $f(n) = \mathcal{O}(n^{\log_b a-\epsilon})$ for some constant $\epsilon > 0$, then:

$$T(n) \in \Theta(n^{\log_b a}).$$

2. If $f(n) = \Theta\left(n^{\log_b a}(\log n)^k\right)$, with $k \geq 0$, then:

$$T(n) \in \Theta\left(n^{\log_b a}(\log n)^{k+1}\right).$$

3. If $f(n) = \Omega(n^{\log_b a+\epsilon})$ with $\epsilon > 0$, and $f(n)$ satisfies the regularity condition ($af(n/b) \leq df(n)$ for some constant $d < 1$, and for all n sufficiently large), then:

$$T(n) \in \Theta(f(n)).$$

For example, for

$$T(n) = \begin{cases} 1 & \text{if } n = 1, \\ T(n/2) + 2^n & \text{if } n > 1, \end{cases}$$

it is possible to find an $\epsilon > 0$ such that the order of $f(n) = 2^n$ will be greater than that of $n^{\log_2 1+\epsilon} = n^\epsilon$. Indeed, in this example any value $\epsilon > 0$ would be valid since the exponential order of 2^n is always greater than the order of a polynomial. Thus, this recurrence relation falls in the third case of the master theorem, which implies that $T(n) \in \Theta(2^n)$.

When $f(n)$ is a polynomial of degree k we can apply the following simpler version of the master theorem:

$$T(n) \in \begin{cases} \Theta(n^k) & \text{if } \frac{a}{b^k} < 1 \\ \\ \Theta(n^k \log n) & \text{if } \frac{a}{b^k} = 1 \\ \\ \Theta(n^{\log_b a}) & \text{if } \frac{a}{b^k} > 1 \end{cases} \tag{3.28}$$

This result can be derived through the expansion method. Consider the following recurrence relation:

$$T(n) = \begin{cases} c & \text{if } n = 1, \\ aT(n/b) + dn^k & \text{if } n > 1, \end{cases} \tag{3.29}$$

where $a \geq 1$, $b > 1$, $c \geq 0$, and $d \geq 0$. The expansion process is:

$$\begin{aligned}
T(n) &= aT(n/b) + dn^k \\
&= a\left[aT(n/b^2) + d(n/b)^k\right] + dn^k \\
&= a^2T(n/b^2) + dn^k(1 + a/b^k) \\
&= a^2\left[aT(n/b^3) + d(n/b^2)^k\right] + dn^k(1 + a/b^k) \\
&= a^3T(n/b^3) + dn^k(1 + a/b^k + a^2/b^{2k}) \\
&\vdots \\
&= a^iT\left(\frac{n}{b^i}\right) + dn^k \sum_{j=0}^{i-1} \left(\frac{a}{b^k}\right)^j.
\end{aligned}$$

The base case is $T(1) = c$, which is reached when $i = \log_b n$. Substituting yields:

$$T(n) = ca^{\log_b n} + dn^k \sum_{j=0}^{\log_b n-1} \left(\frac{a}{b^k}\right)^j = cn^{\log_b a} + dn^k \sum_{j=0}^{\log_b n-1} \left(\frac{a}{b^k}\right)^j,$$

where there are three cases depending on the values of a, b, and k:

1. If $a < b^k$ then $\sum_{j=0}^{\log_b n-1} (a/b^k)^j$ will be a constant term (it cannot be infinity). Notice that the infinite sum $\sum_{i=0}^{\infty} r^i = 1/(1-r)$ is a constant (i.e., it does does not diverge) for $r < 1$. Thus, in this scenario:

$$T(n) = cn^{\log_b a} + dKn^k,$$

for some constant K. In addition, since $a < b^k$ implies that $\log_b a < k$, the highest-order term is n^k. Therefore,

$$T(n) \in \Theta(n^k).$$

2. If $a = b^k$ then $T(n) = cn^k + dn^k \sum_{j=0}^{\log_b n-1} 1 = cn^k + dn^k \log_b n$, which implies that

$$T(n) \in \Theta(n^k \log n).$$

3. If $a > b^k$, then, solving the geometric series:

$$\begin{aligned} T(n) &= cn^{\log_b a} + dn^k \frac{\left(\frac{a}{b^k}\right)^{\log_b n} - 1}{\left(\frac{a}{b^k}\right) - 1} \\ &= cn^{\log_b a} + dn^k \frac{\frac{n^{\log_b a}}{n^k} - 1}{K} \\ &= \left(c + \frac{d}{K}\right) n^{\log_b a} - \frac{d}{K} n^k, \end{aligned}$$

where $K = a/b^k - 1$ is a constant. Finally, since $a > b^k$ implies that $\log_b a > k$, we have:

$$T(n) \in \Theta(n^{\log_b a}).$$

3.3.1.3 Additional examples

A recurrence relation that captures the runtime of Listings 1.2 and 2.4 could be:

$$T(n) = \begin{cases} 1 & \text{if } n = 1, \\ 2T(n/2) + 1 & \text{if } n > 1, \end{cases} \tag{3.30}$$

where we have made the assumptions that multiplicative constants are equal to 1, and that the size of the subproblems is exactly half of the original problem's. The recurrence multiplies $T(n/2)$ times two since the algorithm needs to invoke itself twice in the recursive cases, with different arguments in each call. In addition, it adds the constant 1 since the results of the subproblems need to be added and returned by the method. Lastly, for simplicity, we do not need to specify a base case for $n = 2$. These assumptions will not affect the runtime's order of growth.

The recurrence relation can be solved through the expansion method as follows:

$$\begin{aligned} T(n) &= 2T(n/2) + 1 \\ &= 2[2T(n/4) + 1] + 1 = 4T(n/4) + 2 + 1 \\ &= 4[2T(n/8) + 1] + 2 + 1 = 8T(n/8) + 4 + 2 + 1 \\ &= 8[2T(n/16) + 1] + 4 + 2 + 1 = 16T(n/16) + 8 + 4 + 2 + 1 \\ &\vdots \\ &= 2^i T(n/2^i) + \sum_{j=0}^{i-1} 2^j = 2^i T(n/2^i) + 2^i - 1. \end{aligned}$$

The base case $T(1) = 1$ occurs when $n = 2^i$. Substituting yields:

$$T(n) = n + n - 1 = 2n - 1 \in \Theta(n),$$

where $T(n)$ is a linear function of n. Naturally, this result is in accordance with the master theorem (see (3.28)), since the recurrence relation is a special case of (3.29), where $a = 2$, $b = 2$, and $k = 0$. Therefore, $T(n) \in \Theta(2^{log_2 n}) = \Theta(n)$.

Consider the functions in Listing 1.5. The first two methods decompose the problem by reducing its size by a unit. Thus, the associated recurrence relation could be:

$$T(n) = \begin{cases} 1 & \text{if } n = 1, \\ T(n-1) + 1 & \text{if } n > 1, \end{cases}$$

which is a special case of (3.21), where $T(n) \in \Theta(n)$. Instead, the third method decomposes the original problem in two of half the size, where the subsolutions do not need to be processed any further. Therefore, the corresponding recurrence relation can be identical to the one in (3.30).

Finally, we will study the computational time complexity of Listings 2.2 and 2.3. The associated recurrence relation (ignoring multiplicative constants) is:

$$T(n) = \begin{cases} 1 & \text{if } n < 10, \\ T(n/10) + 1 & \text{if } n \geq 10. \end{cases} \tag{3.31}$$

It is different than the previous recurrences since the base case is defined on an interval. Nevertheless, by assuming that the input will be a power of 10, we can use an alternative definition of the recurrence:

$$T(n) = \begin{cases} 1 & \text{if } n = 1, \\ T(n/10) + 1 & \text{if } n > 1. \end{cases} \tag{3.32}$$

Both functions are equivalent for $n = 10^p$ (and $p \geq 0$).

The second recurrence is a special case of (3.29), where $a = 1$, $b = 10$, and $k = 0$. Thus, the master theorem indicates that its complexity is logarithmic, since $T(n) \in \Theta(n^k \log n) = \Theta(\log n)$.

We can obtain this result by the expansion method as follows:

$$\begin{aligned}
T(n) &= T(n/10) + 1 \\
&= [T(n/100) + 1] + 1 = T(n/10^2) + 2 \\
&= [T(n/1000) + 1] + 2 = T(n/10^3) + 3 \\
&= [T(n/10000) + 1] + 3 = T(n/10^4) + 4 \\
&\vdots \\
&= T(n/10^i) + i.
\end{aligned}$$

The base case is obtained when $n/10^i = 1$, or in other words, when $i = \log_{10} n$. Substituting we obtain:

$$T(n) = T(1) + \log_{10} n = 1 + \log_{10} n \in \Theta(\log n).$$

3.3.2 General method for solving difference equations

The expansion method is effective at solving recurrence relations in which the recursive function appears only once in the recursive definitions. This section describes a powerful approach, denoted in this book as the "general method for difference equations," which allows us to solve recurrences that may "call" themselves several times. In particular, the method can be used to solve recurrences of the following form:

$$T(n) = \underbrace{-a_1T(n-1) - \cdots - a_kT(n-k)}_{\text{“}T\text{ difference” terms}} + \underbrace{P_1^{d_1}(n)b_1^n + \cdots + P_s^{d_s}(n)b_s^n}_{\text{polynomial } \times \text{ exponential terms}}, \quad (3.33)$$

where a_i and b_i are constants, and $P_i^{d_i}(n)$ are polynomials (of n) of degree d_i. Terms involving T may appear several times on the right-hand side of the definition. Moreover, their arguments are necessarily n minus some integer constant. Thus, these recurrence relations are also known as "difference equations." In this book we will call these terms involving T the "T difference" terms, to emphasize that the arguments cannot take the form n/b, where b is a constant (if these terms appear it is necessary to transform the recurrence). In addition, the right-hand side of the definition may contain several terms that consist of a polynomial times a power of the input n (i.e., an exponential).

Instead of providing a general procedure directly, the following subsections will explain the method progressively, starting with simple recurrences, and introducing new elements as they grow in complexity.

3.3.2.1 *Homogeneous recurrences and different roots*

Homogeneous recurrence relations only contain T difference terms:

$$T(n) = -a_1 T(n-1) - \cdots - a_k T(n-k).$$

The first step for solving them consists of passing every term to the left-hand side of the definition:

$$T(n) + a_1 T(n-1) + \cdots + a_k T(n-k) = 0.$$

Subsequently, we define its associated **characteristic polynomial** by applying the change $x^{k-z} = T(n-z)$ for $z = 0, \ldots, k$:

$$x^k + a_1 x^{k-1} + \cdots + a_{k-1} x + a_k.$$

All we have done is replace $T(n)$ with x^k, $T(n-1)$ with x^{k-1}, and so on. The coefficient (a_k) associated with the T difference term that has the smallest argument will be the polynomial's constant. The next step consists of finding the k roots of the characteristic polynomial. If r_i is its i-th root then it can be expressed in factorized form as:

$$(x - r_1)(x - r_2) \cdots (x - r_k).$$

If all of the roots are different then the nonrecursive expression for T will be:

$$T(n) = C_1 r_1^n + \cdots + C_k r_k^n. \tag{3.34}$$

Finally, the values of the constants will depend on the base cases of T, and will be found by solving a system of k linear equations with k unknown variables (the constants C_i). We will need to find k initial values (base cases, for small values of n) of T in order to construct the k equations.

The following example illustrates the approach applied to the basic Fibonacci function defined in (1.2), and coded in Listing 1.3. It can be rewritten as:

$$T(n) = \begin{cases} 0 & \text{if } n = 0, \\ 1 & \text{if } n = 1, \\ T(n-1) + T(n-2) & \text{if } n > 1, \end{cases} \tag{3.35}$$

where we have included an extra base case for $n = 0$ that will be useful shortly. The first step consists of writing each T difference term on the left-hand side of the recursive identity:

$$T(n) - T(n-1) - T(n-2) = 0.$$

It is a homogeneous recurrence since there is a 0 on the right-hand side. We then form the characteristic polynomial:

$$x^2 - x - 1.$$

The roots of this quadratic function are:

$$r_1 = \frac{1+\sqrt{5}}{2}, \qquad \text{and} \qquad r_2 = \frac{1-\sqrt{5}}{2},$$

which are different. Therefore, we can express the recurrence nonrecursively as:

$$T(n) = C_1 r_1^n + C_2 r_2^n = C_1\left(\frac{1+\sqrt{5}}{2}\right)^n + C_2\left(\frac{1-\sqrt{5}}{2}\right)^n. \tag{3.36}$$

The last step consists of finding values for the constants C_i by solving a system of linear equations. Each equation is formed by choosing a small value for n corresponding to a base case, and plugging it in (3.36). The simplest equation is obtained for $n = 0$, which is the reason why we considered the base case $n = 0$ in (3.35). For the second equation we can use $n = 1$. This provides the following system of linear equations:

$$\left.\begin{array}{rcrcccc} C_1 & + & C_2 & = & 0 & = & T(0) \\ \left(\frac{1+\sqrt{5}}{2}\right)C_1 & + & \left(\frac{1-\sqrt{5}}{2}\right)C_2 & = & 1 & = & T(1) \end{array}\right\}.$$

The solutions are:

$$C_1 = \frac{1}{\sqrt{5}}, \qquad \text{and} \qquad C_2 = -\frac{1}{\sqrt{5}}.$$

Therefore, the Fibonacci function can be expressed as:

$$T(n) = F(n) = \frac{1}{\sqrt{5}}\left(\frac{1+\sqrt{5}}{2}\right)^n - \frac{1}{\sqrt{5}}\left(\frac{1-\sqrt{5}}{2}\right)^n \in \Theta\left(\frac{1+\sqrt{5}}{2}\right)^n, \tag{3.37}$$

which is an exponential function. Observe that r_1^n clearly dominates the growth of the function. On the one hand, $r_1 > r_2$. On the other hand, $|r_2| < 1$, which means that r_2^n will approach 0 as n approaches infinity. Finally, despite its complex appearance, the result is obviously an integer.

For the next example consider the mutually recursive functions defined in (1.17) and (1.18). They must be redefined exclusively in terms of themselves in order to apply the method. On the one hand, $B(n) = A(n-1)$ implies that $B(n-1) = A(n-2)$. In addition, $A(2) = 1$. Thus, we can redefine $A(n)$ as:

$$A(n) = \begin{cases} 0 & \text{if } n = 1, \\ 1 & \text{if } n = 2, \\ A(n-1) + A(n-2) & \text{if } n \geq 3. \end{cases} \tag{3.38}$$

On the other hand, since $A(n-1) = B(n)$, and $A(n) = B(n+1)$, we can substitute in (1.17) in order to obtain $B(n+1) = B(n) + B(n-1)$. Furthermore, since $B(2) = 0$, we can define $B(n)$ as:

$$B(n) = \begin{cases} 1 & \text{if } n = 1, \\ 0 & \text{if } n = 2, \\ B(n-1) + B(n-2) & \text{if } n \geq 3. \end{cases} \tag{3.39}$$

Both of these functions have the form in (3.36). Thus, the only difference between them is the value of the constants. In particular:

$$A(n) = \left(\frac{1}{2} - \frac{1}{2\sqrt{5}}\right)\left(\frac{1+\sqrt{5}}{2}\right)^n + \left(\frac{1}{2} + \frac{1}{2\sqrt{5}}\right)\left(\frac{1-\sqrt{5}}{2}\right)^n, \tag{3.40}$$

and

$$B(n) = \left(-\frac{1}{2} + \frac{3}{2\sqrt{5}}\right)\left(\frac{1+\sqrt{5}}{2}\right)^n + \left(-\frac{1}{2} - \frac{3}{2\sqrt{5}}\right)\left(\frac{1-\sqrt{5}}{2}\right)^n, \tag{3.41}$$

where it is possible to assume that $A(0) = 1$, and $B(0) = -1$. Lastly, it is not hard to see that $A(n) + B(n)$ are Fibonacci numbers $F(n)$, since adding (3.40) and (3.41) yields (3.37).

3.3.2.2 Roots with multiplicity greater than one

The previous section showed that if the multiplicity of a root r_i is 1, corresponding to a term $(x - r_i)^1$ in the factorization of the characteristic polynomial, then the nonrecursive version of $T(n)$ will contain the term $C_i r_i^n$. In contrast, when the multiplicity m of a root r is greater than 1, resulting from $(x - r)^m$, then $T(n)$ will incorporate m terms of the form: constant × polynomial × r^n. In particular, the polynomials will be different powers of n, ranging from 1 to n^{m-1}. For example, a term

$(x-2)^4$ in the factorization of the characteristic polynomial would lead to the four following terms in the nonrecursive version of $T(n)$:

$$C_1 2^n + C_2 n 2^n + C_3 n^2 2^n + C_4 n^3 2^n,$$

for some constants C_i. Therefore, $T(n)$ can be expressed in general as:

$$T(n) = C_1 P_1(n) r_1^n + \cdots + C_k P_k(n) r_k^n, \tag{3.42}$$

where $P_i(n)$ is a polynomial of the form n^c, for some c.

For example, consider the following recurrence:

$$T(n) = 5T(n-1) - 9T(n-2) + 7T(n-3) - 2T(n-4),$$

with $T(0) = 0$, $T(1) = 2$, $T(2) = 11$, and $T(3) = 28$. It can be expressed as $T(n) - 5T(n-1) + 9T(n-2) - 7T(n-3) + 2T(n-4) = 0$, leading to following the characteristic polynomial:

$$x^4 - 5x^3 + 9x^2 - 7x + 2.$$

It can be factorized (for example, by using Ruffini's rule) as:

$$(x-1)^3(x-2),$$

which implies that the recurrence will have the following form:

$$\begin{aligned} T(n) &= C_1 \cdot 1 \cdot 1^n + C_2 \cdot n \cdot 1^n + C_3 \cdot n^2 \cdot 1^n + C_4 \cdot 1 \cdot 2^n \\ &= C_1 + C_2 n + C_3 n^2 + C_4 2^n, \end{aligned}$$

where there are three terms associated with root $r = 1$. Finally, the constants can be recovered by solving the following system of linear equations:

$$\left.\begin{array}{rcl} C_1 \qquad\qquad\qquad + \ C_4 &=& 0 = T(0) \\ C_1 + C_2 + C_3 + 2C_4 &=& 2 = T(1) \\ C_1 + 2C_2 + 4C_3 + 4C_4 &=& 11 = T(2) \\ C_1 + 3C_2 + 9C_3 + 8C_4 &=& 28 = T(3) \end{array}\right\}.$$

The solutions are $C_1 = -1$, $C_2 = -2$, $C_3 = 3$, and $C_4 = 1$ (Listing 3.2 shows how to solve the system of linear equations, expressed as $\mathbf{Ax} = \mathbf{b}$, with the NumPy package). Finally,

$$T(n) = -1 - 2n + 3n^2 + 2^n \in \Theta(2^n),$$

whose order of growth is exponential.

Listing 3.2 Solving a system of linear equations, $\mathbf{Ax} = \mathbf{b}$, in Python.

```
1 import numpy as np
2
3 A = np.array([[1, 0, 0, 1], [1, 1, 1, 2],
4               [1, 2, 4, 4], [1, 3, 9, 8]])
5 b = np.array([0, 2, 11, 28])
6 x = np.linalg.solve(A, b)
```

3.3.2.3 Nonhomogeneous recurrences

A nonhomogeneous recurrence contains nonrecursive terms on its right-hand side. We will be able to solve these recurrences when such terms consist of a polynomial times an exponential, as shown in (3.33). The procedure is exactly the same, but for every term $P_i^{d_i}(n)b_i^n$ we will need to include $(x - b_i)^{d_i+1}$ in the characteristic polynomial, where d_i is the degree of the polynomial $P_i(n)$.

Consider the following recurrence relation:

$$T(n) = 2T(n-1) - T(n-2) + 3^n + n3^n + 3 + n + n^2.$$

As in the previous examples, the first step consists of moving the T difference terms to the left-hand side of the recurrence:

$$T(n) - 2T(n-1) + T(n-2) = 3^n + n3^n + 3 + n + n^2.$$

For the next step it is useful to express the terms on the right-hand side as the product of a polynomial times an exponential. Naturally, if a term only contains an exponential then it can be multiplied by 1, which is a polynomial. Similarly, we can multiply polynomials times 1^n. Therefore, the recurrence can be written as:

$$T(n) - 2T(n-1) + T(n-2) = 1 \cdot 3^n + n \cdot 3^n + 3 \cdot 1^n + n \cdot 1^n + n^2 \cdot 1^n.$$

Furthermore, even though a particular exponential may appear several times, we must consider that it multiplies a single polynomial. Thus, the recurrence should be regarded as:

$$T(n) - 2T(n-1) + T(n-2) = (1+n) \cdot 3^n + (3+n+n^2) \cdot 1^n.$$

The next step consists of determining the characteristic polynomial. From the left-hand side we have $(x^2 - 2x + 1) = (x-1)^2$. From the term $(1+n) \cdot 3^n$ we need to include $(x-3)^2$, where the 3 is the base of the exponential, and 2 is the degree of the polynomial (1) plus one. Similarly, $(3+n+n^2)\cdot 1^n$ provides the new term $(x-1)^3$. Therefore, the characteristic polynomial is;

$$(x-1)^2(x-3)^2(x-1)^3 = (x-1)^5(x-3)^2,$$

and $T(n)$ has the following form:

$$T(n) = C_1 + C_2 n + C_3 n^2 + C_4 n^3 + C_5 n^4 + C_6 3^n + C_7 n 3^n.$$

The next example illustrates the approach with the following non-homogeneous recurrence relation:

$$T(n) = \begin{cases} 1 & \text{if } n = 0, \\ 2T(n-1) + n + 2^n & \text{if } n > 0. \end{cases}$$

In this case, we can write the recursive case as $T(n) - 2T(n-1) = n\cdot 1^n + 1\cdot 2^n$. The corresponding factorized characteristic polynomial is:

$$\underbrace{(x-2)}_{\text{from } T(n)-2T(n-1)} \cdot \underbrace{(x-1)^2}_{\text{from } n\cdot 1^n} \cdot \underbrace{(x-2)}_{\text{from } 1\cdot 2^n} = (x-1)^2(x-2)^2,$$

which implies that the recurrence has the following form:

$$T(n) = C_1 + C_2 n + C_3 2^n + C_4 n 2^n.$$

Since there are four unknown constants we need the values of T evaluated at four different inputs. Staring with the base case $T(0) = 1$, we can compute $T(1)$, $T(2)$, and $T(3)$ by using $T(n) = 2T(n-1) + n + 2^n$. In particular, $T(1) = 5$, $T(2) = 16$, and $T(3) = 43$. With this information we can build the following system of linear equations:

$$\left.\begin{array}{rcrcrcrcrcl} C_1 & & & + & C_3 & & & = & 1 & = & T(0) \\ C_1 & + & C_2 & + & 2C_3 & + & 2C_4 & = & 5 & = & T(1) \\ C_1 & + & 2C_2 & + & 4C_3 & + & 8C_4 & = & 16 & = & T(2) \\ C_1 & + & 3C_2 & + & 8C_3 & + & 24C_4 & = & 43 & = & T(3) \end{array}\right\}.$$

The solutions are $C_1 = -2$, $C_2 = -1$, $C_3 = 3$, and $C_4 = 1$. Therefore, the nonrecursive expression of $T(n)$ is:

$$T(n) = -2 - n + 3\cdot 2^n + n2^n \in \Theta(n2^n).$$

3.3.2.4 Fractional inputs

The general method for difference equations can only be applied when the inputs of terms involving T are differences of the form $n-b$, where b is some integer constant. Nevertheless, it is possible to solve some recurrence relations that contain terms where the inputs to T are fractions of the form n/b, for some constant b. The key resides in transforming the fraction into a difference through the change of variable $n = b^k$, and therefore assuming that n is a power of b.

Consider the recurrence relation in (3.30), whose recursive case is:

$$T(n) = 2T(n/2) + 1,$$

and where we can assume that the input is a power of two. Since the argument in the T term of the right-hand side is $n/2$ we need to apply the change of variable $n = 2^k$ in order to obtain:

$$T(2^k) = 2T(2^k/2) + 1 = 2T(2^{k-1}) + 1.$$

In this new definition $T(2^k)$ is a function of k. Therefore, we can substitute it with the term $t(k)$:

$$t(k) = 2t(k-1) + 1,$$

which we can solve through the general method for difference equations. In particular, the expression can be written as $t(k) - 2t(k-1) = 1 \cdot 1^n$. Therefore, the characteristic polynomial is $(x-2)(x-1)$, and the function will have the following form:

$$t(k) = C_1 + C_2 2^k. \tag{3.43}$$

By undoing the change of variable we obtain a general solution to the recurrence in terms of the original variable:

$$T(n) = C_1 + C_2 n. \tag{3.44}$$

The last step consists of determining the constants C_1 and C_2. This can be done through either (3.43) or (3.44). For T we can use the base cases $T(1) = 1$ and $T(2) = 3$. The analogous base cases for t are: $t(0) = T(2^0) = T(1) = 1$, and $t(1) = T(2^1) = T(2) = 3$. Either way, the system of linear equations is:

$$\left.\begin{array}{rcccl} C_1 + C_2 & = & 1 & = & T(1) = t(0) \\ C_1 + 2C_2 & = & 3 & = & T(2) = t(1) \end{array}\right\}.$$

The solutions are $C_1 = -1$ and $C_2 = 2$, and the nonrecursive expression of $T(n)$ is:

$$T(n) = 2n - 1 \in \Theta(n).$$

In the next recurrence relation we can assume that the input is a power of four. In particular, let:

$$T(n) = \begin{cases} 1 & \text{if } n = 1, \\ 2T(n/4) + \log_2 n & \text{if } n > 1. \end{cases}$$

In this case applying the change of variable $n = 4^k$ leads to:

$$T(4^k) = 2T(4^k/4) + \log_2 4^k = 2T(4^{k-1}) + k \log_2 4 = 2T(4^{k-1}) + 2k.$$

The next step consists of performing the substitution $t(k) = T(4^k)$:

$$t(k) = 2t(k-1) + 2k.$$

We will call this operation a "change of function" in order to emphasize that only the T terms are replaced in the expression (by the t terms). Therefore, the change does not affect the term $2k$. The recurrence can be rewritten as $t(k) - 2t(k-1) = 2k \cdot 1^k$, where the associated characteristic polynomial is $(x-2)(x-1)^2$. Thus, the recurrence has the form:

$$t(k) = C_1 2^k + C_2 + C_3 k.$$

In order to undo the change variable we can use $k = \log_4 n$, but we need an expression for 2^k. In particular, notice that $n = 4^k = (2^2)^k = (2^k)^2$. Therefore, $2^k = \sqrt{n}$, which leads to:

$$T(n) = C_1\sqrt{n} + C_2 + C_3 \log_4 n.$$

Finally, from the base case $T(1) = 1$ we can compute $T(4) = 2T(1) + \log_2 4 = 4$, and $T(16) = 2T(4) + \log_2 16 = 12$. This allows us to compute the constants C_i by solving the following system of linear equations:

$$\left.\begin{array}{rcrcrcrcr} C_1 & + & C_2 & & & = & 1 & = & T(1) \\ 2C_1 & + & C_2 & + & C_3 & = & 4 & = & T(4) \\ 4C_1 & + & C_2 & + & 2C_3 & = & 12 & = & T(16) \end{array}\right\}.$$

The solutions are $C_1 = 5$, $C_2 = -4$, and $C_3 = -2$, and the nonrecursive expression of $T(n)$ is:

$$T(n) = 5\sqrt{n} - 4 - 2\log_4 n \in \Theta(n^{1/2}).$$

When the argument of T in the recursive expression is a square root, we can use the change of variable $n = 2^{(2^k)}$. Consider the following recurrence relation:

$$T(n) = 2T(\sqrt{n}) + \log_2 n \tag{3.45}$$

where $T(2) = 1$ and $n = 2^{(2^k)}$ (note that $2^{(2^k)} \neq 4^k$). This last restriction on n guarantees that the recursive procedure will terminate at the base case for $n = 2$. Applying the change of variable, we have:

$$T(2^{(2^k)}) = 2T(2^{(2^k/2)}) + \log_2 2^{(2^k)} = 2T(2^{(2^{k-1})}) + 2^k.$$

By using the change of function $t(k) = T(2^{(2^k)})$, the recurrence is $t(k) = t(k-1) + 2^k$, whose characteristic polynomial is $(x-2)(x-2)$. Therefore, the new recurrence will have the form:

$$t(k) = C_1 2^k + C_2 k 2^k.$$

In order to undo the change of variable we can use $k = \log_2(\log_2 n)$, and $2^k = \log_2 n$. The recurrence as a function of n is therefore:

$$T(n) = C_1 \log_2 n + C_2(\log_2(\log_2 n))\log_2 n. \tag{3.46}$$

Finally, we can use the base cases $T(2) = 1$ and $T(4) = 4$ in order to find the constants. In particular, we need to solve the following system of linear equations:

$$\left.\begin{array}{rcccl} C_1 & & = & 1 & = & T(2) \\ 2C_1 & + 2C_2 & = & 4 & = & T(4) \end{array}\right\}.$$

The solutions are $C_1 = C_2 = 1$. Therefore, the final nonrecursive formula for $T(n)$ is:

$$T(n) = \log_2 n + (\log_2(\log_2 n))\log_2 n \in \Theta((\log(\log n))\log n). \tag{3.47}$$

3.3.2.5 *Multiple changes of variable or function*

The previous recurrence in (3.45) can also be solved by applying two consecutive changes of variable (and function). Firstly, the change $n = 2^k$ leads to:

$$T(2^k) = 2T(\sqrt{2^k}) + \log_2 2^k = T(2^{k/2}) + k.$$

Replacing $T(2^k)$ with $t(k)$, we obtain the following recurrence:

$$t(k) = 2t(k/2) + k,$$

which still cannot be solved through the method since it is not a difference equation. However, we can apply a new change of variable in order to transform it into one. With the change $k = 2^m$ we obtain:

$$t(2^m) = 2t(2^{m-1}) + 2^m,$$

which can be written as a difference equation by performing the change of function $u(m) = t(2^m)$:

$$u(m) = 2u(m-1) + 2^m.$$

Its characteristic polynomial is $(x-2)^2$, which implies that the recurrence has the following form:

$$u(m) = C_1 2^m + C_2 m 2^m.$$

Undoing the changes of variable leads to:

$$t(k) = C_1 k + C_2(\log_2(k))k,$$

and finally:

$$T(n) = C_1 \log_2 n + C_2(\log_2(\log_2 n)) \log_2 n,$$

which is identical to (3.46). Thus, the solution is provided in (3.47).

The strategy of using several changes of variables or functions can be used to solve more complex recurrences, such as the following nonlinear one:

$$T(n) = \begin{cases} 1/3 & \text{if } n = 1, \\ n\,[T(n/2)]^2 & \text{if } n > 1, \end{cases}$$

where n is a power of two. Firstly, we can apply the change of variable $n = 2^k$, which leads to:

$$T(2^k) = 2^k\left[T(2^k/2)\right]^2 = 2^k\left[T(2^{k-1})\right]^2.$$

In addition, with the change of function $t(k) = T(2^k)$ we obtain:

$$t(k) = 2^k\left[t(k-1)\right]^2,$$

which we still cannot solve. However, we can take logarithms on both sides of the definition:

$$\log_2 t(k) = \log_2 \left(2^k \left[t(k-1)\right]^2\right) = k + 2\log_2 t(k-1),$$

and apply the more complex change of function $u(k) = \log_2 t(k)$, which leads to:

$$u(k) = k + 2u(k-1),$$

and that we can solve through the method. In particular, its characteristic polynomial is $(x-2)(x-1)^2$, which implies that $u(k)$ has the following form:

$$u(k) = C_1 2^k + C_2 + C_3 k.$$

Undoing the changes, we first have:

$$t(k) = 2^{C_1 2^k + C_2 + C_3 k},$$

and finally:

$$T(n) = 2^{C_1 n + C_2 + C_3 \log_2 n}.$$

The last step consists of determining the constants. Using the initial recurrence with $T(1) = 1/3$, we obtain $T(2) = 2[T(1)]^2 = 2/9$, and $T(4) = 4[T(2)]^2 = 4(2/9)^2 = 16/81$. We can use these values to build the following system of (nonlinear) equations:

$$\left.\begin{array}{rcrcl} 2^{C_1+C_2} & = & 1/3 & = & T(1) \\ 2^{2C_1+C_2+C_3} & = & 2/9 & = & T(2) \\ 2^{4C_1+C_2+2C_3} & = & 16/81 & = & T(4) \end{array}\right\},$$

which can be transformed into a system of linear equations by taking logarithms on both sides of the equations:

$$\left.\begin{array}{rcr} C_1 + C_2 & = & -log_2 3 \\ 2C_1 + C_2 + C_3 & = & 1 - \log_2 9 = 1 - 2\log_2 3 \\ 4C_1 + C_2 + 2C_3 & = & 4 - \log_2 81 = 4 - 4\log_2 3 \end{array}\right\}.$$

The solutions are: $C_1 = 2 - \log_2 3 = \log_2(4/3)$, $C_2 = -2$, and $C_3 = -1$. Therefore, $T(n)$ can be expressed as:

$$T(n) = 2^{(\log_2(4/3))n - 2 - \log_2 n} = \left[2^{\log_2(4/3)}\right]^n \cdot 2^{-2} \cdot 2^{\log_2(1/n)}$$

$$= \left[\frac{4}{3}\right]^n \cdot \frac{1}{4n} \in \Theta\left(\left[\frac{4}{3}\right]^n \cdot \frac{1}{n}\right).$$

3.4 EXERCISES

Exercise 3.1 — Prove the following identity:

$$\left(\frac{a}{b^k}\right)^{\log_b n} = \frac{n^{\log_b a}}{n^k}$$

Exercise 3.2 — By using limits, show that $\log n! \in \Theta(n \log n)$. Hint: when n approaches infinity, $n!$ can be substituted by "Stirling's approximation":

$$n! \sim \sqrt{2\pi n}\left(\frac{n}{e}\right)^n.$$

Exercise 3.3 — Show that $n \log n \in \mathcal{O}(n^{1+a})$, where $a > 0$. Use limits, and L'Hopital's rule.

Exercise 3.4 — Let m and n be some integers. Determine:

$$\sum_{i=m}^{n} 1.$$

Exercise 3.5 — Write the sum of the first n odd integers in sigma notation, and simplify it (the result should be a well-known polynomial).

Exercise 3.6 — Use the following identity:

$$\sum_{i=1}^{n} \frac{i(i+1)}{2} = \sum_{i=1}^{n} i(n-i+1)$$

to provide an expression for the sum of the first n squared integers $(1^2 + 2^2 + \cdots + n^2)$.

Exercise 3.7 — Show that a general partial sum of n terms of an arithmetic sequence ($s_i = s_{i-1} + d$, for some initial s_0) follows:

$$\sum_{i=0}^{n-1} s_i = \frac{n}{2}\left(s_0 + s_{n-1}\right).$$

Exercise 3.8 — An algorithm processes some of the bits of the binary representations of numbers from 1 to $2^n - 1$, where n is the number of bits of each number. In particular, the algorithm processes the least significant bits of each number (from right to left), until it finds a bit set to 1. Given n, determine, using sums, the total number of bits that the algorithm processes. For example, for $n = 4$ the algorithm processes the 26 shaded bits in Figure 3.11.

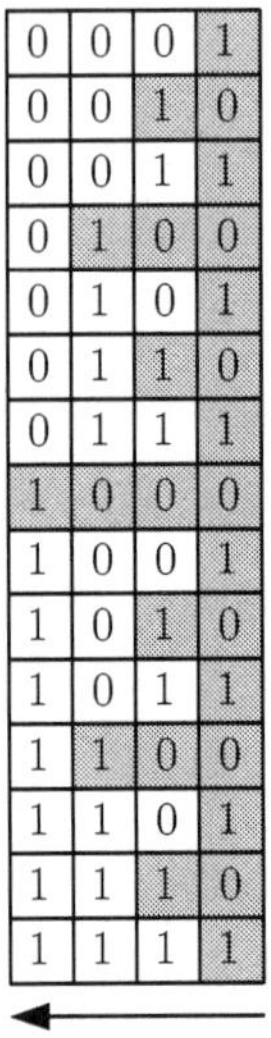

0	0	0	1
0	0	1	0
0	0	1	1
0	1	0	0
0	1	0	1
0	1	1	0
0	1	1	1
1	0	0	0
1	0	0	1
1	0	1	0
1	0	1	1
1	1	0	0
1	1	0	1
1	1	1	0
1	1	1	1

Figure 3.11 An algorithm processes the shaded bits of numbers from 1 to $2^n - 1$ (for $n = 4$).

Exercise 3.9 — Specify recurrence relations that describe the runtime of algorithms that implement functions (1.14), (1.15), (1.16), and (1.19).

Exercise 3.10 — Solve the following recurrences, but without computing multiplicative constants (C_i). Thus, in this exercise the base cases are not needed.

a) $T(n) = 4T(n-1) - 5T(n-2) + 2T(n-3) + n - 3 + 5n^2 \cdot 2^n$

b) $T(n) = T(n-1) + 3n - 3 + n^3 \cdot 3^n$

c) $T(n) = 5T(n-1) - 8T(n-2) + 4T(n-3) + 3 + n^2 + n2^n$

Exercise 3.11 — Define a recurrence relation for the runtime of Listing 2.6, then solve it through the expansion method and the general method for difference equations. Finally, determine its order of growth.

Exercise 3.12 — Solve the following recurrence relation through the expansion method:

$$T(n) = \begin{cases} 1 & \text{if } n = 0, \\ 1 + \sum_{j=0}^{n-1} T(j) & \text{if } n > 0. \end{cases}$$

Exercise 3.13 — Solve the following recurrence relation through the expansion method:

$$T(n) = \begin{cases} 1 & \text{if } n = 1, \\ 2T(n/4) + \log_2 n & \text{if } n > 1. \end{cases}$$

Exercise 3.14 — Solve the recurrence relation in (3.18) through the general method for difference equations.

Exercise 3.15 — Define a recurrence relation for the runtime of Listing 2.9, solve it through the master theorem, the expansion method, and the general method for difference equations. Finally, determine its order of growth.

Exercise 3.16 — Solve the following recurrence relation through the master theorem, the expansion method, and the general method for difference equations. Finally, determine its order of growth.

$$T(n) = \begin{cases} 1 & \text{if } n = 1, \\ 3T(n/2) + n & \text{if } n > 1, \end{cases}$$

where n is a power of 2.

Exercise 3.17 — Solve the recurrence relation in (3.32) through the general method for difference equations.

Exercise 3.18 — Solve the following recurrence relations:

a) $T(n) = 2T(n-1) + 3n - 2$, with $T(0) = 0$.

b) $T(n) = T(n/2) + n$, with $T(1) = 1$, and n is a power of 2.

c) $T(n) = T(n/\alpha) + n$, with $T(1) = 0$, and where $\alpha \geq 2$ is an integer.

d) $T(n) = T(n/3) + n^2$, with $T(1) = 1$, and n is a power of 3.

e) $T(n) = 3T(n/3) + n^2$, with $T(1) = 0$, and n is a power of 3.

f) $T(n) = 2T(n/4) + n$, with $T(1) = 1$, and n is a power of 4.

g) $T(n) = T(n/2) + \log_2 n$, with $T(1) = 1$, and n is a power of 2.

h) $T(n) = 4T(n/2) + n$, with $T(1) = 1$, and n is a power of 2.

i) $T(n) = 2T(n/2) + n\log_2 n$, with $T(1) = 1$, and n is a power of 2.

j) $T(n) = \frac{3}{2}T(n/2) - \frac{1}{2}T(n/4) - \frac{1}{n}$, with $T(1) = 1$, $T(2) = 3/2$, and n is a power of 2.

CHAPTER 4

Linear Recursion I: Basic Algorithms

Do the difficult things while they are easy and do the great things while they are small. A journey of a thousand miles must begin with a single step.

— Lao Tzu

RECURSIVE methods can be categorized according to the number of times they invoke themselves in the recursive cases. This chapter examines methods that not only call themselves just once, but also process the output of the recursive call before producing or returning their own result. This common type of recursion is known as **linear recursion**, where the single recursive call is not the last operation carried out by the method. This chapter will present numerous problems and respective linear-recursive solutions that we will design by relying on the concepts and methodology introduced in Chapters 1 and 2. Finally, if the single recursive call is the method's last action, we say we are using **tail recursion**. This special type of recursion will be covered in Chapter 5.

Linear recursion is the simplest type of recursion, and can be viewed as an alternative way to implement iterative loops that carry out a certain number of operations repeatedly. In this regard, although it is fairly straightforward to solve the problems discussed in this chapter iteratively, linear recursion provides the clearest examples of how to think and program recursively. The chapter will therefore present examples of how to apply recursive concepts (i.e., problem decomposition and induction), which the reader should comprehend before tackling more complex problems and types of recursion.

4.1 ARITHMETIC OPERATIONS

This section presents recursive solutions to several elemental arithmetic computations. We will examine them for illustration purposes, since most correspond to simple operations that can be implemented through basic commands or expressions.

4.1.1 Power function

A classical problem used to illustrate recursion is the power function. The goal is to calculate b to the power of n:

$$b^n = \prod_{i=1}^{n} b = \underbrace{b \times b \times \cdots b \times b}_{n \text{ times}},$$

where the base b is a real number, and the exponent n is a nonnegative integer (in Python, powers can be obtained through the ** operator.) The following subsections examine algorithms that can compute the power in linear and logarithmic time.

4.1.1.1 Powers in linear time

According to the methodology presented in Figure 2.1, the first step consists of determining the size of the problem, which is related to the input parameters. For this problem, only the exponent is relevant for determining the number of operations that algorithms need to solve it. Clearly, the runtime of the algorithms increases as n grows, since the result would require more multiplications. Thus, the exponent n is the size of the problem. The base case corresponds to trivial instances of the problem, with minimum size. Since n is a nonnegative integer, the smallest instance is b^0, which is equal to one.

In the decomposition step we have to consider smaller and simpler self-similar subproblems that stem from reducing the size of the problem. In this section we will simply decrement it by a single unit, which leads to the following diagram:

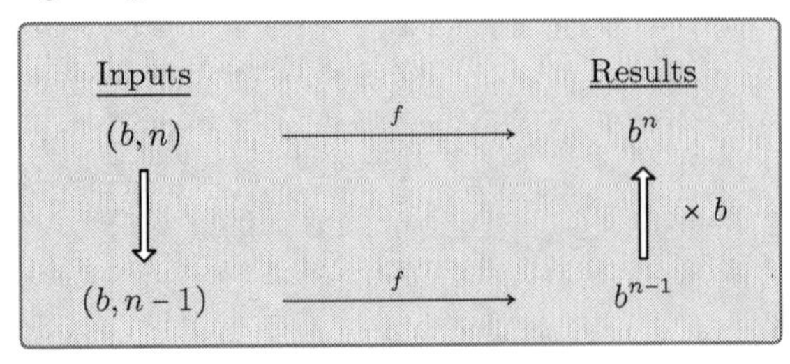

Listing 4.1 Power function in linear time for nonnegative exponents.

```
1 def power_linear(b, n):
2     if n == 0:
3         return 1
4     else:
5         return b * power_linear(b, n - 1)
```

where $f(b,n) = b^n$ is naturally a function of the two parameters b and n. It is apparent that if we assume to know the solution to $f(b,n-1) = b^{n-1}$, then all we need to do in the recursive case is multiply it times b in order to obtain b^n. Together with the base case, the function can be implemented as shown in Listing 4.1.

It is an example of linear recursion since there is only one function call in the recursive case, and the recursive call is not the last operation that the algorithm performs in the recursive case. Even though the function call is the last element in the expression (in line 5), the method first evaluates it, and then it processes it, multiplying the result times b. Thus, the function is not tail-recursive. It is also important to note that the function does not need a base case when $n = 1$, since it would be redundant. Finally, the computational complexity of this algorithm is linear with respect to n. In particular, the runtime can be defined through:

$$T(n) = \begin{cases} 1 & \text{if } n = 0, \\ T(n-1) + 1 & \text{if } n > 0, \end{cases} \tag{4.1}$$

where $T(n) = n + 1 \in \Theta(n)$.

If the problem statement allowed negative integer exponents as well, the size of the problem would be the absolute value of n. In this variant the base case also occurs when $n = 0$, and the recursive case is identical if $n > 0$. When n is negative the decomposition needs to reduce the size of the problem, ensuring that the corresponding subproblem gets closer to the base case. Thus, for this recursive case we need to increment n. The diagram that illustrates the thought process would be:

Listing 4.2 Power function in linear time.

```
1 def power_general_linear(b, n):
2     if n == 0:
3         return 1
4     elif n > 0:
5         return b * power_general_linear(b, n - 1)
6     else:
7         return power_general_linear(b, n + 1) / b
```

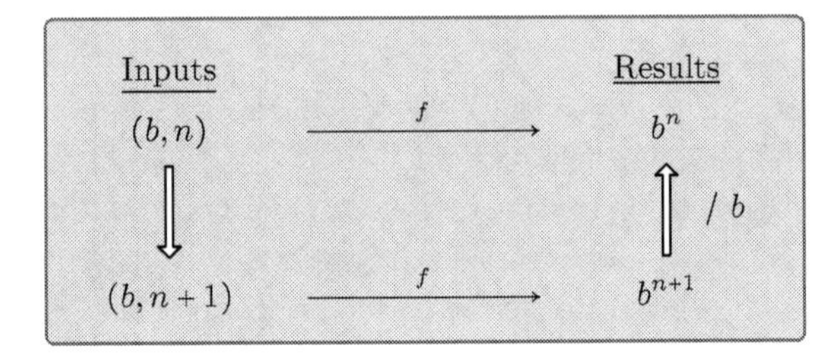

Thus, the result of the subproblem must be divided by b in order to obtain the desired power b^n. The recursive case is therefore $f(b,n) = f(b,n+1)/b$, as shown in Listing 4.2.

4.1.1.2 Powers in logarithmic time

It is possible to design more efficient algorithms that calculate powers, which run in logarithmic time, by considering a subproblem of half the size in the decomposition stage. Assuming that n is a nonnegative integer, we need two diagrams in order to account for the parity of n. If n is even, the recursive design thought process can be visualized as follows:

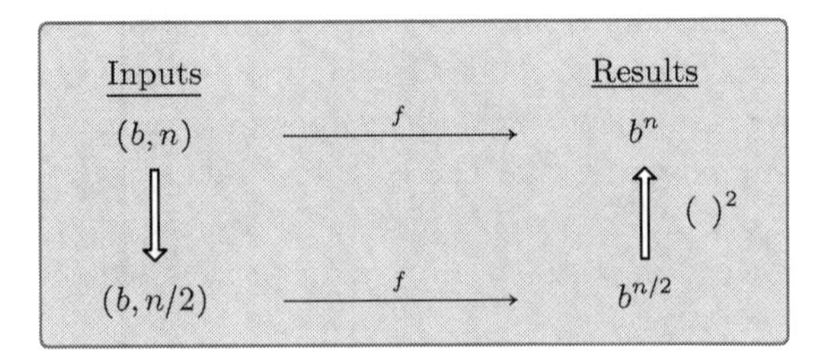

Thus, the result of the subproblem must be squared in order to obtain b^n. Instead, if n is odd, we can consider using a subproblem of size $(n-1)/2$ (which is an integer). The diagram would be:

Listing 4.3 Power function in logarithmic time for nonnegative exponents.

```
def power_logarithmic(b, n):
    if n == 0:
        return 1
    elif n % 2 == 0:
        return power_logarithmic(b, n // 2)**2
    else:
        return b * (power_logarithmic(b, (n - 1) // 2)**2)
```

Listing 4.4 Inefficient implementation of the power function that runs in linear time.

```
def power_alt(b, n):
    if n == 0:
        return 1
    elif n % 2 == 0:
        return power_alt(b, n // 2) * power_alt(b, n // 2)
    else:
        return (power_alt(b, (n - 1) // 2)
                * power_alt(b, (n - 1) // 2) * b)
```

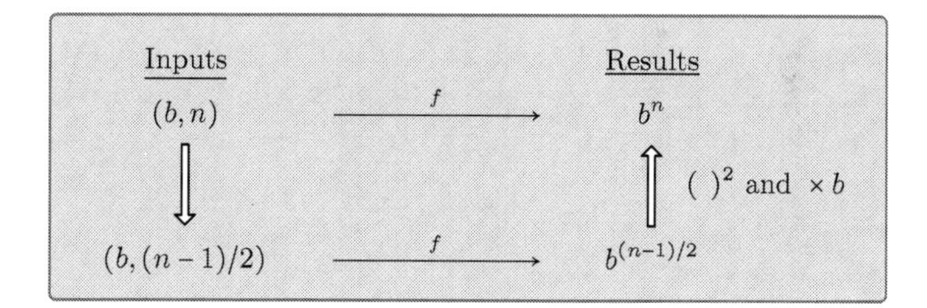

In this case, the output of the subproblem also needs to be squared, but a final multiplication times the base b is also required. With these recursive cases, and the trivial base case, the recursive function is:

$$b^n = \begin{cases} 1 & \text{if } n = 0, \\ \left(b^{n/2}\right)^2 & \text{if } n > 0 \text{ and } n \text{ is even}, \\ \left(b^{(n-1)/2}\right)^2 \cdot b & \text{if } n > 0 \text{ and } n \text{ is odd}. \end{cases}$$

A proper implementation of the function is shown in Listing 4.3, which again uses linear recursion. The use of the integer division (`//`) instead of a real division `/` is optional, but included in order to emphasize that the second input parameter is an integer. In addition, when n is odd `(n-1)//2` is equivalent to `n//2`. The code uses the former expression since it resembles the mathematical definition of the function more closely.

Its runtime can be defined through:

$$T(n) = \begin{cases} 1 & \text{if } n = 0, \\ T(n/2) + 1 & \text{if } n > 0, \end{cases}$$

where $T(n) = 2 + \log_2 n$ for $n > 0$. Thus, $T(n) \in \Theta(\log n)$. The superior performance stems from dividing the size of the problem by two in the decomposition stage. However, the function must make a single recursive call in each recursive case. For example, the code in Listing 4.4 does not run in logarithmic time even though the decomposition divides the problem size by two. The issue is that it calculates the result of the same subproblem twice by using two identical recursive calls, which is obviously unnecessary. The runtime cost for this function is:

$$T(n) = \begin{cases} 1 & \text{if } n = 0, \\ 2T(n/2) + 1 & \text{if } n > 0, \end{cases}$$

where $T(n) = n + 1 \in \Theta(n)$. Notice that it is analogous to the runtime of the function in Listing 4.1, whose runtime is described through (4.1). Thus, the extra performance that could have been gained by halving the size of the input is lost by making two identical recursive calls in the recursive cases.

4.1.2 Slow addition

This problem consists of adding two nonnegative integers a and b, where we cannot use the general arithmetic commands such as addition, subtraction, multiplication, or division. Instead, we are only allowed to add or subtract single units from the numbers (a constant amount of times). This simple problem will help illustrate how the choice of the size can affect the way we decompose a problem, which leads to different recursive algorithms.

In consonance with the template in Figure 2.1, the first step consists of determining the size of the problem. Since we are only allowed to add or subtract 1 from a and b, it is reasonable to think that we will have to perform these simple operations $a + b$ times until a base case condition is met (we will see shortly that it can be solved in only a or b operations). If we choose $a + b$ to be the size of the problem, then a base case occurs

Listing 4.5 Slow addition of two nonnegative integers.

```
1 def slow_addition(a, b):
2     if a == 0:
3         return b
4     elif b == 0:
5         return a
6     else:
7         return slow_addition(a - 1, b) + 1
```

when $a = b = 0$, which corresponds to the instance of smallest size. In addition, we have to consider other possible base cases where we can provide a result easily, without the need to carry out a recursive call. For this problem, when $a = 0$ the result is obviously b, and when $b = 0$ the output should be a. Furthermore, with these two base cases it is not necessary to include the one for $a = 0$ *and* $b = 0$. Lastly, the base cases guarantee that both a and b will be positive in the recursive case.

We could proceed in the decomposition step by reducing the size of the problem by a unit. Since our choice for the size of the problem is $a + b$, we could subtract a unit from either a or b. The recursive diagram related to decrementing a would be:

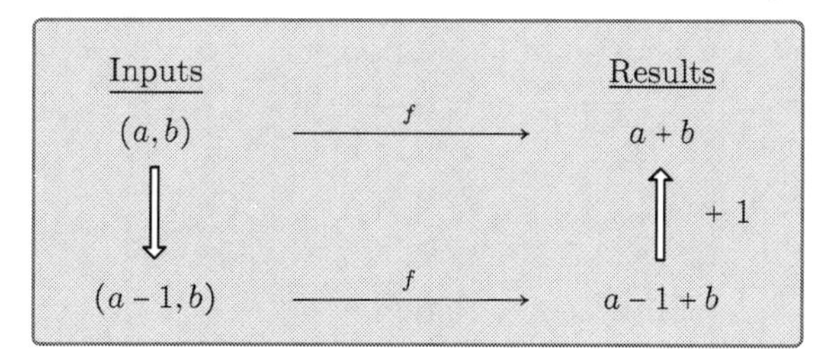

If the function we are implementing is f, then clearly $f(a,b) = f(a-1,b)+1$ defines the recursive case, and it can be expressed mathematically as:

$$f(a,b) = \begin{cases} b & \text{if } a = 0, \\ a & \text{if } b = 0, \\ f(a-1,b)+1 & \text{if } a > 0,\ b > 0. \end{cases} \quad (4.2)$$

Listing 4.5 shows an implementation of the function in Python. Had we chosen a to be the size of the problem the two base cases would also appear in the definition of the function, and we would have also decremented the value of a when decomposing the problem. Therefore,

Listing 4.6 Quicker slow addition of two nonnegative integers.

```
def quicker_slow_addition(a, b):
    if a == 0:
        return b
    elif b == 0:
        return a
    elif a < b:
        return quicker_slow_addition(a - 1, b) + 1
    else:
        return quicker_slow_addition(a, b - 1) + 1
```

the resulting function would be identical. Alternatively, choosing b as the size of the problem would imply replacing the recursive case by $f(a,b) = f(a,b-1)+1$, since we would have decremented b instead of a.

The function in Listing 4.5 can be slow if a is large. Alternatively, we can build a more efficient algorithm by considering that the size of the problem is the smallest input parameter, i.e., $\min(a,b)$. In this scenario the two base cases $f(0,b) = b$, and $f(a,0) = a$, correspond to the smallest instances of the problem ($f(0,0)$ would be redundant). For the recursive cases we must find a decomposition that guarantees reducing the size of the problem. A first approach consists of decrementing the smallest input parameter, where the size of the subproblem would be $\min(a,b)-1$. For example, if $a < b$ the recursive rule would be $f(a,b) = f(a-1,b)+1$, while if $a \geq b$ we would apply $f(a,b) = f(a,b-1)+1$. Thus, the mathematical function would be:

$$f(a,b) = \begin{cases} b & \text{if } a = 0, \\ a & \text{if } b = 0 \text{ (and } a \neq 0), \\ f(a-1,b)+1 & \text{if } a < b \text{ (and } a \neq 0,\ b \neq 0), \\ f(a,b-1)+1 & \text{if } b \leq a \text{ (and } a \neq 0,\ b \neq 0), \end{cases}$$

which can be coded as shown in Listing 4.6.

A second idea consists in decrementing both parameters, which also ensures reducing the size of the problem. In particular, note that $\min(a-1,b-1) = \min(a,b)-1$. For this decomposition the recursive diagram would be:

Listing 4.7 Alternative quicker slow addition of two nonnegative integers.

```
1 def quicker_slow_addition_alt(a, b):
2     if a == 0:
3         return b
4     elif b == 0:
5         return a
6     else:
7         return quicker_slow_addition_alt(a - 1, b - 1) + 1 + 1
```

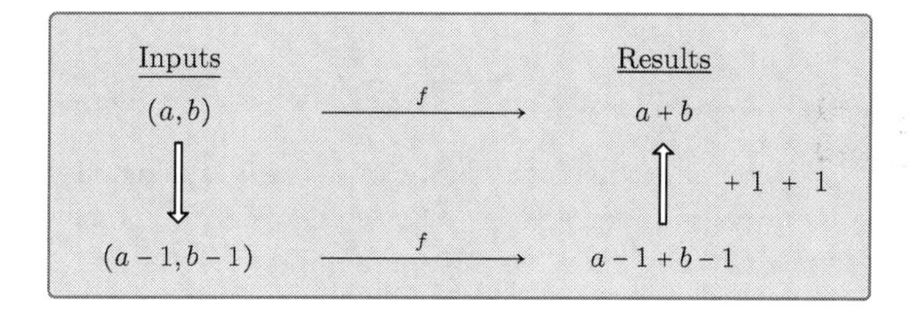

The recursive rule is therefore: $f(a,b) = f(a-1,b-1)+1+1$, for $a>0$ and $b>0$. According to the problem statement, this is valid if we assume that we can perform increments a *constant* number of times (in this case, $f(a-1,b-1)$ is incremented twice), which does not depend on the inputs. Listing 4.7 shows the associated code.

4.1.3 Double sum

The next example involves calculating the particular double sum:

$$f(m,n) = \sum_{i=1}^{m}\sum_{j=1}^{n}(i+j). \tag{4.3}$$

By using the properties of sums it can be simplified to a simple expression:

$$f(m,n) = \sum_{i=1}^{m}\sum_{j=1}^{n}(i+j) = \sum_{i=1}^{m}\left(in + \frac{n(n+1)}{2}\right) = \frac{nm(m+1)}{2} + \frac{mn(n+1)}{2}$$
$$= \frac{mn(m+n+2)}{2}.$$

However, we will develop a recursive solution that uses two recursive functions (one for each sum).

Firstly, the outer sum is a function of the parameters m and n, whose size is m. The function returns 0 in its base case when m is smaller than the lower index (i.e., when $m \leq 0$). The general diagram for the recursive case, assuming that the size of the subproblem is $m-1$, is:

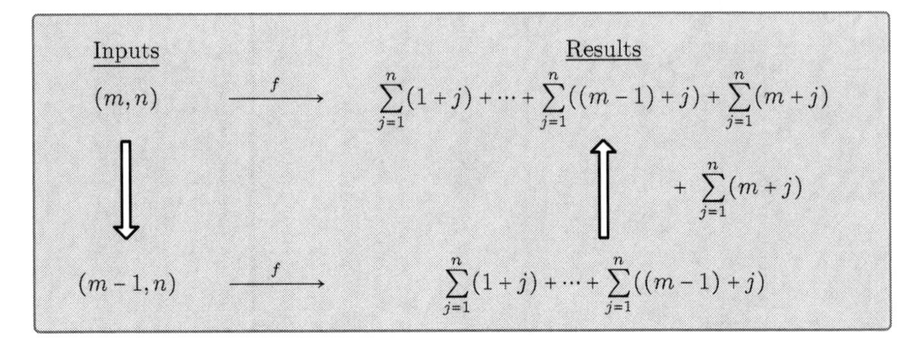

Clearly, all we have to do to the result of $f(m-1,n)$, in order to obtain $f(m,n)$, is add the inner sum $\sum_{j=1}^{n}(m+j)$, which is a function g that depends on n and m. Therefore, we can define f as follows:

$$f(m,n) = \begin{cases} 0 & \text{if } m \leq 0, \\ f(m-1,n) + g(n,m) & \text{if } m > 0. \end{cases}$$

We can also define $g(n,m)$ recursively. Firstly, we need to determine its size, which is n, since it adds n terms. In accordance with other sums, it returns 0 in the base case, which occurs when $n \leq 0$. The general diagram for its recursive case, again decrementing the size of the problem by a unit, is:

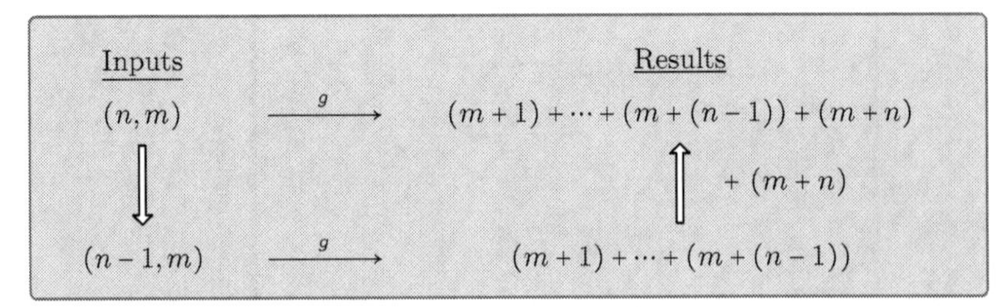

In this case, we need to add $(m+n)$ to the result of the subproblem in order to obtain $g(n,m)$. Thus, the function is:

$$g(n,m) = \begin{cases} 0 & \text{if } n \leq 0, \\ g(n-1,m) + (m+n) & \text{if } n > 0. \end{cases}$$

Finally, Listing 4.8 shows an implementation of both sums.

Listing 4.8 Recursive functions that compute the double sum in (4.3).

```
def inner_sum(n, m):
    if n <= 0:
        return 0
    else:
        return inner_sum(n - 1, m) + (m + n)


def outer_sum(m, n):
    if m <= 0:
        return 0
    else:
        return outer_sum(m - 1, n) + inner_sum(n, m)
```

4.2 BASE CONVERSION

Numbers can be represented in different ways, depending on a particular **base** or radix. Typically, we use a base-10 numeral system where a sequence of digits, say 142, represents the number $1 \cdot 10^2 + 4 \cdot 10^1 + 2 \cdot 10^0$. In general, for a particular base b, the value of a number x can be expressed as a unique sequence of m digits $d_{m-1} \cdots d_0$, where:

$$x = \sum_{i=0}^{m-1} d_i b^i, \tag{4.4}$$

with $0 \le d_i < b$, and $d_{m-1} \neq 0$ (i.e., we omit writing leading zeros). Therefore, different bases lead to distinct sequences of digits that represent the same number. Regarding notation, the base can be specified through a subscript, which we usually omit when it is 10. For example, $142_{10} = 142$, but $142_5 = 1 \cdot 5^2 + 4 \cdot 5^1 + 2 \cdot 5^0 = 25 + 20 + 2 = 47$. In this section we will examine algorithms for converting numbers expressed in some base to another one.

4.2.1 Binary representation of a nonnegative integer

In this example the goal consists of developing a recursive function that, given a particular nonnegative integer n (in base 10), returns a new integer that contains the binary representation of n, which is the sequence of base-2 digits (i.e., bits) associated with the value of n. The output will also be a number in base 10, but its digits will only be either zero or one. Thus, we can interpret that the digits actually correspond to bits.

Since we need to create the sequence of "bits," the size of the problem is the number of bits in the binary representation of n. Mathematically, this quantity is $\lfloor \log_2 n \rfloor + 1$ (for $n > 0$). However, we do not need the formula in order to design the recursive algorithm. All we require is a clear definition of the size of the problem that will enable us to define base cases and decompose the problem. In particular, the smallest instances of the problem correspond to numbers that contain a single bit, which are 0 and 1. Thus, the base case occurs when $n < 2$, where the output is simply n.

For the recursive case we need to decide how we can reduce the size of the problem. The simplest way consists of decrementing the number of bits by a single unit, which is accomplished by performing an integer division of n by two (this shifts the bits one place to the right, where the least significant bit is discarded). We can start the analysis with concrete instances. For example:

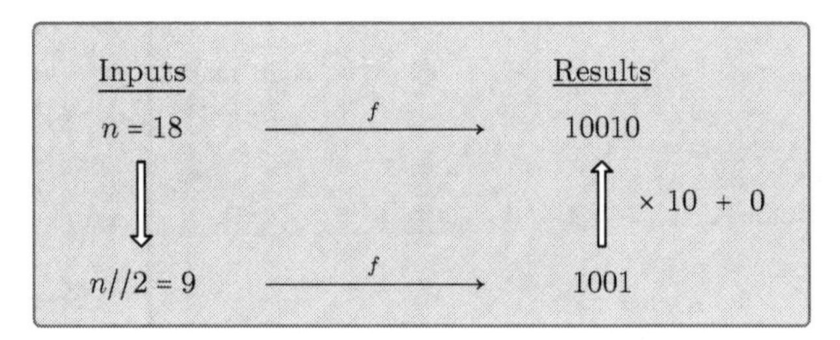

and

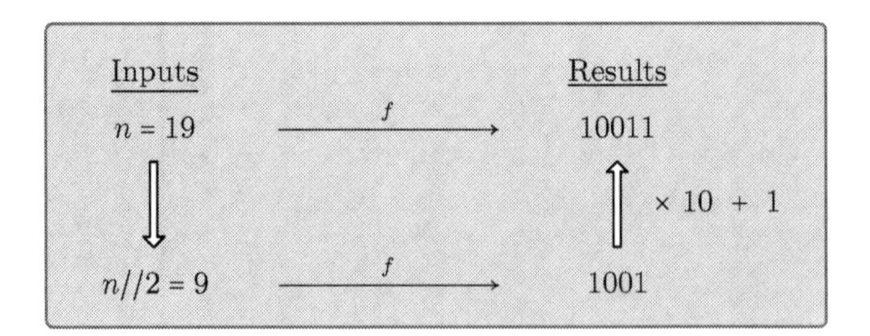

Note that the output of $f(18)$ is ten thousand and ten, since it is expressed in base 10. The diagrams illustrate that we can obtain the desired result by multiplying the output of the subproblem by 10, and then adding 1 if n is odd. Therefore, the recursive case seems to be $f(n) = 10f(n//2) + n\%2$.

We can proceed more rigorously through the following general diagram of the recursive thought process:

Listing 4.9 Binary representation of a nonnegative integer.

```
1 def decimal_to_binary(n):
2     if n < 2:
3         return n
4     else:
5         return 10 * decimal_to_binary(n // 2) + (n % 2)
```

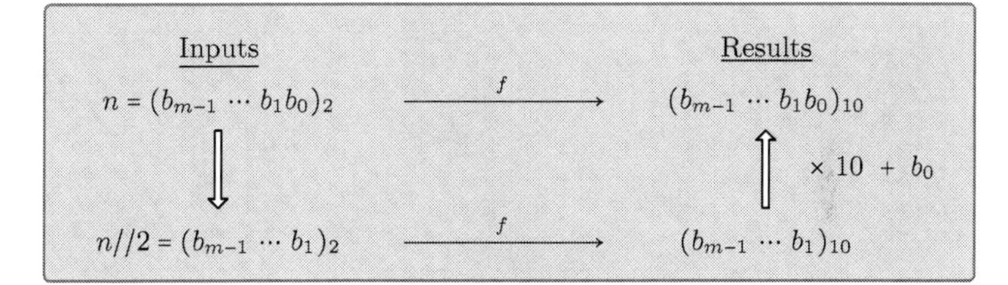

where b_i represents a bit. In particular, $f(n//2)$ is the decimal number that represents the sequence of "bits" of n except for the least significant one (b_0), whose value is $n\%2$ (or $n\&1$, where & denotes the bitwise AND operator). In order to append the discarded bit we need to multiply $f(n//2)$ by 10 (which shifts the digits one place to the left, and appends a zero), and add it to the result. Therefore, the recursive case is indeed $f(n) = 10f(n//2)+n\%2$, which we obtained by analyzing concrete instances. The complete function is:

$$f(n) = \begin{cases} n & \text{if } n < 2, \\ 10f(n//2) + n\%2 & \text{if } n \geq 2. \end{cases}$$

Listing 4.9 shows the corresponding code. Finally, the binary representation of $142 = 142_{10}$ is $10001110_2 = 128 + 8 + 4 + 2$.

4.2.2 Decimal to base b conversion

The problem in Section 4.2.1 is a particular case of the more general problem that consists of converting a decimal number into another in another base b, where $b \geq 2$. It can be tackled in a similar way by using diagrams. Figure 4.1 shows the steps of a general algorithm that solves the problem. In particular, it illustrates how to obtain the base-5 representation of 142, which is 1032 (i.e., $142_{10} = 1032_5$). The full problem is shown in (a), where the idea consists of performing successive integer divisions of n by b until reaching 0. At each step the remainder of the

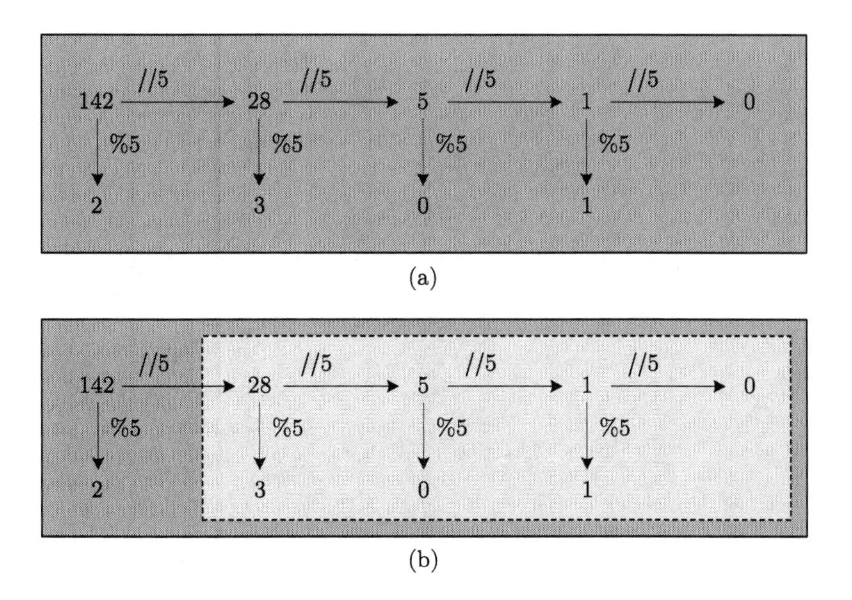

Figure 4.1 Conversion of 142 into its representation (1032) in base 5.

Listing 4.10 Conversion of a nonnegative integer n into its representation in base b.

```
def decimal_to_base(n, b):
    if n < b:
        return n
    else:
        return 10 * decimal_to_base(n // b, b) + (n % b)
```

division by b is also calculated, and will constitute a digit in the new representation in base b, whose position is controlled by a variable associated with a power of 10. The diagram in (b) highlights a subproblem within the original, where $28 = 103_5$.

In order to develop a recursive solution we need to establish the size of the problem. For this problem it is the number of times it is necessary to divide n until reaching 0, which is essentially the number of digits of the input in base b. The base case occurs when n can be represented by a single digit in base b. In other words, when $n < b$, where the result is simply n.

The recursive case requires identifying a subproblem within the original, as shown in Figure 4.1(b). Note that the subproblem is self-similar to the original, but starts at 28, which is 142//5. Thus, the decomposition consists of performing the integer division $n//b$, which reduces the size of the problem by a unit. Subsequently, we must determine how it is possible to modify the solution to the subproblem (103), in order to obtain the original (1032). The solution consists of multiplying the result of the subproblem times 10, and adding $n\%b$ (which is 2 in the example). Thus, the function can be coded as shown in Listing 4.10.

4.3 STRINGS

This section analyzes two problems involving strings, which are essentially sequences of characters, and constitute a fundamental data type in many programming languages.

4.3.1 Reversing a string

Consider the problem of reversing a string. In particular, we will develop a function f that receives an input string and returns its reverse. For example, f(`'abcd'`) =`'dcba'`.

The size of the problem is the length of the input string. A base case occurs when the input string is empty, where the function obviously returns an empty string. In the recursive case we need to discard a character from the input string in order to reduce the size of the problem. The first and last characters of the string are clear candidates. Omitting the first one leads to the following diagram where the input string **s** is written as the sequence $s_0 s_1 \cdots s_{n-2} s_{n-1}$ of characters:

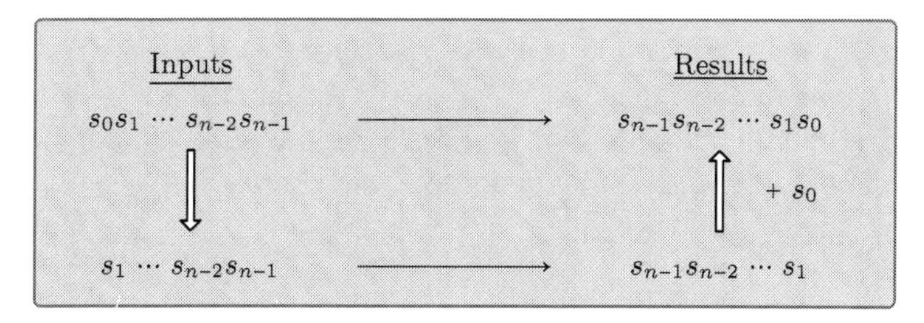

Thus, the function simply has to concatenate the first character to the result of the subproblem associated with $s_1 \cdots s_{n-2} s_{n-1}$ (the + symbol represents string concatenation). Together with the base case, the recursive function in Python is shown in Listing 4.11.

Listing 4.11 Conversion of a nonnegative integer n into its representation in base b.

```
1 def reverse_string(s):
2     if s == '':
3         return ''
4     else:
5         return reverse_string(s[1:]) + s[0]
```

4.3.2 Is a string a palindrome?

The next problem consists of determining whether a string is a palindrome, which is a sequence of characters that reads the same forward as it does backward (e.g., “radar”). Its size is the length of the string, since it determines the number of operations needed in order to obtain a true result. There are two base cases for this problem: (a) when the string is empty, and (b) when it comprises a single character. In both situations the output is `True`. Similarly to the function in Listing 2.6, the second base case is necessary since we will decompose the problem by reducing its size by two units. In particular, we will consider the subproblem that discards the first and last characters of the original string of length $n \geq 2$. In addition, the output of the function for some string $\mathbf{s} = s_0 s_1 \cdots s_{n-2} s_{n-1}$ of length n is:

$$(s_0 = s_{n-1}) \wedge (s_1 = s_{n-2}) \wedge \cdots \wedge (s_{\lfloor \frac{n}{2} \rfloor - 1} = s_{n - \lfloor \frac{n}{2} \rfloor}),$$

where $\wedge$ denotes logical AND (conjunction), and s_i represents the character from $\mathbf{s}$ at position i (where the first character of the string is located at position 0). Observe that the indices of the characters that need to be compared add up to $n-1$. Furthermore, we only need to compare $\lfloor n/2 \rfloor$ pairs of characters. The expression can be written more compactly as:

$$\bigwedge_{i=0}^{\lfloor n/2 \rfloor - 1} (s_i = s_{n-i-1})$$

similarly to “sigma” notation. The corresponding recursive diagram is:

Listing 4.12 Function that determines if a string is a palindrome.

```
1 def is_palindrome(s):
2     n = len(s)
3     if n <= 1:
4         return True
5     else:
6         return (s[0] == s[n - 1]) and is_palindrome(s[1:n - 1])
```

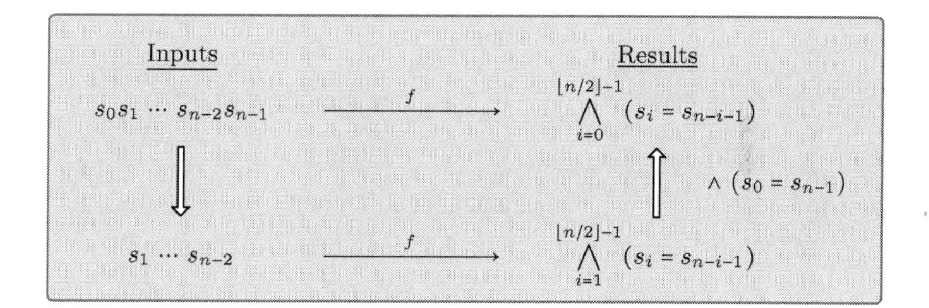

Thus, the Boolean function can be defined as:

$$f(\mathbf{s}) = \begin{cases} \text{true} & \text{if } n < 2, \\ (s_0 = s_{n-1}) \ \wedge \ f(\mathbf{s}_{1..n-2}) & \text{if } n \geq 2, \end{cases}$$

where $\mathbf{s}_{1..n-2}$ simply denotes the substring $s_1 \cdots s_{n-2}$. Finally, Listing 4.12 shows the corresponding code, whose time complexity is $\Theta(n)$.

4.4 ADDITIONAL PROBLEMS

This section contains several classical problems that can be solved elegantly through recursion.

4.4.1 Selection sort

The selection sort algorithm is one of the simplest strategies for sorting the elements of a list. Given an input list $\mathbf{a}$ of n numbers, assume the method must sort it in ascending order. The algorithm begins by placing the smallest element of the list in the first position of the output list. In the second step, the method searches for the minimum element in $\mathbf{a}_{1..n-1}$, and places it in the second position of the output list. This procedure is carried out $n-1$ times until the array is finally sorted. Clearly, at the end of the k-th step the smallest k numbers are sorted correctly, and after $n-1$ the entire list is completely sorted.

We will now show a linear-recursive version of the algorithm. Firstly, the size of the problem is the length of the list (n). The base case occurs when the list contains one element, where the output is the input list since it is obviously sorted. In addition, the algorithm can also return an empty list if the input is empty. Thus, the method can return **a** if $n \leq 1$.

For the recursive case we can decompose the problem by reducing its size by a unit, where the main idea is to discard the smallest element from the list. This can be done in two ways. Firstly, the element can be swapped with the first one on the list. This naturally does not change the result (the sorted list), but allows us to discard the first element of the list after having swapped the elements. Therefore, if the input list is $[7, 5, 3, 8, 4]$, it can be modified to $[3, 5, 7, 8, 4]$ by swapping the 3 and the 7. In that case, the input to the subproblem of size $n - 1$ would be $[5, 7, 8, 4]$, and a recursive diagram with this particular example could be:

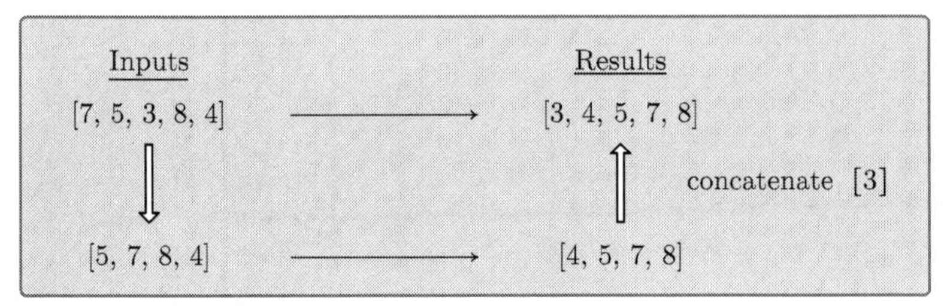

In order to solve the original problem, the recursive rule must concatenate the discarded smallest element ([3]) with the output list of the subproblem (note that the results are sorted lists). Listing 4.13 shows an implementation of the method that uses the function `min` to determine the smallest value in a list, the method `index` that returns the location of an element in a list, and the `+` operator to concatenate lists. An important detail regarding this algorithm is that it needs to make a new copy of the input list (in line 5) in order to not alter it when calling the method. In particular, without this copy the method would swap the first and smallest elements of the list.

Another possibility for reducing the size of the problem consists of discarding the smallest element of the list directly, by calling the `remove` method. In this case, the recursive diagram would be:

Listing 4.13 Recursive selection sort algorithm.

```
1 def select_sort_rec(a):
2     if len(a) <= 1:
3         return a
4     else:
5         b = list(a)
6         min_index = b.index(min(b))
7         aux = b[min_index]
8         b[min_index] = b[0]
9         b[0] = aux
10
11        return [aux] + select_sort_rec(b[1:])
```

Listing 4.14 Recursive variant of the selection sort algorithm.

```
1 def select_sort_rec_alt(a):
2     if len(a) <= 1:
3         return a
4     else:
5         b = list(a)
6         m = min(b)
7         b.remove(m)
8
9         return [m] + select_sort_rec_alt(b)
```

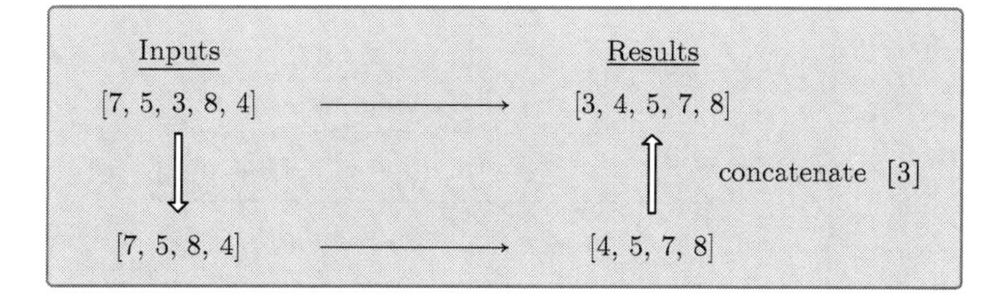

The only difference with respect to the previous diagram is the order of the elements in the input to the subproblem. This does not affect the recursive rule, where we must also concatenate the discarded smallest element ([3]) with the output list of the subproblem. Listing 4.14 shows a possible implementation of the function, which relies on the `remove` method. Again, the fifth line makes a copy of the input list in order to keep it unaltered when calling the method.

Finally, the runtime cost of these algorithms is characterized by (3.24), since finding (and removing) the smallest element of a list requires on the order of n operations. Thus, the algorithms run in $\Theta(n^2)$.

4.4.2 Horner's method for evaluating polynomials

In this problem the goal consists of evaluating a polynomial of degree n:

$$P(x) = c_n x^n + c_{n-1}x^{n-1} + \cdots + c_1 x + c_0, \tag{4.5}$$

at some real value x. The sum contains powers of x that are multiplied by the coefficients c_i. A naive algorithm that computes each power independently would require on the order of n^2 multiplications. Instead, Horner's method only needs to perform n products. Its clever idea is based on expressing the polynomial as:

$$P(x) = c_0 + x(c_1 + x(c_2 + \cdots + x(c_{n-1} + c_n x))).$$

The size of the problem is clearly the degree n. Thus, the base case occurs when $n = 0$, where the result is obviously c_0. In practice, $\mathbf{c}$ will be a list (or a similar data structure such as an array) of $n+1$ elements that represents the polynomial. Therefore, the base case is reached when the length of $\mathbf{c}$ is one.

In order to apply recursion we need to detect a self-similar subproblem of smaller size. The decomposition of the problem, decrementing its size by a unit, is:

$$P(x) = \underbrace{c_0 + x\big(\overbrace{c_1 + x(c_2 + \cdots + x(c_{n-1} + c_n x))\cdots}^{\text{subproblem}}\big)}_{\text{full problem}}.$$

By discarding coefficient c_0 and the first multiplication times x, the resulting subproblem has exactly the same structure as the original problem. If $p(\mathbf{c}, x)$ is the function that evaluates the polynomial defined by $\mathbf{c}$ at x, the recursive formula would be:

$$p(\mathbf{c}, x) = c_0 + x \cdot p(\mathbf{c}_{1..n}, x),$$

where the list $\mathbf{c}_{1..n}$ is simply $\mathbf{c}$ without its first element. The full function is:

$$p(\mathbf{c}, x) = \begin{cases} c_0 & \text{if } n = 1, \\ c_0 + x \cdot p(\mathbf{c}_{1..n}, x) & \text{if } n > 1, \end{cases}$$

Listing 4.15 Horner's method for evaluating a polynomial.

```
def horner(c, x):
    if len(c) == 1:
        return c[0]
    else:
        return c[0] + x * horner(c[1:], x)
```

and the corresponding code is shown in Listing 4.15.

It is important to note that discarding c_n instead of c_0 does not lead to Horner's method. In that case the recursive formula would be:

$$p(\mathbf{c}, x) = p(\mathbf{c}_{0..n-1}, x) + c_n x^n,$$

which would evaluate the polynomial as in (4.5). If the power x^n is computed in linear or logarithmic time the method would require $\Theta(n^2)$ or $\Theta(n \log n)$ operations, respectively. However, Horner's method is more efficient since it runs in linear time ($\Theta(n)$).

4.4.3 A row of Pascal's triangle

Pascal's triangle is the triangular arrangement of binomial coefficients shown in Figure 4.2, where the triangle in (a) shows the particular binomial coefficients, and the one in (b) shows the actual integer values associated with them. The following problem consists of determining a list containing the binomial coefficients of the n-th row:

$$\binom{n}{0} \binom{n}{1} \cdots \binom{n}{n-1} \binom{n}{n},$$

where the first and last elements are always 1. The problem can be viewed from a recursive point of view by considering the definition in 3.2. Graphically, it means that a binomial coefficient is the sum of the two immediately above it in the previous row of Pascal's triangle. For example, $\binom{4}{3} = \binom{3}{2} + \binom{3}{3}$. Therefore, it is possible to define a row of Pascal's triangle if we know the values of the previous row. Figure 4.3 shows the relationship between a problem (n-th row) and a subproblem ($(n-1)$-th row).

The size of the problem is n. Thus, the base case corresponds to $n = 0$, where the output is simply a list containing a 1 (`[1]`). For $n = 1$ we cannot apply the recursive rule in 3.2. Thus, initially it appears that we may need an additional base case for $n = 1$, where the output is

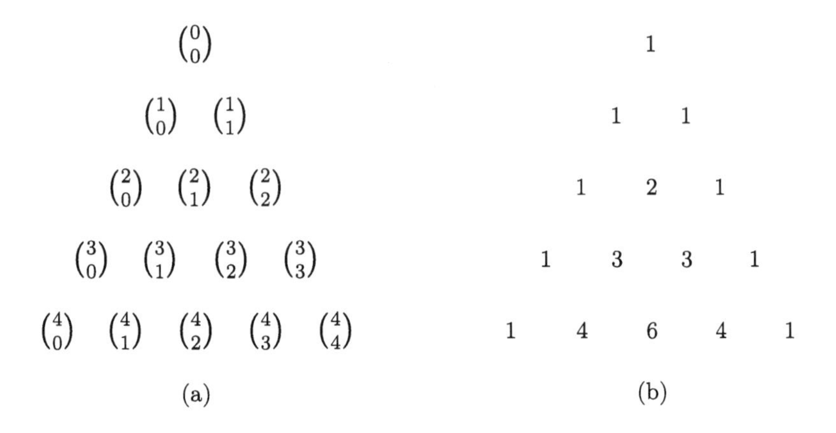

Figure 4.2 Pascal's triangle.

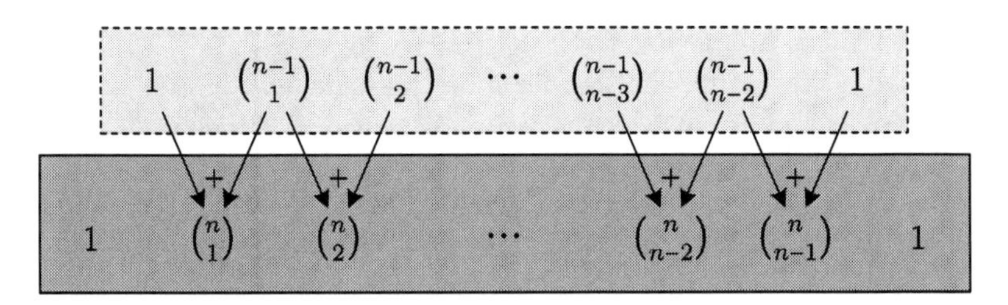

Figure 4.3 Decomposition and recovery of a row of Pascal's triangle.

Listing 4.16 Function that generates the n-th row of Pascal's triangle.

```
1  def pascal(n):
2      if n == 0:
3          return [1]
4      else:
5          row = [1]
6          previous_row = pascal(n - 1)
7          for i in range(len(previous_row) - 1):
8              row.append(previous_row[i] + previous_row[i + 1])
9          row.append(1)
10     return row
```

(`[1,1]`). However, since all of the rows for $n > 0$ begin and end with a 1, these elements can be incorporated in the recursive case by default. Thus, in this scenario a special base case for $n = 1$ would be unnecessary.

In particular, in the recursive case we assume that we know the solution to the subproblem, which allows us to compute every element

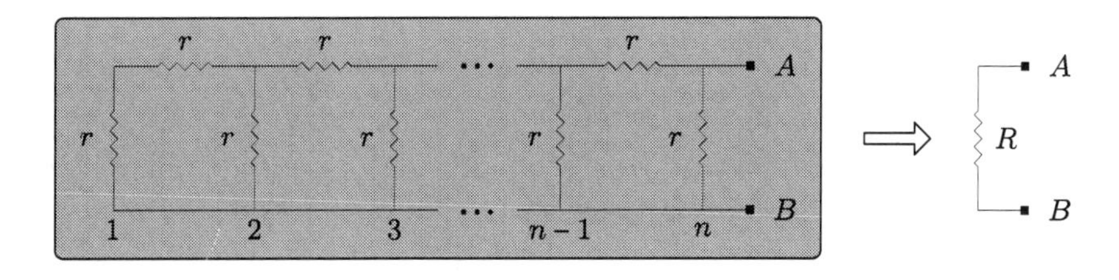

Figure 4.4 Ladder of resistors problem.

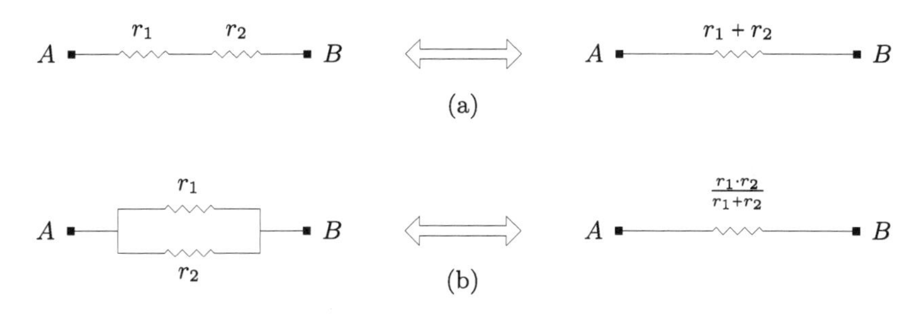

Figure 4.5 Equivalence of circuits with resistors.

of the solution except for the ones at the extremes (these elements can be appended at some point during the process). Listing 4.16 shows a possible solution to the problem. The recursive case starts by inserting a 1 in a list (`row`) that will contain the result. Line 6 computes the result of the subproblem of size $n-1$, and in lines 7 and 8 the loop adds consecutive integers of the subsolution (as shown in Figure 4.3), appending the sum to the result. Finally, a 1 is inserted at the end, which completes the solution. Exercise 4.14 proposes replacing the loop by a recursive function.

4.4.4 Ladder of resistors

In the next problem the goal is to simplify the electrical circuit in Figure 4.4 resembling a ladder that contains several layers of resistors whose resistance is r. In particular, it is possible to substitute the entire circuit by an equivalent one composed of a single resistor. Thus, the objective consists of determining the value of the resistance R (as a function of the resistance r) that will yield equivalent circuits. Figure 4.5 shows how to transform a circuit with two resistors with resistances r_1 and r_2 when they are connected in series, as shown in (a), or in parallel, as illustrated in (b). When they are connected in series the resulting resistance

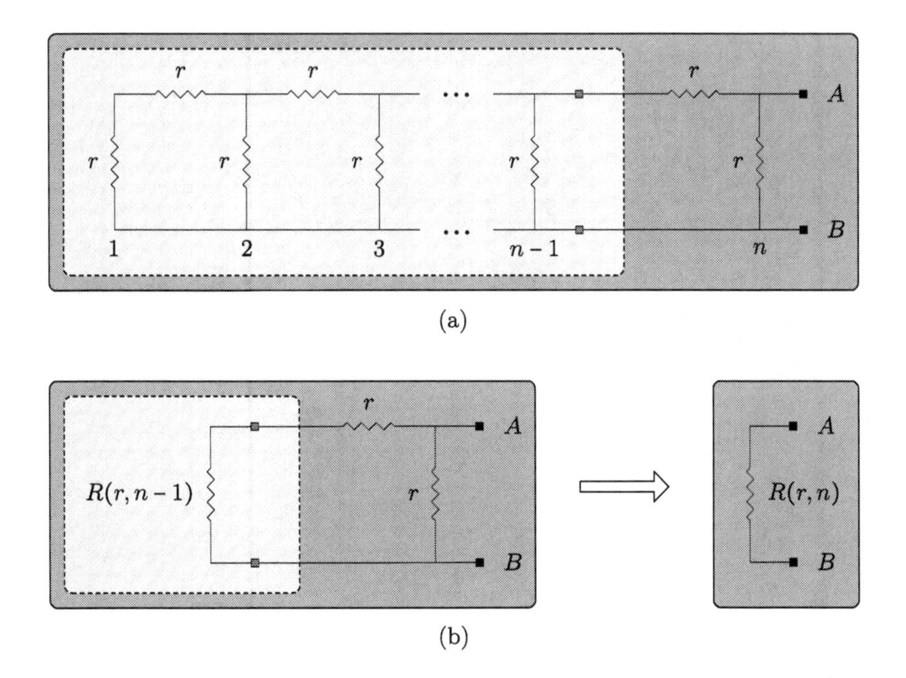

Figure 4.6 Decomposition of the ladder of resistors problem, and derivation of the recursive case through induction.

is $r_1 + r_2$. Instead, when they are connected in parallel the new resistance is $r = (r_1 \cdot r_2)/(r_1 + r_2)$. Alternatively, it can be expressed as:

$$\frac{1}{r} = \frac{1}{r_1} + \frac{1}{r_2}. \tag{4.6}$$

These rules can be applied successively to pairs of resistors until we obtain a circuit that contains a single resistor. However, this process is tedious. Instead, recursion provides a succinct and elegant solution.

The problem has two input parameters: the resistance (r) and the number of rungs (n) of the ladder. The size of the problem is clearly n (r plays a role in the final value, but is not responsible for the runtime of the algorithm). Let $R(r, n)$ denote the recursive function. The base case occurs when $n = 1$, where the initial ladder would only contain a single resistor. Therefore, in that case $R(r, 1) = r$. For the recursive case we need to find a subproblem within the original with exactly the same structure. Figure 4.6(a) shows the decomposition of the problem by decrementing its size by a unit. The circuit associated with the subproblem can be replaced by a single resistor of resistance $R(r, n - 1)$,

Listing 4.17 Function that solves the ladder of resistors problem.

```
1 def circuit(n, r):
2     if n == 1:
3         return r
4     else:
5         return 1 / (1 / r + 1 / (circuit(n - 1, r) + r))
```

as shown in (b), where we can assume that we know its value by using induction. Finally, it is fairly straightforward to simplify the resulting circuit that contains only three resistors. Firstly, the left and top resistors are connected in series. Thus, they can be merged to form a resistor with resistance $R(r, n-1) + r$. Finally, this new resistor is connected in parallel to the right resistor. By applying (4.6), $R(r, n)$ can be defined through:

$$\frac{1}{R(r,n)} = \frac{1}{r} + \frac{1}{R(r,n-1)+r}.$$

The recursive function is therefore:

$$R(r,n) = \begin{cases} r & \text{if } n = 1, \\ 1/\left(\frac{1}{r} + \frac{1}{R(r,n-1)+r}\right) & \text{if } n > 1. \end{cases}$$

Listing 4.17 shows the associated code. Finally, as a curious note, it is possible to show that $R(r, n) = rF(2n-1)/F(2n)$, where F is the Fibonacci function.

4.5 EXERCISES

Exercise 4.1 — Listing 2.6 contains a linear-recursive Boolean function that determines whether a nonnegative integer n is even, where the decomposition reduces the size of the problem (n) by two units. Define and code an alternative method based on decrementing the size of the problem by a single unit.

Exercise 4.2 — Implement a recursive function that computes the power b^n in logarithmic time, for a real base b, and an integer n, which can be negative.

Exercise 4.3 — Implement a recursive function that computes the n-th power of a square matrix, which is another square matrix of the same dimensions. Use the NumPy package and its methods:

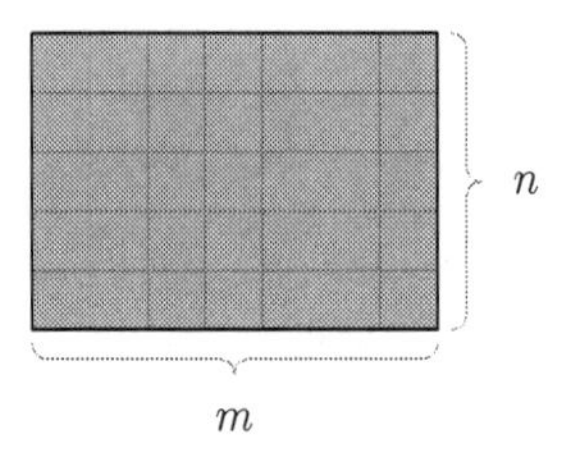

Figure 4.7 The product of two nonnegative integers n and m can be represented as the number of unit squares that form an $n \times m$ rectangle.

- `identity`: returns the identity matrix.
- `shape`: indicates the dimensions of a matrix.
- `dot`: multiplies matrices.

Finally, compute matrices:

$$\begin{bmatrix} 1 & 1 \\ 1 & 0 \end{bmatrix}^n$$

for $n = 1, \dots, 10$, and print the element in their first row and second column. What are these numbers?

Exercise 4.4 — Implement recursive functions that compute a "slow" product of two nonnegative integers m and n. The algorithms are allowed to add and subtract numbers, but they may not use the multiplication (`*`) operator. Use decompositions that decrease one or both of the inputs by a single unit. In addition, illustrate the recursive design thought process through diagrams (for example, by using the general diagram in Figure 2.5; or through diagrams that represent the product of n and m as the area of a rectangle of height n and width m, where the result is the number of square blocks of unit area needed to build an $n \times m$ rectangle, as shown in Figure 4.7).

Exercise 4.5 — Implement a more efficient "slow" product recursive function by using the decomposition that divides both input parameters by two. Use rectangular diagrams like the one described in Figure 4.7.

Exercise 4.6 — In Python and other programming languages methods can also be parameters to other methods. Define and code a general recursive function that computes the sum:

$$g(m, n, f) = \sum_{i=m}^{n} f(i) = f(m) + f(m+1) + \cdots + f(n-1) + f(n),$$

where m and n are integers, and f is a function. Use it in order to calculate and print:

$$\sum_{i=1}^{n} i^3,$$

for $n = 0, \ldots, 4$.

Exercise 4.7 — Define and code a function that computes the number of digits of a nonnegative integer n.

Exercise 4.8 — Define and code a function that, given a decimal number n whose digits are either zero or one, returns the number whose binary representation is precisely the sequence of zeros and ones in n. For example, if $n = 10110_{10}$ the function returns 22, since $10110_2 = 22$.

Exercise 4.9 — Derive the recursive case of the "decimal to base b conversion" problem in Section 4.2.2 by using the general diagram for thinking about recursive cases in Figure 2.5.

Exercise 4.10 — Recall the problem in 3.8 that processes (binary) numbers of n bits (from right to left). Derive and code a recursive algorithm that, given a particular number x of n bits, determines the position of its least significant bit set to 1. In particular, consider that the least-significant bit is located at position 1. For example, for $x = 01110\underline{1}00$ the rightmost bit set to 1 is located at position 3.

Exercise 4.11 — Write a recursive method that uses the function developed in Exercise 4.10 in order to solve Exercise 3.8 computationally. Print the solutions for numbers expressed with $n = 1, \ldots, 5$ bits.

Exercise 4.12 — Code a recursive function that returns the number of vowels in a given string.

Exercise 4.13 — A binomial coefficient $\binom{n}{m}$ is a function of two integer parameters n and m. Define three linear-recursive functions for calculating binomial coefficients by considering three decompositions: one that decrements n, another that decrements m, and one that decrements both. Use the definition in (3.1) to develop the recursive cases.

Exercise 4.14 — Replace the loop in Listing 4.16 with a recursive function. It should receive a row of Pascal's triangle, and return a list with the sum of its consecutive terms. For example, if the input is `[1, 3, 3, 1]` the result should be `[4, 6, 4]`.

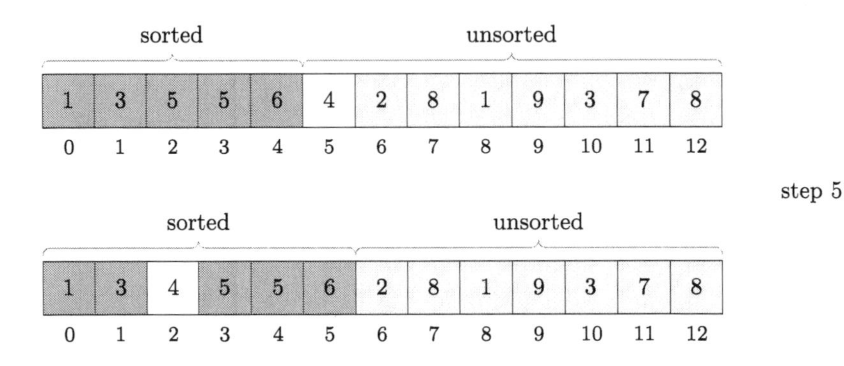

Figure 4.8 Step of the insertion sort algorithm.

Exercise 4.15 — Consider a nonnegative integer n whose digits always appear in ascending order from left to right, such as 24667. In other words, if $d_{m-1} \cdots d_1 d_0$ represents the sequence of (m) digits of n, then $d_i \leq d_j$ for $i < j$. Given an additional digit $0 \leq x \leq 9$, write a function that returns the integer that results from inserting x in n, such that its digits also appear in ascending order from left to right. For instance, if $n = 24667$ and $x = 5$, the function should return 245667. Avoid redundant base cases.

Exercise 4.16 — This exercise is similar to Exercise 4.15. Given a sorted list of numbers in ascending order (**a**), and a number x, implement a function that inserts x in a location in **a** such that the resulting list will remain sorted.

Exercise 4.17 — The "insertion sort" method sorts a list by repeatedly applying the procedure shown in Figure 4.8. At the i-th step the elements from indices 0 to $i-1$ will be sorted, while the rest of the list will not be (necessarily) sorted. In order to proceed, the method inserts the element at index i into the sorted sublist in a location where the resulting list (from index 0 to i) will remain sorted. If this operation is performed from index (step) 1 (or 0) until $n-1$ the list will be sorted after the last step. Implement a recursive variant of the method that uses the solution to Exercise 4.16 to sort a list.

CHAPTER 5

Linear Recursion II: Tail Recursion

You are only as beautiful as your last action.

— Stephen Richards

TAIL recursion, also known as "final" recursion, is a form of linear recursion in the sense that the recursive cases only invoke the method once. However, in tail recursion a recursive call is the last action performed by the method. For example, the function in (1.15) is tail-recursive since it does not manipulate the result of the function call in the recursive case. This implies that it will return a value directly obtained in a base case. In addition, functions may require more parameters in order to store intermediate results that will be used in the base cases for providing the final output. The tail-recursive methods seen so far include Listings 2.3, 2.5, and 2.6.

In the previous chapter all of the methods were based on decrementing the size of the problem by one or two units, except for the ones in Section 4.1.1.2. In contrast, this chapter will present more algorithms that divide the size of the problem by two. In these cases the recursive calls will solve a single problem of half the size of the original's, since the methods only invoke themselves once. Some authors refer to this strategy as "divide and conquer." However, we will only use this term when the algorithms need to solve *several* independent subproblems whose size is a fraction of that of the original's. These algorithms will be covered in Chapter 6.

Finally, tail recursion is also special due to its relationship with iteration. Chapter 11 will analyze this connection.

5.1 BOOLEAN FUNCTIONS

This introductory section examines how the choice of base cases can lead to linear or tail-recursive algorithms for some Boolean functions.

5.1.1 Does a nonnegative integer contain a particular digit?

Given a nonnegative integer n, and a digit $0 \leq d \leq 9$ (which is also an integer), in this problem the goal consists of determining whether d is present in n. The recursive method will therefore be a Boolean function with parameters n and d. Additionally, we will assume that n does not contain any leading zeros (for example, 358 cannot be written as 0358). However, we will consider that the number zero contains the digit 0.

5.1.1.1 Linear-recursive algorithm

The first step consists of establishing the size of the problem, which is the number (m) of digits of n. A base case occurs when n only contains a single digit (i.e., when $n < 10$), where the result is simply whether $n = d$. Let us assume for the time being that this is the only base case.

For the recursive case, it is cumbersome to express the value of a digit in terms of n. Therefore, we can use an alternative notation, as discussed in Section 2.5.3. In particular, let $d_{m-1} \cdots d_1 d_0$ represent the sequence of m digits of n in base 10. We can form the following diagram when decrementing the size of the problem by a unit (i.e., using $n//10$, which discards the least significant digit of n):

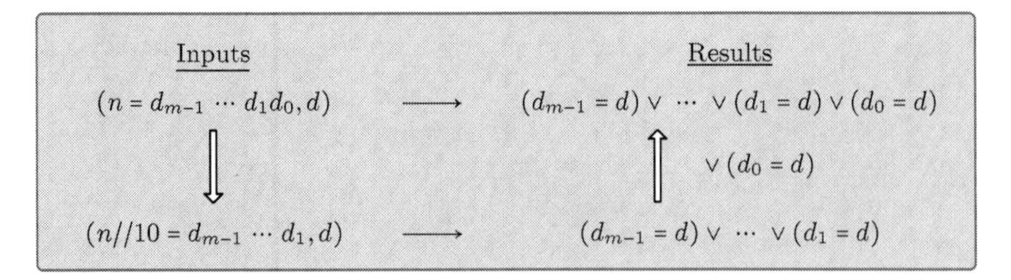

where $\vee$ denotes logical OR (disjunction). The diagram clearly illustrates that if d is a digit of n, then it is either present in $n//10$ (this is the result of the recursive call), or it should be equal to d_0.

Other notations are also possible. We can simply choose words or expressions, even if they are not technically correct logically or mathematically. For instance,

Listing 5.1 Linear-recursive Boolean function that determines if a non-negative integer contains a digit.

```
1 def contains_digit(n, d):
2     if n < 10:
3         return n == d
4     else:
5         return (n % 10 == d) or contains_digit(n // 10, d)
```

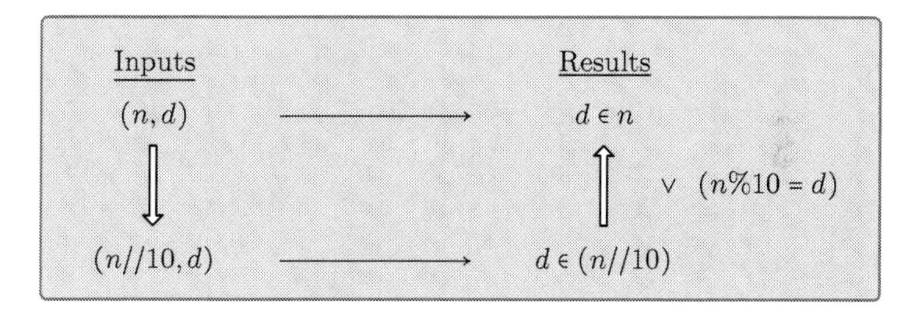

introduces a clear abuse of notation, since the membership symbol $\in$ should be used with a set. Thus, $d \in n$ actually represents whether d belongs to the set of digits that form n. Nevertheless, the diagram serves its purpose, accurately indicating the function's recursive case.

Listing 5.1 shows the code that implements the function, where $n\%10$ is the least significant digit of n. Observe that the method is linear-recursive, since the function invokes itself once, and needs to process the output of the subproblem (i.e., compute the result of the expression involving the `or` operator). Additionally, if the condition `(n%10==d)` is `True` then a compiler or interpreter that uses "short-circuit" evaluation would automatically return a true value, avoiding the call to `contains_digit(n//10,d)`. In particular, short-circuit evaluation allows us to avoid unnecessary computations by considering the properties: `True` $\vee\ b =$ `True`, and `False` $\wedge\ b =$ `False`, where b is some Boolean expression. Since these hold regardless of the value of b, there is no need to evaluate b.

5.1.1.2 *Tail-recursive algorithm*

Most programming languages support short-circuit evaluation. However, we can force it by considering an additional base case, which leads to a tail-recursive algorithm. In particular, for this problem an algorithm can return `True` as soon as it finds a digit in n that is equal to d. Therefore, we can return `True` if the last digit of n is equal to d (i.e., if $n\%10 = d$).

Listing 5.2 Tail-recursive Boolean function that determines if a nonnegative integer contains a digit.

```
1 def contains_digit_tail(n, d):
2     if n < 10:
3         return n == d
4     elif n % 10 == d:
5         return True
6     else:
7         return contains_digit_tail(n // 10, d)
```

By considering this base case we can be sure that $n\%10 = d_0 \neq d$ in the recursive case. Therefore, the recursive diagram is now:

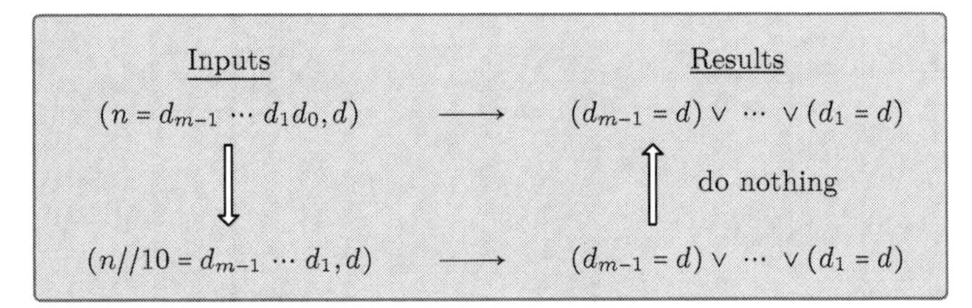

This implies that the result of the subproblem is exactly the output of the original problem, and the function can simply return the result of the recursive call, without processing it. Together with the base cases, this leads to a tail-recursive algorithm that can be coded as shown in Listing 5.2.

Lastly, it is important to understand that although the chosen decomposition divides the input by a constant (10), the size of the problem is only reduced by a unit. The time complexity for the algorithm is logarithmic with respect to the input n, but it is linear with respect to the number of digits (m) of n.

Finally, this problem has analogous counterparts that rely on data structures such as lists, arrays, etc., since these also represent sequences of elements. For example, it is very similar to deciding if a string contains a character, or if an element is present in a list. Although the codes may be different, the underlying reasoning is essentially identical.

5.1.2 Equal strings?

The following problem consists of determining whether two strings are equal (of course, in Python we can simply use the == operator). A

Boolean function that solves the problem will therefore have two string input parameters. If their lengths are different the algorithm can return `False` immediately in a base case. Thus, the challenge lies in solving the problem when they have the same length, which would constitute its size.

5.1.2.1 Linear-recursive algorithm

The smallest instance therefore corresponds to two empty strings, where the result is obviously `True`. In this section we will assume that there are no additional base cases, which leads to a linear-recursive function. In particular, the decomposition can decrease the size of the problem by a unit. This implies discarding one character (located at the same position) from each input string. Omitting the first one leads to the following recursive diagram, where **s** and **t** are two input strings of length n:

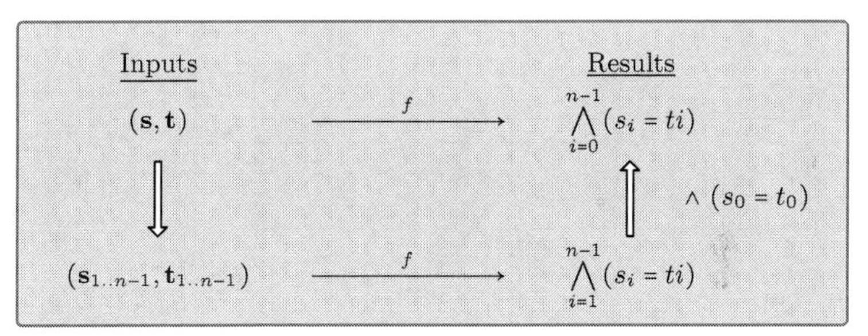

Clearly, the method has to check that the first characters are the same, *and* that the remaining substrings are also identical through a recursive call. The Boolean function is therefore:

$$f(\mathbf{s},\mathbf{t}) = \begin{cases} \text{false} & \text{if length}(\mathbf{s}) \neq \text{length}(\mathbf{t}), \\ \text{true} & \text{if } n = 0, \\ (s_0 = t_0) \wedge f(\mathbf{s}_{1..n-1},\mathbf{t}_{1..n-1}) & \text{if } n > 0, \end{cases}$$

Listing 5.3 shows the corresponding code, which describes a linear-recursive function.

5.1.2.2 Tail-recursive algorithm

Similarly to the example in Section 5.1.1, we can force a short-circuit evaluation by considering an additional base case. In particular, an algorithm can return `False` as soon as it detects two different characters

Listing 5.3 Linear-recursive function that determines if two strings are identical.

```
1 def equal_strings(s, t):
2     if len(s) != len(t):
3         return False
4     elif s == '':
5         return True
6     else:
7         return s[0] == t[0] and equal_strings(s[1:], t[1:])
```

Listing 5.4 Tail-recursive function that determines if two strings are identical.

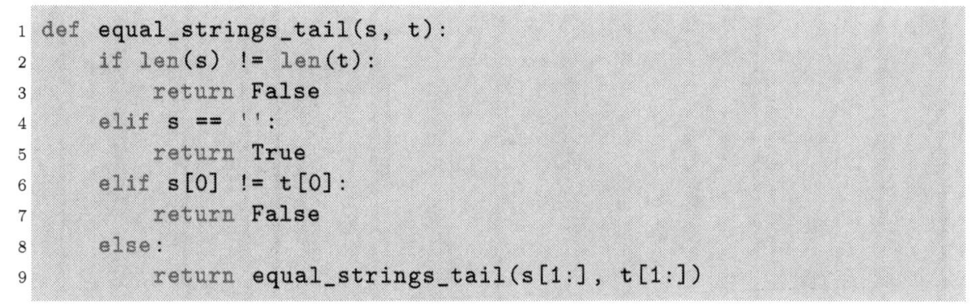

```
1 def equal_strings_tail(s, t):
2     if len(s) != len(t):
3         return False
4     elif s == '':
5         return True
6     elif s[0] != t[0]:
7         return False
8     else:
9         return equal_strings_tail(s[1:], t[1:])
```

at the same position in the strings. Thus, we can incorporate a base case that checks if $s_0 \neq t_0$. In that case, the algorithm would automatically return `False`. With this base case we can be sure that $s_0 = t_0$ in the recursive case. Therefore, the new recursive diagram would be:

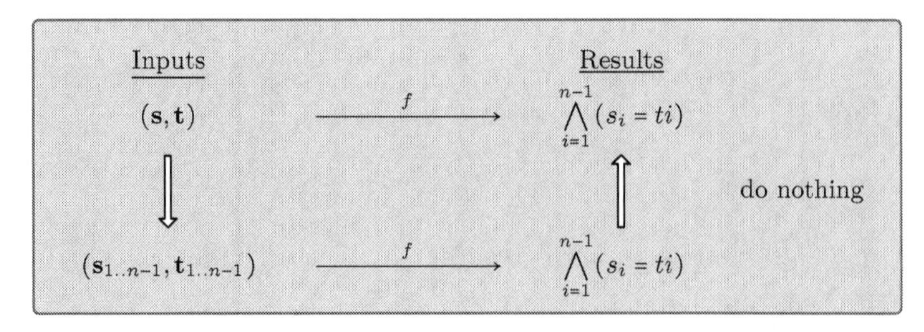

Again, the result of the subproblem is exactly the output of the original problem, which leads to a tail-recursive algorithm. Finally, together with the base cases, the function can be coded as in Listing 5.4.

Listing 5.5 Tail-recursive linear search of an element in a list.

```
def linear_search_tail(a, x):
    n = len(a)
    if a == []:
        return -1
    elif a[n - 1] == x:
        return n - 1
    else:
        return linear_search_tail(a[:n - 1], x)
```

5.2 SEARCHING ALGORITHMS FOR LISTS

This section will cover algorithms that solve the classical problem of finding the position i of an element x within a list $\mathbf{a} = [a_0, a_1, \ldots, a_{n-1}]$. In other words, given $\mathbf{a}$ and x, the goal is to find an index i for which $a_i = x$. If the element x appears several times in the list the methods that solve the problem can return any valid index that indicates a location where x is found in the list.

For a list of length n, a method that solves the problem will return an integer index ranging from 0 until $n-1$, when the element is present in the list. In contrast, if the list does not contain x the algorithms must return some value outside of that interval. Some Python implementations simply return a Boolean false value. However, this option is not valid in many other languages (C, Java, Pascal, etc.), where methods can only return values of a single data type. Since the indices are necessarily integers, in these languages the value that indicates that an element is not contained in the list must also be an integer. Thus, we will use the integer value -1 to indicate that an element is not present in a list, which is compatible with other programming languages.

5.2.1 Linear search

Firstly, assume that the list $\mathbf{a}$ is not (necessarily) sorted. Any algorithm would require n comparisons in the worst case, which occurs when the element does not belong to the list. Note that any method would have to analyze every single element of the list in order to conclude that the element is missing. In this situation the algorithms would perform a "linear search" for an element in a list, which runs in $\mathcal{O}(n)$.

The size of the problem is clearly n. If the list is empty the method can trivially return -1. Another base case can return the location of

Listing 5.6 Linear-recursive linear search of an element in a list.

```
def linear_search_linear(a, x, n):
    if a == []:
        return -n - 1
    elif a[0] == x:
        return 0
    else:
        return 1 + linear_search_linear(a[1:], x, n)


def linear_search_linear_wrapper(a, x):
    return linear_search_linear(a, x, len(a))
```

the element if it is found. This base case will depend on the choice of decomposition. If we reduce the size of the problem by discarding the last element of the list, then the base case will need to check whether x is located at the last position. If indeed $a_{n-1} = x$, then the method can simply return $n - 1$. Otherwise, it will need to carry out a recursive call in order to solve the subproblem of size $n - 1$, and the final result will be exactly that of the subproblem's, which leads to a tail-recursive solution. Listing 5.5 shows a possible implementation in Python of the described linear search function.

Finally, it is worth noticing that with the previous decomposition the locations of the elements in the sublist are the same as those in the original list. Alternatively, if the decomposition omitted the first element of the list, all of the indices in the sublist would be decremented by a unit with respect to those in the original list. For example, if the list were $[2, 8, 5]$, the 8 is also located at position 1 in $[2, 8]$, but it appears at location 0 in $[8, 5]$. This leads to more complex methods that require an additional parameter. For example, Listing 5.6 shows a solution that adds a unit to the result of each recursive call. The method trivially returns 0 if $x = a_0$, but the base case for an empty list is more complicated. Note that if that base case is reached the algorithm will have added n units in the n previous recursive calls. Thus, it needs to return $-n - 1$, where n is the length of the initial list (not the particular length of the input argument since it is 0 in that case) in order to return -1, indicating that x does not appear in the list. Since n cannot be obtained from an empty list it has to be passed as an extra parameter in every function call in order to be recovered in the base case. The code

Listing 5.7 Alternative tail-recursive linear search of an element in a list.

```
def linear_search_tail_alt(a, x, index):
    if a == []:
        return -1
    elif a[0] == x:
        return index
    else:
        return linear_search_tail_alt(a[1:], x, index + 1)


def linear_search_alt_tail_wrapper(a, x):
    return linear_search_tail_alt(a, x, 0)
```

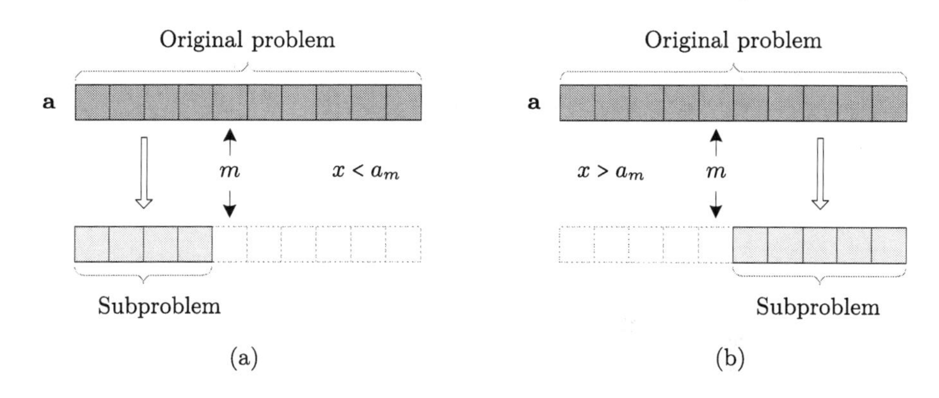

Figure 5.1 Decomposition related to the binary search algorithm.

therefore requires a wrapper method, where the third parameter of the recursive function is initialized to n.

The last decomposition implies that the algorithm will search for the element starting at index 0, and will progressively advance until reaching index $n-1$. Another solution consists of explicitly specifying the index of the element that will be compared to x each time the method invokes itself. This introduces a new parameter that can be viewed as an accumulator variable that is incremented with each function call, and leads to the tail-recursive function in Listing 5.7, which correctly returns -1 if the list is empty. The function must be called with the accumulator `index` parameter set to 0 from a wrapper method. Lastly, observe that the result is stored in that parameter.

Listing 5.8 Binary search of an element in a list.

```
def binary_search(a, x, lower, upper):
    if lower > upper:  # empty list
        return -1
    else:
        middle = (lower + upper) // 2

        if x == a[middle]:
            return middle
        elif x < a[middle]:
            return binary_search(a, x, lower, middle - 1)
        else:
            return binary_search(a, x, middle + 1, upper)


def binary_search_wrapper(a, x):
    return binary_search(a, x, 0, len(a) - 1)
```

5.2.2 Binary search in a sorted list

When the input list is sorted it is possible to search for an element by using a faster approach. Assume that the list **a** is sorted in ascending order, where $a_i \leq a_{i+1}$, for $i = 0, \dots, n-2$. The problem can be solved in $\mathcal{O}(\log n)$ running time by "binary search" algorithms. The main idea behind these methods (there are several variants) consists of dividing the size of the problem progressively by two. Figure 5.1 illustrates the decomposition of the problem. Firstly, the algorithms compute the middle index m. If $x = a_m$ they can terminate in a base case simply returning m. Otherwise, since the list is sorted, it is possible to keep searching for x in only one of the sublists to the right or left of m. In particular, if $x < a_m$ then x cannot appear in the sublist $[a_m, \dots, a_{n-1}]$, and the methods can continue to search for x in $[a_0, \dots, a_{m-1}]$, as shown in (a). Analogously, if $x > a_m$ then x can only appear in $[a_{m+1}, \dots, a_{n-1}]$, as illustrated in (b). This approach essentially divides the size of the problem (n) by two. Therefore, the runtime will be characterized by (3.20), which implies that the methods will run in logarithmic time with respect to n in the worst case. Lastly, the functions will return -1 (at a base case) when the list is empty.

Similarly to the linear search functions in Listings 5.6 and 5.7, the binary search methods need to include additional input parameters in order to indicate the original indices of the list. For example, consider searching for an 8 in the list [1,3,3,5,6,8,9]. If we simply pass the sublist $[6, 8, 9]$ to the function in a recursive call, the 8 is now at index 1, while it was located at position 5 in the original list. Thus, we need to indicate that the sublist's first index should really be 5. In other words, we need to specify an extra parameter that contains the position in the original list where the sublist begins. Listing 5.8 shows an implementation of the method that not only includes this parameter (`lower`), but also uses an additional one (`upper`), which is the index of the original list in which a sublist ends. Thus, `lower` and `upper` simply specify the boundaries of a sublist within `a`. The `upper` parameter is not strictly necessary in Python since we can use the length of the list in order to determine the sublist's upper index (Exercise 5.4 proposes implementing a method that only uses the `lower` parameter). However, the code in Listing 5.8 is simpler in the sense that it requires less arithmetic operations, does not invoke itself with sublists (the recursive calls use the entire list `a`), and shows how the method can be implemented in other programming languages where the length of lists is not directly available. In particular, if `lower` is greater than `upper` then the list is empty and the method returns -1. Otherwise, the function checks the base case associated with the condition $x = a_m$. If x is not at index m, the method carries out one of the two recursive calls, on the appropriate sublist, depending on the condition $x < a_m$. Finally, the wrapper method is used to initialize the `lower` and `upper` indices to 0 and $n - 1$, respectively.

5.3 BINARY SEARCH TREES

A binary search tree is a data structure used for storing data items, each of which is associated with a certain key (we will consider that it has to be unique), similarly to the entries in a Python dictionary. It is an important data structure that allows us to carry out certain operations efficiently, such as searching, inserting, or removing elements. The data in each node of the binary tree is a pair (key, item), and the nodes are arranged according to the values of the keys, which can be compared through the `<` operator, or some equivalent function that allows us to determine if a key is less than another one (i.e., a total preorder binary relation). In particular, given any node in the tree, every key in its left subtree must be less than its key, while all of the keys in its right subtree must be greater than its key. This is also known as the "binary search tree property," which implies that every subtree of a binary search tree is also a binary search tree.

For example, this data structure can be used to store the information of a birthday calendar. Figure 5.2 shows a binary tree of seven persons that allows us to retrieve their birthdays according to their names, which are strings that we can assume are unique (naturally, we can include last names or tags in order to identify each person uniquely). Keys can be strings since they can be compared, for example, according to the lexicographic order (in Python we can simply use the < operator). Therefore, the names in the binary search tree are sorted as they would appear in a regular dictionary. For instance, consider the root node associated with "Emma". Observe that all of the names contained in its left subtree would appear before "Emma" in a dictionary, and all of the names in the right subtree would be found after "Emma". Furthermore, this property holds for every node of the binary search tree.

In order to implement a binary search tree, we first have to decide how to code a binary tree. There are several possibilities (one of the most common approaches consists of using object oriented features in order to declare a class associated with the nodes of the tree), but in this book we will simply use lists. In particular, every node of the tree will consist of a list of four elements: the key, the item, the left subtree, and the right subtree, where the subtrees are also binary search trees. Thus, the binary search tree in Figure 5.2 would correspond to the following list:

```
['Emma', '2002/08/23',
 ['Anna', '1999/12/03', [], []],
 ['Paul', '2000/01/13',
  ['Lara', '1987/08/23',                    (5.1)
   ['John', '2006/05/08', [], []],
   ['Luke', '1976/07/31', [], []]],
  ['Sara', '1995/03/14', [], []]]]
```

which is illustrated graphically in Figure 5.3. Observe that the left and right subtrees are lists, which are empty at the leaf nodes.

5.3.1 Searching for an item

The goal of the next problem is to search for some item with a known key k in a binary search tree, and retrieve its associated item. We can assume that the size of the problem is the height of the tree. A trivial base case occurs when the list that represents the binary tree is empty. In that case the algorithm can simply return `None`. There is another situation where the algorithm can provide a result without carrying out a recursive call. If the key of the root node is k, the method will have

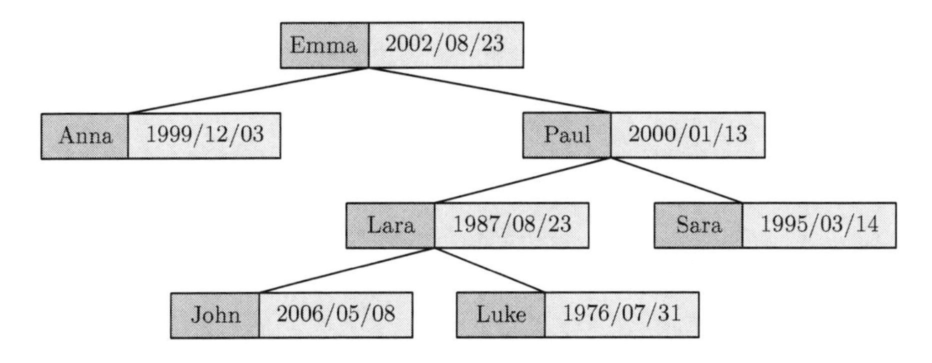

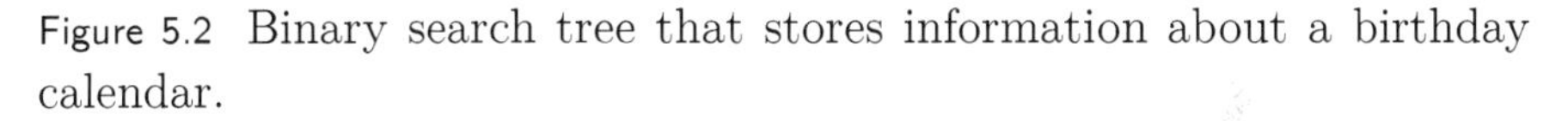
Figure 5.2 Binary search tree that stores information about a birthday calendar.

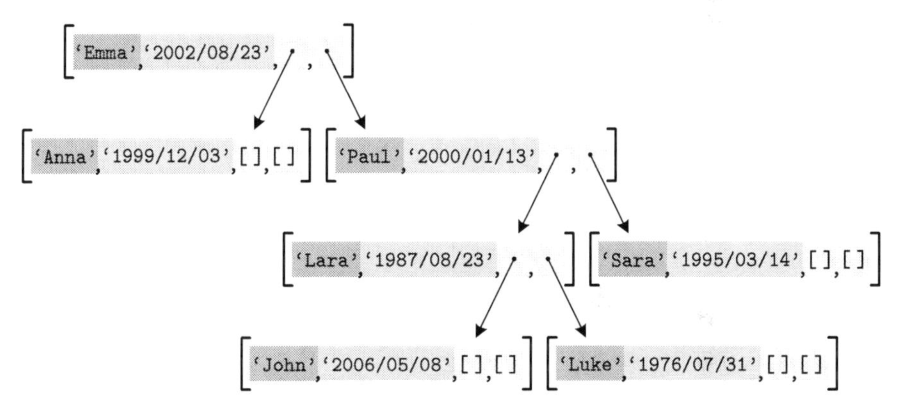

Figure 5.3 Binary search tree in Figure 5.2 and (5.1), where each node is a list of four elements: name (string), birthday (string), left subtree (list), and right subtree (list).

found the item, and can therefore return it. Thus, in the recursive cases we can be sure that the root node does not contain the searched item.

In the next step our goal is to find an appropriate decomposition of the problem that reduces its size. We have already seen that trees are composed recursively of subtrees. Thus, we could consider searching for the item in the two subtrees of the binary search tree. This guarantees reducing the size of the problem by a unit, since it discards the root node. Nevertheless, it is easy to see that in this problem we can also avoid searching in an entire subtree. If the key k is less than the key of the root node (k_{root}), then we can be sure that the item we are looking for will not be in the right subtree, due to the binary tree search property.

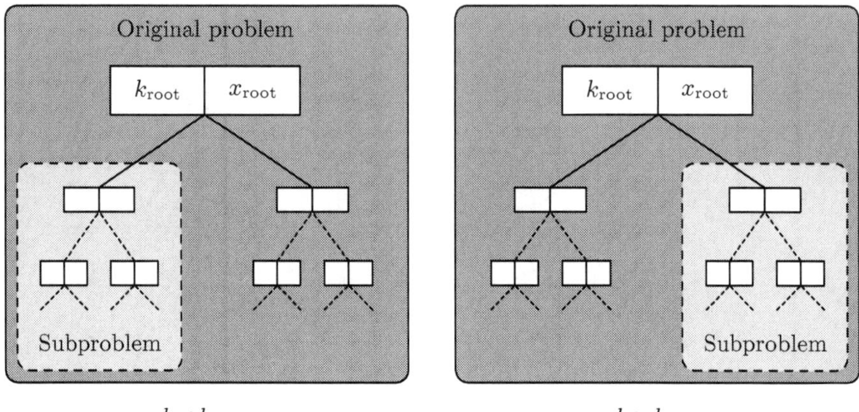

Figure 5.4 Decomposition associated with several algorithms related to binary search trees.

Listing 5.9 Algorithm for searching an item with a particular key in a binary search tree.

```
1 def bst_search(T, key):
2     if T == []:
3         return None
4     elif T[0] == key:
5         return T[1]   # return the root item
6     elif key < T[0]:
7         return bst_search(T[2], key)  # search in left subtree
8     else:
9         return bst_search(T[3], key)  # search in right subtree
```

Analogously, if $k > k_{\text{root}}$, the item will not appear in the left subtree. Figure 5.4 illustrates this idea, where x_{root} is the item stored in the root node. Clearly, for this particular problem there are two recursive cases. If $k < k_{\text{root}}$ the method must keep searching in the left subtree through a recursive call, while if $k > k_{\text{root}}$ it will search for the item in the right subtree. Listing 5.9 shows an implementation of the searching algorithm, where each node of the binary tree is coded as a list of four components, as described in Section 5.3.

Finally, the height of the binary tree determines the cost of the algorithm in the worst case. If the tree is balanced (i.e., it has approximately the same number of nodes on the left and right subtrees of nodes that

Listing 5.10 Procedure for inserting an item with a particular key in a binary search tree.

```
1  def insert_binary_tree(x, T):
2      if T == []:
3          T.extend([x[0], x[1], [], []])
4      else:
5          if x[0] < T[0]:
6              if T[2] == []:
7                  T[2] = [x[0], x[1], [], []]
8              else:
9                  insert_binary_tree(x, T[2])
10         else:
11             if T[3] == []:
12                 T[3] = [x[0], x[1], [], []]
13             else:
14                 insert_binary_tree(x, T[3])
```

appear on the same level) it runs in $\Theta(\log n)$, where n is the number of nodes in the tree, since it would discard approximately half of the nodes with each recursive call. However, the tree could have a linear structure, in which case the algorithm would run in $\mathcal{O}(n)$.

5.3.2 Inserting an item

The goal in this problem is to insert an item with a given key k in a binary search tree, such that the resulting tree also satisfies the binary search tree property. The method will be a procedure that receives two parameters: (1) a tuple x containing the key and the item, and (2) the binary search tree T. In particular, it will extend the list (T) by adding a new leaf to the tree.

Again we can consider that the size of the problem is the height of the tree. The simplest base case occurs when the tree is empty. In that situation the algorithm must extend the tree in order to incorporate a single node, which obviously will not have any children. The diagrams in Figure 5.4 also illustrate the decomposition that we will follow for solving this problem. Clearly, if the tree is not empty, then the procedure will insert a node in the left subtree if k is smaller than the key of the root node. Otherwise, it will insert it in the right subtree. Firstly, assume that $k < k_{\text{root}}$. There are two possible scenarios that lead to a base case and a recursive case. If the root node does not have a left subtree, then the procedure can simply replace the associated empty list by the node

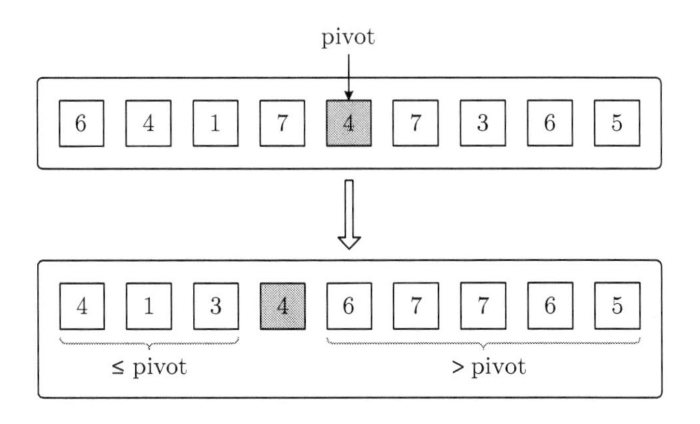

Figure 5.5 Partitioning of a list used in the quicksort and quickselect algorithms.

containing the key and the item (and two empty subtrees). However, if the root node does have a nonempty subtree then the method must insert the new node in that subtree, which naturally leads to a recursive call. The reasoning regarding the right subtree is analogous. Finally, Listing 5.10 shows a possible implementation of the procedure.

5.4 PARTITIONING SCHEMES

A well-known problem in computer science consists of partitioning a list in the following manner:

1. Choose a "pivot" element from the list. Typical choices include the first or middle element of the list, or simply some element selected at random.

2. Construct a new list with the same elements as the input list, or simply permute the input list, where the first elements are smaller than or equal to the pivot, and the last ones are greater than the pivot.

Figure 5.5 illustrates the idea through a concrete example. The problem is relevant since the partitioning is key in other eminent algorithms such as quickselect (see Section 5.5) or quicksort (see Section 6.2.2). There are several well-known and efficient algorithms, also denoted as "partitioning schemes," which solve the problem. The most popular is Hoare's partitioning method, developed by Tony Hoare, which we will analyze shortly. Nevertheless, first we will examine a simpler approach for the sake of introducing more intuitive recursive algorithms.

5.4.1 Basic partitioning scheme

A simple idea consists of scanning the entire input list in order to build: (1) a new list that contains the elements that are smaller than or equal to the pivot, and (2) another list formed by the elements that are greater than the pivot. Afterwards, they can be concatenated together with the pivot element, which would obviously be placed in between them.

Since the problems are very similar we will explain only the first one. In particular, the inputs to the problem are a list **a** of length n, and some value x that plays the role of the pivot. Its size is clearly the length of the list. The base case occurs when the input list is empty, where algorithms can simply return an empty list. For the recursive case we can decompose the problem by discarding the first element of the list, which reduces its size by a unit. In that case, we can form the following recursive diagrams, depending on whether the first element of the list is smaller than or equal to the pivot. If it is, a concrete diagram could be:

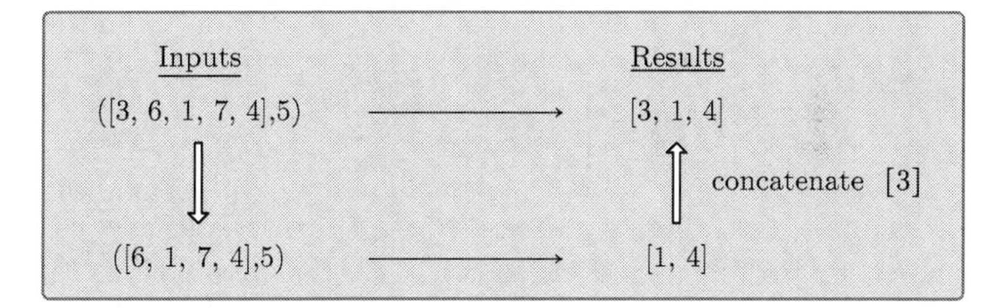

Clearly, the algorithm must concatenate the first element of the list to the output of the subproblem. In contrast, if the first element is greater than x, the diagram is:

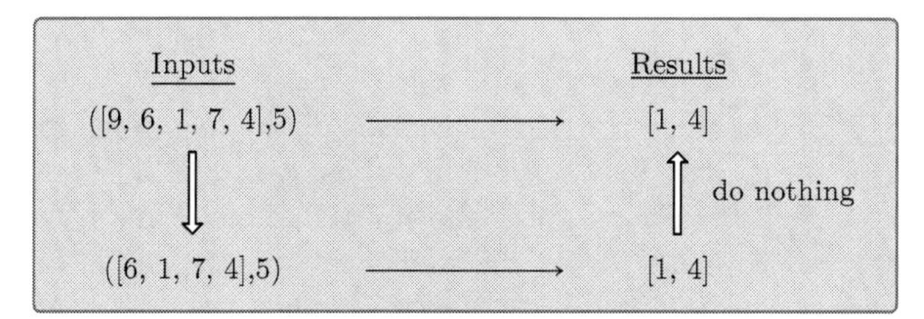

In this case, the algorithm simply has to return the solution (list) to the subproblem. Listing 5.11 shows the linear-recursive codes that solve

Listing 5.11 Auxiliary methods for partitioning a list.

```
def get_smaller_than_or_equal_to(a, x):
    if a == []:
        return []
    elif a[0] <= x:
        return [a[0]] + get_smaller_than_or_equal_to(a[1:], x)
    else:
        return get_smaller_than_or_equal_to(a[1:], x)


def get_greater_than(a, x):
    if a == []:
        return []
    elif a[0] > x:
        return [a[0]] + get_greater_than(a[1:], x)
    else:
        return get_greater_than(a[1:], x)
```

both problems (notice that in the recursive cases that perform the concatenation the function call is not the last action of the method). The runtime cost of the methods can be characterized by:

$$T(n) = \begin{cases} 1 & \text{if } n = 0, \\ T(n-1) + 1 & \text{if } n > 0, \end{cases}$$

which implies that their runtime is linear with respect to n. Lastly, the functions can be substituted with Python expressions that implement "filters." In particular, `get_smaller_than_or_equal_to(a,x)` and `get_greater_than(a,x)` return the same lists as the expressions `[y for y in a if y <= x]` and `[y for y in a if y > x]`, respectively.

5.4.2 Hoare's partitioning method

Listing 5.12 contains an iterative version of Hoare's partitioning algorithm. In particular, it partitions a sublist within the input list (**a**), which is specified through lower and upper indices. Firstly, the method chooses a pivot (in this case, the element in the middle of the sublist) and swaps it with the first element of the sublist (lines 3–6). Afterwards, it sets a left index parameter at the second position of the sublist, and a right index at the last location of the sublist (lines 8 and 9). Subsequently the method enters the main loop of the algorithm. Inside it, it

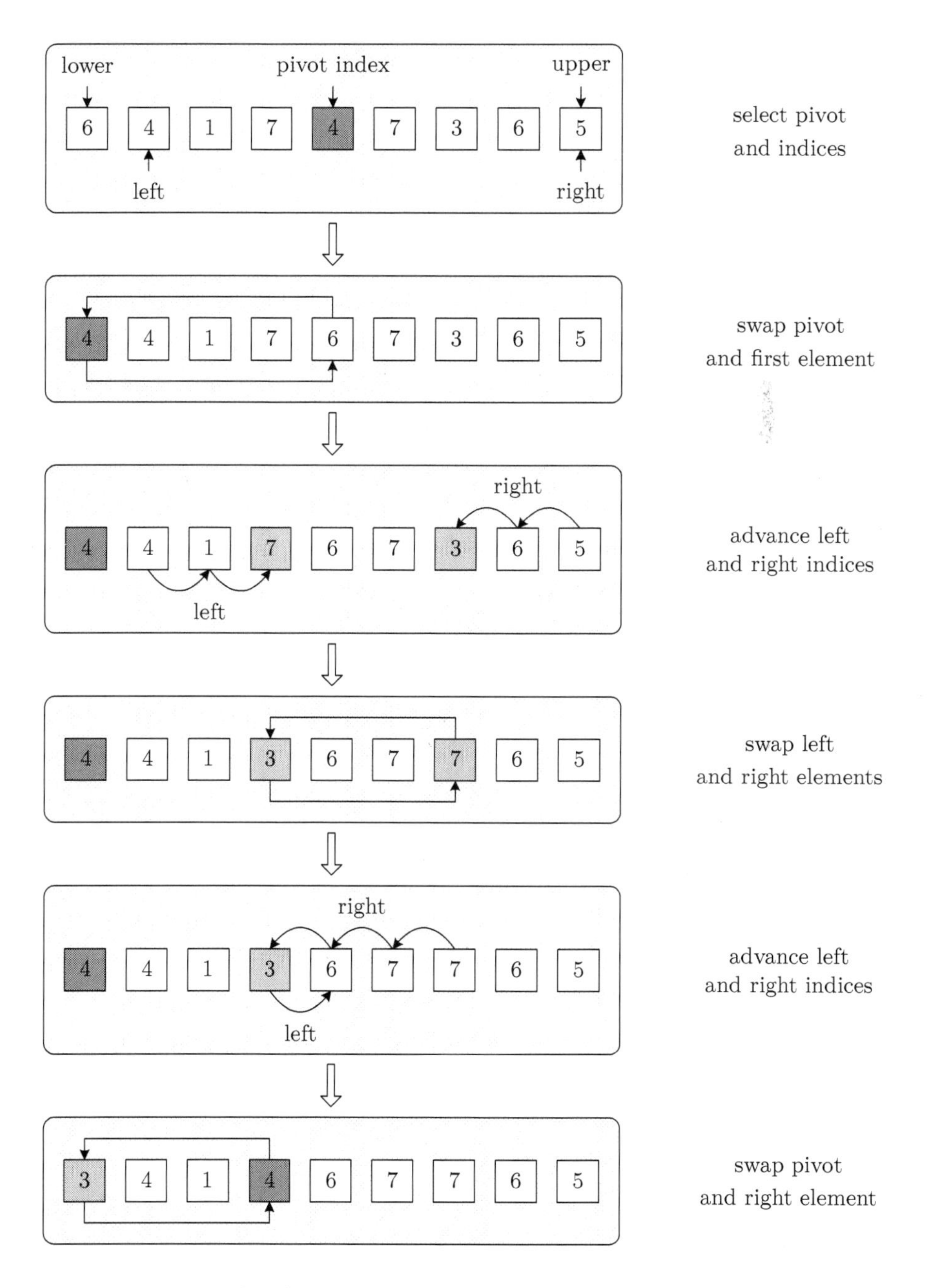

Figure 5.6 Example of Hoare's partition method.

Listing 5.12 Hoare's iterative partitioning algorithm.

```
1  def partition_Hoare(a, lower, upper):
2      if upper >= 0:
3          middle = (lower + upper) // 2
4          pivot = a[middle]
5          a[middle] = a[lower]
6          a[lower] = pivot
7
8          left = lower + 1
9          right = upper
10
11         finished = False
12         while not finished:
13             while left <= right and a[left] <= pivot:
14                 left = left + 1
15
16             while a[right] > pivot:
17                 right = right - 1
18
19             if left < right:
20                 aux = a[left]
21                 a[left] = a[right]
22                 a[right] = aux
23
24             finished = left > right
25
26         a[lower] = a[right]
27         a[right] = pivot
28
29         return right
```

iteratively increases the left index until it references an element that is larger than the pivot (lines 13 and 14). Analogously, it decreases the right index until it references an element that is less than or equal to the pivot (lines 16 and 17). The algorithm then swaps the elements referenced by the left and right index (if the left index is smaller than the right one) in lines 19–22. This process is repeated until the indices cross (i.e., until the left index is greater than the right one, which is checked in line 24). At the end of the loop the elements located up to the right index will be smaller than or equal to the pivot, and the elements from the left index until the end of the sublist will be greater than the pivot. Finally, the procedure completes the partition by swapping the element referenced by the right index with the pivot (lines 26 and 27), and re-

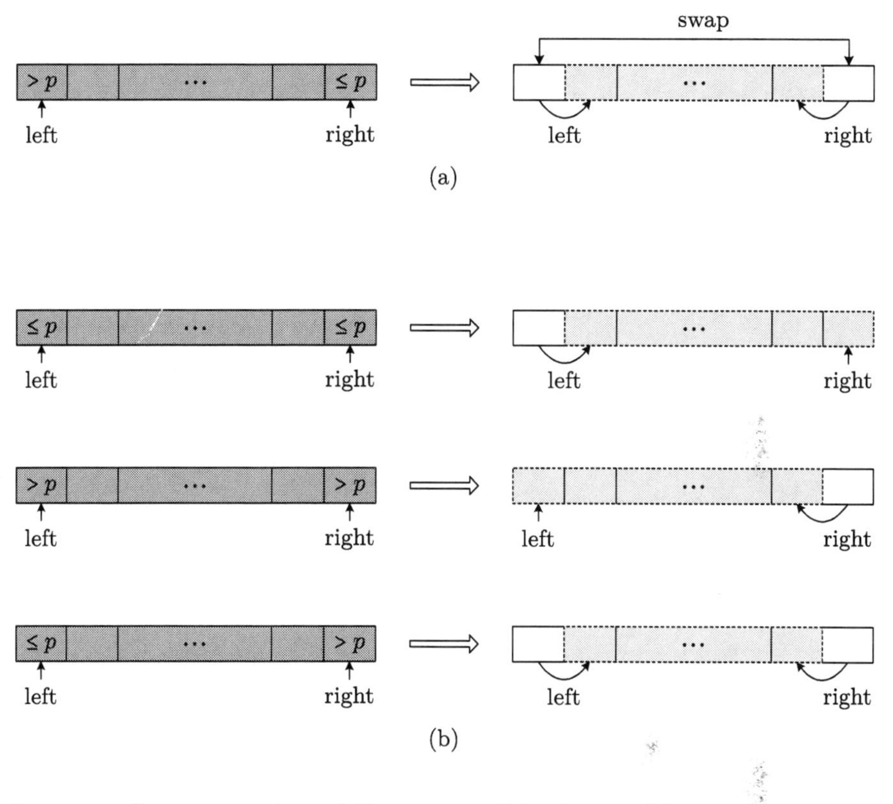

Figure 5.7 Decomposition of Hoare's partitioning problem.

turns the final location of the pivot. Figure 5.6 shows a concrete example of the partitioning method.

We will now see a tail-recursive algorithm that can substitute the main loop in Hoare's partitioning scheme. Given the input list **a**, initial left and right indices, as well as the value of the pivot, the method should partition the list analogously as the loop in Hoare's method, returning the final location of the right index. The size of the problem is the difference between the left and right indices, since it determines the number of increment/decrement operations to be performed on the indices until they cross. The base case occurs precisely when they cross (when the left index is greater than the right one), where the method can simply return the right index.

For the recursive case we must reduce the size of the problem by incrementing the left index and/or decrementing the right one. There are two different scenarios, illustrated in Figure 5.7, where p denotes

Listing 5.13 Alternative recursive version of Hoare's partitioning scheme.

```
1  def partition_Hoare_rec(a, left, right, pivot):
2      if left > right:
3          return right
4      else:
5          if a[left] > pivot and a[right] <= pivot:
6              aux = a[left]
7              a[left] = a[right]
8              a[right] = aux
9              return partition_Hoare_rec(a, left + 1,
10                                         right - 1, pivot)
11         else:
12             if a[left] <= pivot:
13                 left = left + 1
14             if a[right] > pivot:
15                 right = right - 1
16             return partition_Hoare_rec(a, left, right, pivot)
17
18
19 def partition_Hoare_wrapper(a, lower, upper):
20     if upper >= 0:
21         middle = (lower + upper) // 2
22         pivot = a[middle]
23         a[middle] = a[lower]
24         a[lower] = pivot
25
26         right = partition_Hoare_rec(a, lower + 1, upper, pivot)
27
28         a[lower] = a[right]
29         a[right] = pivot
30
31         return right
```

the pivot. If $a_{\text{left}} > p$ and $a_{\text{right}} \leq p$ the method first needs to swap the elements referenced by the indices, and afterwards can perform a function call by advancing both indices, as shown in (a). This leads to a first recursive case that reduces the size of the problem by two units. If the previous condition is `False`, then the method will not swap elements, and will advance at least one of the indices (thus, reducing the size of the problem), as shown in (b). If $a_{\text{left}} \leq p$ it will increment the left index, and if $a_{\text{right}} > p$ it must decrement the right one. Subsequently, the method can invoke itself with the updated indices. Listing 5.13 shows the corresponding code, together with a wrapper method that completes the partitioning algorithm. Note that it is identical to Listing 5.12, where

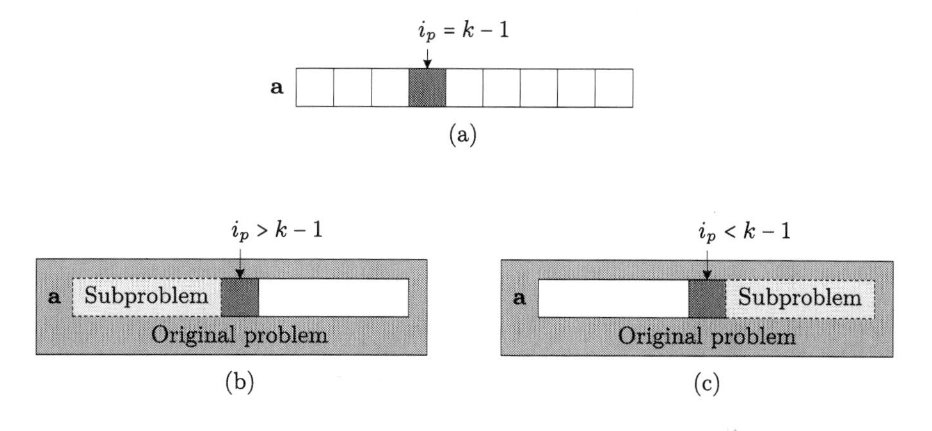

Figure 5.8 Base case and problem decomposition used by the quickselect algorithm.

the code associated with the loop has been substituted by a call to the recursive function. Lastly, the method runs in $\mathcal{O}(n)$ since the runtime of the recursive method can be characterized by the recurrence $T(n) = T(n-1) + 1$ in the worst case.

5.5 THE QUICKSELECT ALGORITHM

The quickselect algorithm, also developed by Tony Hoare, is a selection method based on Hoare's partitioning scheme (see Section 5.4.2). In particular, it finds the k-th smallest number (also called the k-th order statistic) in a list that is not sorted. Assume that an instance of the problem is defined through a numerical input list **a**, lower and upper indices defining a sublist within it, and a positive integer k. The algorithm will therefore search for the k-th smallest number in the specified sublist. The size of the problem is the length of the sublist. The smallest instances of the problem correspond to lists that contain a single element. Therefore, the base case occurs when the lower and upper indices are identical, where the method can simply return the element of the list.

For larger lists the method first applies Hoare's partition scheme, which arranges the list in three parts. A first sublist contains the elements of **a** that are less than or equal to a chosen "pivot" element from the list. This list is followed by another one that only contains the pivot. Finally, a third list contains the elements of **a** that are greater than the pivot.

Listing 5.14 Tail-recursive quickselect algorithm.

```
1  def quickselect(a, lower, upper, k):
2      if lower == upper:
3          return a[lower]
4      else:
5          pivot_index = partition_Hoare_wrapper(a, lower, upper)
6
7          if pivot_index == k - 1:
8              return a[pivot_index]
9          elif pivot_index < k - 1:
10             return quickselect(a, pivot_index + 1, upper, k)
11         else:
12             return quickselect(a, lower, pivot_index - 1, k)
```

Let i_p denote the index of the pivot. Since the function that implements Hoare's partition scheme returns i_p, we can check whether it is equal to $k-1$ (since indices start at location 0, the j-th element appears at position $j-1$). If it is, then the pivot will be the k-th smallest element in **a**, and the method can terminate at this base case. This scenario is illustrated in Figure 5.8(a).

In the recursive case the algorithm reduces the size of the problem by considering either the sublist to the left, or to the right, of the pivot. If $i_p > k-1$, then the location of the searched element will be smaller than i_p. Thus, the algorithm can focus on the sublist to the left of the pivot, as shown in Figure 5.8(b). Instead, if $i_p < k-1$ then the algorithm will proceed by solving the subproblem related to the list to the right of the pivot, as illustrated in (c). Finally, Listing 5.14 contains an implementation of the tail-recursive function.

The runtime cost of the algorithm depends on where the pivot is located after performing the decomposition. If it is always located at the middle position of the list, the running time can be characterized by:

$$T(n) = \begin{cases} 1 & \text{if } n \leq 1, \\ T(n/2) + cn & \text{if } n > 1, \end{cases}$$

since the algorithm needs to solve a subproblem of approximately half the size of the original's, while the methods that perform the partition run in linear time. This is the best case scenario, where $T(n) \in \Theta(n)$. However, if the pivot is always located at an extreme of the list, then the runtime cost is determined by:

$$T(n) = \begin{cases} 1 & \text{if } n \le 1, \\ T(n-1) + cn & \text{if } n > 1, \end{cases}$$

which is a quadratic function (i.e, $T(n) \in \Theta(n^2)$). This situation corresponds to the worst case scenario for the algorithm.

5.6 BISECTION ALGORITHM FOR ROOT FINDING

Consider a continuous real-valued function $f(x)$ defined over some interval $[a, b]$. The bisection algorithm (also known as the interval halving, binary search, or dichotomy method) is a strategy for finding an approximation $\hat{z}$ to a root of $f(x)$ in the (a, b) interval. Recall that a root, or zero, of $f(x)$ is a value r for which $f(r) = 0$.

The inputs to the problem are the extremes of the interval a and b, where $a \le b$, the function f, and a small value $\epsilon > 0$ that determines the accuracy of the approximation. In particular, the output of the function $\hat{z}$ will satisfy $|\hat{z} - r| \le \epsilon$. Thus, the smaller ϵ is, the better the approximation will be. In addition, another prerequisite over the inputs is that $f(a)$ and $f(b)$ must have opposite signs. This guarantees that there will exist a root inside the interval (a, b). Lastly, the approximation $\hat{z}$ is defined as the midpoint between a and b. In other words, $\hat{z} = (a + b)/2$.

The method works by progressively dividing the interval by two where the root must exist. Figure 5.9 illustrates a few steps of the process. In the initial setting the root is located to the left of the approximation ($r < \hat{z}$). Therefore, in the next step b is updated to $\hat{z}$, which halves the interval. Similarly, at step 1, $\hat{z} < r$, and a is assigned the value of $\hat{z}$ in the following step. This procedure is applied repeatedly until the interval is small enough, guaranteeing that $\hat{z}$ meets the required accuracy.

The size of the problem depends on the length of the interval $[a, b]$, and the base case occurs when $b - a \le 2\epsilon$, where the method returns $\hat{z} = (a + b)/2$. Observe that this condition guarantees $|\hat{z} - r| \le \epsilon$, as illustrated in Figure 5.10. In particular, if $\hat{z}$ is the midpoint of the interval $[a, b]$, then the distance between $\hat{z}$ and r cannot be greater than ϵ. In addition, the method can also return $\hat{z}$ trivially if $f(\hat{z}) = 0$ (or if $f(\hat{z})$ is sufficiently small).

The decomposition of the problem involves dividing its size by two. This is accomplished by replacing one of the extremes of the interval $[a, b]$ by its midpoint $\hat{z}$. Since the signs of the new extremes must be

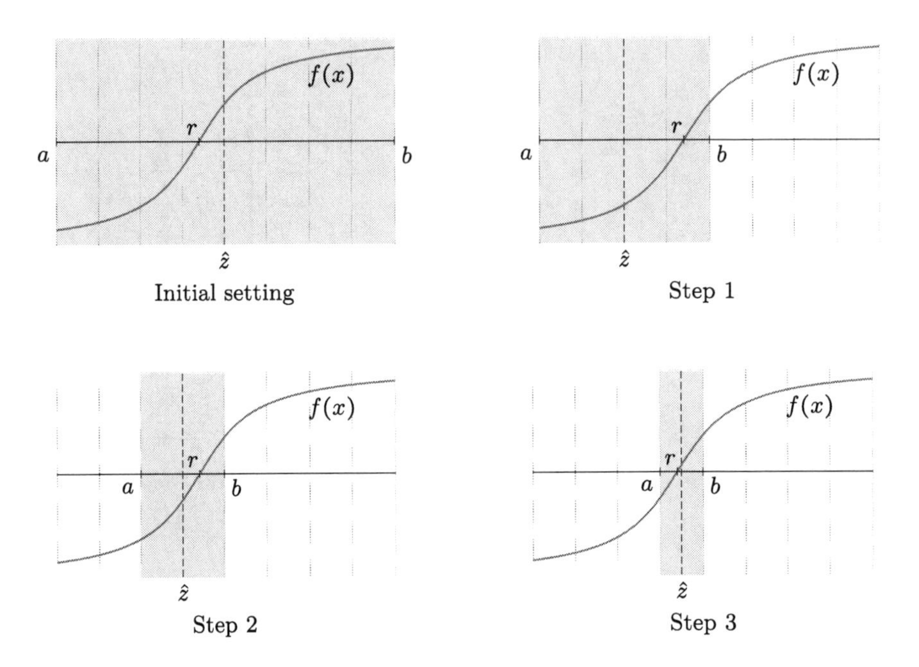

Figure 5.9 Steps of the bisection algorithm.

different, $\hat{z}$ should replace b when $f(a)$ and $f(z)$ have opposite signs. Otherwise, $\hat{z}$ will replace a. This leads to the tail-recursive method in Listing 5.15, which contains two recursive cases. The code uses the function $f(x) = x^2 - 2$ that allows us to find an approximation of $\sqrt{2}$, since it is a root of $f(x)$. Note that the initial interval $[0,4]$ contains $\sqrt{2}$. Finally, the error of the approximation will be less than or equal to 10^{-10} (the result will be accurate up to nine decimal digits to the right of the decimal point).

5.7 THE WOODCUTTER PROBLEM

In this problem a woodcutter needs to collect w units of wood from a forest composed of n trees. He has a special machine that is able to cut the trees at some height h, where he would gather the wood from the trees above h. Assume that w is an integer (not greater than all of the wood in the forest), and that the heights of the n trees, which are also integers, are specified through a list **t**. The goal of the problem is to find the largest value of h that will allow him to collect at least w units of wood. Figure 5.11 shows an instance of the problem for $n = 20$

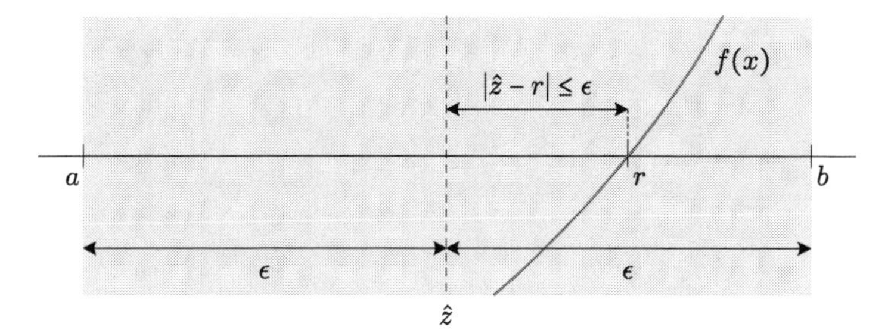

Figure 5.10 Base case of the bisection algorithm ($b - a \leq 2\epsilon$).

Listing 5.15 Bisection algorithm.

```
1  def f(x):
2      return x * x - 2
3
4
5  def bisection(a, b, f, epsilon):
6      z = (a + b) / 2
7
8      if f(z) == 0 or b - a <= 2 * epsilon:
9          return z
10     elif (f(a) > 0 and f(z) < 0) or (f(a) < 0 and f(z) > 0):
11         return bisection(a, z, f, epsilon)
12     else:
13         return bisection(z, b, f, epsilon)
14
15
16 # Print an approximation of the square root of 2
17 print(bisection(0, 4, f, 10**(-10)))
```

trees, where the tallest tree has height $H = 12$. If the goal is to collect 10 units of wood, then the woodcutter should set the height of the cutting machine to $h = 8$, where the total wood collected would be exactly 10 units. If the goal were to collect 7, 8, or 9 units of wood the optimal height h would also be 8. Even though the woodcutter would obtain more wood than initially required, h cannot be higher since cutting at height 9 only provides 6 units of wood.

The problem is interesting from an algorithmic point of view since it can be solved in several ways. For example, the trees can be initially sorted in decreasing order by their height, and subsequently processed from the highest to the lowest until obtaining the optimal height. This

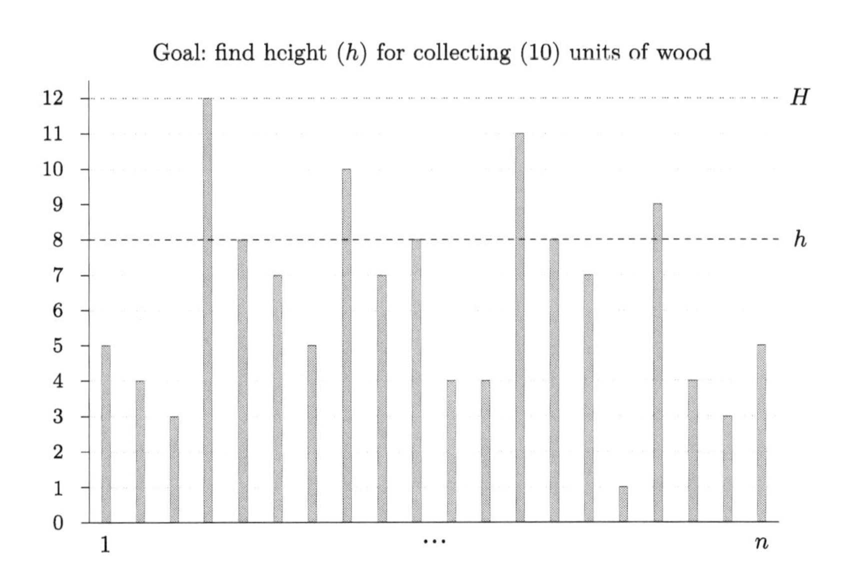

Figure 5.11 Instance of the woodcutter problem.

approach would run in $O(n \log n)$ if we apply general sorting algorithms like merge sort or quicksort (see Chapter 6). Since the heights are integers it is possible to apply linear time sorting algorithms like the counting sort method (see Exercise 5.3), which would lead to a solution that runs in $O(n+H)$. Thus, if H were small this approach could be more efficient.

We will now present a "binary search" solution that runs in $O(n \log H)$, which does not rely on sorting the heights of the trees. Firstly, note that a basic approach consists of starting with $h = H - 1$, and then decreasing h unit by unit until obtaining the required amount of wood.

Listing 5.16 Function that computes the amount of wood collected when cutting trees at height h.

```
1 def compute_wood(t, h):
2     if t == []:
3         return 0
4     else:
5         if t[0] > h:
6             return t[0] - h + compute_wood(t[1:], h)
7         else:
8             return compute_wood(t[1:], h)
```

For simplicity, assume that the method calculates the amount of wood obtained at each new height independently of the amount gathered for other heights. This implies that computing the amount of wood collected from n trees requires on the order of n operations. Therefore, in total, the algorithm for solving the problem would require $O(nH)$ operations in the worst case (at each of the possible H heights the algorithm would need to carry out n computations). In particular, the amount of wood collected at a certain height h can be obtained through the linear-recursive function in Listing 5.16 (Exercise 11.2 proposes implementing a tail-recursive version of the function). The method returns zero in the base case if the list of tree heights is empty. The recursive case processes the first tree height from the list. If it is greater than h the tree will be cut, and the function must return the difference between the height of the tree t_0 and h, plus the wood collected from the rest of the trees (through the recursive call). If the tree's height is less than or equal to h, the tree will not be cut, and the method can simply return the total wood gathered from the remaining trees.

Instead of decreasing h one unit at a time, the following algorithm uses a strategy similar to a binary search of an element in a list, or to the bisection method. The idea is to start searching for the height h in the middle of the interval $[0, H]$, and progressively halving it until a decision can be taken. Thus, the method can receive the tree heights and specified wood as parameters, together with lower and upper limits that indicate where the solution h can be found.

The size of the problem is the difference between the upper and lower limits (which is initially H). A first base case occurs if they share the same value. In that case, the output height h will be precisely that value. Otherwise the algorithm will compute a "middle" height, as the (integer) average of the lower and upper limits. If the total amount of wood that would be gathered from cutting the trees at this middle height is equal to the goal w, then the method can return such middle height in another base case.

Figure 5.12 shows the decomposition of the problem used by the algorithm. The original problem is shown in the top image, given the lower and upper indices. After computing the wood that would be collected at the middle height (h_m), there are two possible scenarios. Firstly, if it is greater than w we can clearly rule out all of the heights below h_m. Thus, we can continue searching for the optimal height (through a recursive call) by replacing the lower limit with h_m. It is important to note that h_m could still be the solution, since the wood gathered for $h_m + 1$ may

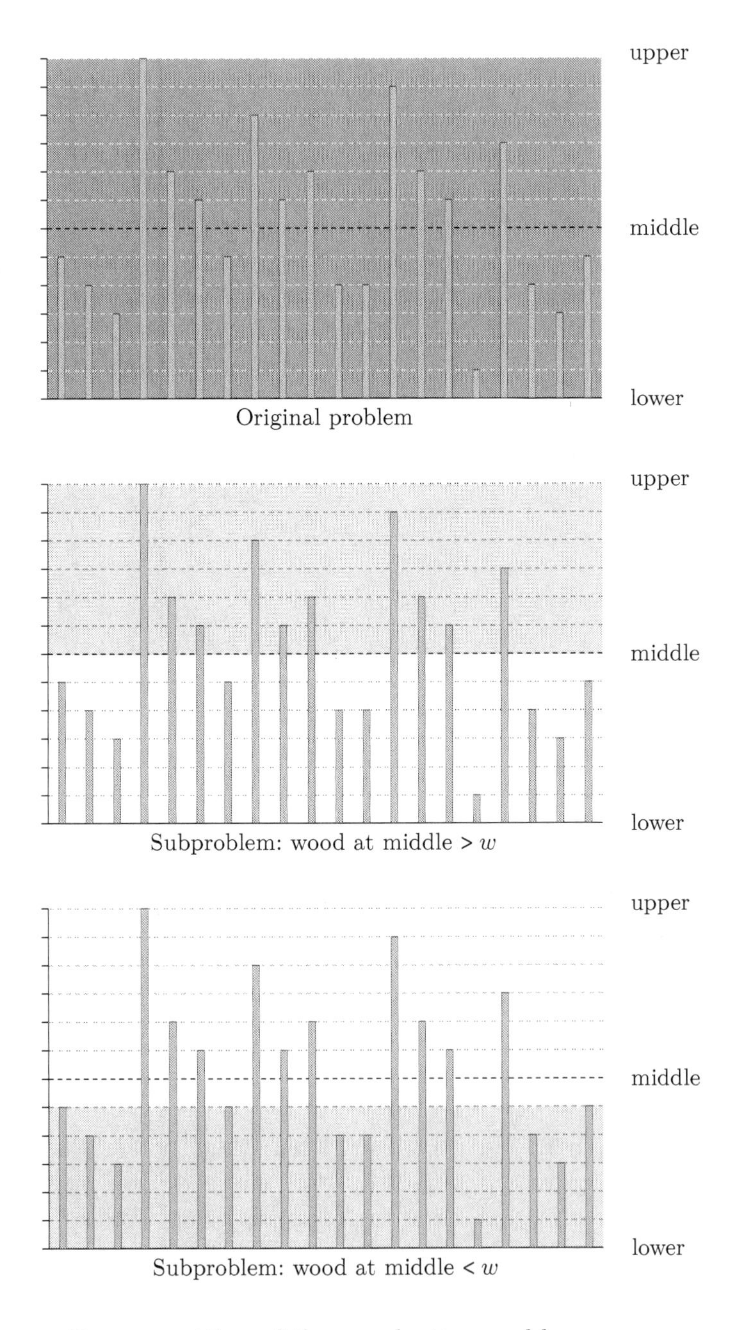

Figure 5.12 Decomposition of the woodcutter problem.

Listing 5.17 Binary search algorithm for the woodcutter problem.

```
1  def collect_wood(t, wood, lower, upper):
2      middle_h = (lower + upper) // 2
3      wood_at_middle = compute_wood(t, middle_h)
4
5      if wood_at_middle == wood or lower == upper:
6          return middle_h
7      elif lower == upper - 1:
8          if compute_wood(t, upper) >= wood:
9              return upper
10         else:
11             return lower
12     elif wood_at_middle > wood:
13         return collect_wood(t, wood, middle_h, upper)
14     else:
15         return collect_wood(t, wood, lower, middle_h - 1)
```

be less than the required amount w. A second scenario occurs when the wood collected at h_m is less than w. In that case we can discard all of the heights that are greater than or equal to h_m, since the woodcutter would not obtain enough wood. The associated recursive case would therefore invoke the method by replacing the upper limit with $h_m - 1$.

Lastly, in the previous decomposition, when the upper limit is just a unit greater than the lower one, the middle height is equal to the lower one. In that situation the first recursive case would not reduce the size of the problem (the limits would not vary). Thus, the algorithm needs an additional base case in order to work properly. When this situation occurs the method has to return either the lower or the upper limit. In particular, if the amount of wood associated with the upper limit is greater than or equal to w, then the upper limit is the correct result. Otherwise, it will be the lower one.

Listing 5.17 shows the code related to the mentioned base and recursive cases. Lastly, Figure 5.13 shows the steps it takes in order to solve the instance in Figure 5.11. Step 1 shows an initial situation where the lower limit is zero and the upper one is $H = \max\{t_i\}$, for $i = 1, \ldots, n$. Step 2 and 3 are associated with applying the first and second recursive cases, respectively. Finally, step 4 applies the additional base case since the upper limit (8) is just a unit over the lower one (7). The solution is $h = 8$, since for that height the woodcutter obtains the requested amount of wood (10 units).

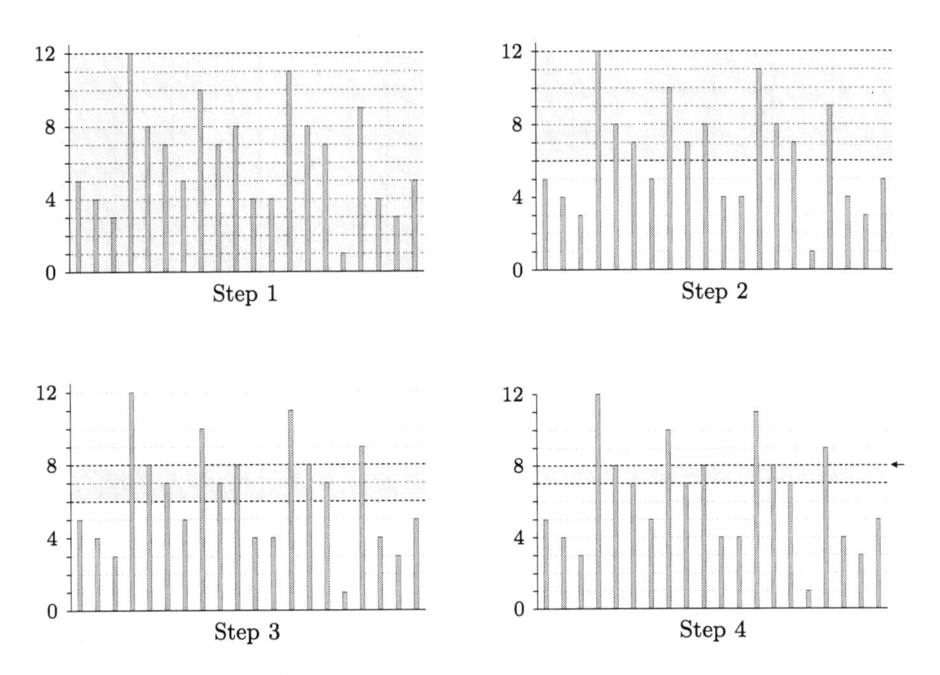

Figure 5.13 Steps of the binary search algorithm related to an instance of the woodcutter problem, for $w = 10$.

5.8 EUCLID'S ALGORITHM

One of the first algorithms developed throughout history is known as Euclid's algorithm, named after the ancient Greek mathematician Euclid, who first described it in his Elements (c. 300 BC). Its purpose is finding the greatest common divisor (gcd) of two nonnegative integers m and n (where both cannot be zero), which is the largest positive integer k that divides both m and n without a remainder (i.e., m/k and n/k are integers). For example, the greatest common divisor of 20 and 24 is 4. It can also be understood as the product of the common prime factors of m and n. For $m = 20 = 2 \cdot 2 \cdot 5$, and $n = 24 = 2 \cdot 2 \cdot 2 \cdot 3$, the product of common factors is $2 \cdot 2 = 4$.

There are several versions of the method. The original one corresponds to the following function:

$$gcd_1(m,n) = \begin{cases} n & \text{if } m = 0, \\ gcd_1(n,m) & \text{if } m > n, \\ gcd_1(m, n-m) & \text{otherwise,} \end{cases} \tag{5.2}$$

Listing 5.18 Euclid's algorithm for computing the greatest common divisor of two nonnegative integers.

```
def gcd1(m, n):
    if m == 0:
        return n
    elif m > n:
        return gcd1(n, m)
    else:
        return gcd1(m, n - m)
```

which can be coded as shown in Listing 5.18. Firstly, it is tail-recursive since it returns the value of a function call in the recursive cases. In addition, observe that the method returns the value of a parameter in the base case, which is common in many tail-recursive functions. In particular, the greatest common divisor of $n > 0$ and 0 is clearly n, since n divides both n and 0 without a remainder. The first recursive case simply switches the order of the arguments. This guarantees that the first one will be larger than or equal to the second one when reaching the second recursive case. In this last recursive case reducing the size of the problem (which depends on m and n) by a unit, or dividing it by two, does not work. Instead, the recursive case decreases the second argument by subtracting m (observe that $n - m \geq 0$, since $n \geq m$). It implements the mathematical property:

$$gcd(m, n) = gcd(m, n - m), \tag{5.3}$$

which is not immediately obvious. It can be shown to be true as follows. Assume $n \geq m$, and let $m = az$, and $n = bz$, where $z = gcd(m, n)$ represents the product of the common prime factors in n and m, and $a \leq b$. This implies that a and b do not share common prime factors. In addition, let $b = a + c$. In that case we can express n and m as:

$$n = bz = (a + c)z = (a_1 \cdots a_k + c_1 \cdots c_l)z, \tag{5.4}$$

$$m = (a_1 \cdots a_k)z, \tag{5.5}$$

where $a_1, \ldots a_k$ and $c_1, \ldots c_l$ are the prime factors of a and c, respectively. The key to the proof is the fact that a and c cannot share prime factors (i.e., $a_i \neq c_j$), because if they did they could be pulled out of the parenthesis in (5.4), implying that a and b share those common prime factors, which is a contradiction. If a and c do not share prime factors then $z = gcd(az, cz)$, and we can conclude that:

$$gcd(az, bz) = z = gcd(az, cz) \Rightarrow gcd(m, n) = gcd(m, n - m).$$

In addition, it can also be shown that the algorithm is guaranteed to reach the base case, since the values of the arguments decrease until one of them is 0. For $m = 20$ and $n = 24$ the method carries out the following recursive calls:

$$\begin{aligned} & gcd_1(20, 24) \\ = \; & gcd_1(20, 4) \\ = \; & gcd_1(4, 20) \\ = \; & gcd_1(4, 16) \\ = \; & gcd_1(4, 12) \\ = \; & gcd_1(4, 8) \\ = \; & gcd_1(4, 4) \\ = \; & gcd_1(4, 0) \\ = \; & gcd_1(0, 4) \; = \; 4. \end{aligned} \tag{5.6}$$

Since tail-recursive functions do not manipulate the result of recursive calls, all of them return the same value, which is obtained at the base case ($gcd_1(0, 4) = 4$). Therefore, tail-recursive functions essentially define relationships between sets of arguments. For example, in (5.6) the pairs $(20, 24)$, $(20, 4)$, and so on, are all related to the value 4. Lastly, a more efficient version that is still used today is:

$$gcd_2(m, n) = \begin{cases} n & \text{if } m = 0, \\ gcd_2(n\%m, m) & \text{if } m \neq 0. \end{cases} \tag{5.7}$$

The proof of the property in the recursive case is similar to the one for (5.3). Assume $n \geq m$, and let $m = az$, and $n = bz$, where z is the product of the common prime factors in n and m, and $a \leq b$. This implies that a and b do not share common prime factors. In addition, let $b = qa + r$, where q and r are the quotient and remainder of the division b/a. The key in this proof is that a and r cannot share common prime factors, since otherwise they would also be prime factors of b, and this is not possible since a and b do not share common prime factors. This implies that $z = gcd(rz, az)$. Thus, we can conclude that:

$$gcd(az, bz) = z = gcd(rz, az) \Rightarrow gcd(m, n) = gcd(n\%m, m).$$

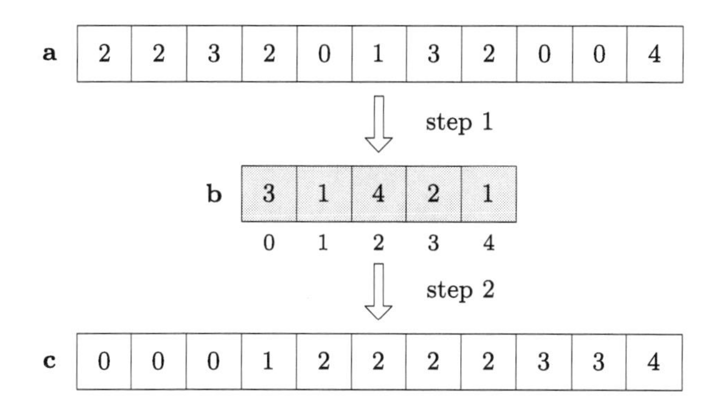

Figure 5.14 Steps in the counting sort algorithm.

5.9 EXERCISES

Exercise 5.1 — Define and code Boolean linear and tail-recursive functions that determine whether a nonnegative integer n contains an odd digit.

Exercise 5.2 — It is more pleasant to display polynomials in a formatted style rather than through a list. Write a method that receives a list of length n that represents a polynomial of degree $n-1$, and prints it in a formatted style. For example, given the input list $\mathbf{p} = [3, -5, 0, 1]$, corresponding to $x^3 - 5x^1 + 3$ (i.e., p_i is the coefficient associated with the term x^i), the method should print a line similar to: `+ 1x^3 - 5x^1 + 3`. Assume that $p_{n-1} \neq 0$, except if the polynomial is the constant 0. Finally, specify its asymptotic computational cost.

Exercise 5.3 — The "counting sort" algorithm is a method for sorting a list $\mathbf{a}$ of n integers in the range $[0, k]$, where k is usually small. The method runs in $\mathcal{O}(n+k)$ time, which implies that it runs in $\mathcal{O}(n)$ time if $k \in \mathcal{O}(n)$.

Given $\mathbf{a}$, the method creates a new list $\mathbf{b}$ that contains the number of occurrences of each integer in $\mathbf{a}$, as shown in step 1 of Figure 5.14. For example, $b_2 = 4$, since the integer 2 appears four times in $\mathbf{a}$. Implement a tail-recursive procedure that receives lists $\mathbf{a}$ and $\mathbf{b}$ (initialized with zeros), and fills list $\mathbf{b}$ with the occurrences of the integers in $\mathbf{a}$.

In addition, implement a linear-recursive function that receives the list of integer occurrences $\mathbf{b}$, and returns a new list ($\mathbf{c}$) that is the sorted version of $\mathbf{a}$, as shown in step 2 of Figure 5.14.

Finally, implement a function that calls the previous two methods in order to implement a version of the counting sort algorithm, and specify its asymptotic computational cost.

Exercise 5.4 — Implement an alternative version of Listing 5.8 that does not use the `upper` parameter. Finally, specify its asymptotic computational cost.

Exercise 5.5 — Implement a Boolean "binary search" function in Python that simply indicates whether an element x appears in a list $\mathbf{a}$. The function will have only two input parameters: x and $\mathbf{a}$, and will be based on decomposing the problem into another one of half its size. Finally, specify its asymptotic computational cost.

Exercise 5.6 — Write a function that searches for the item with the smallest key in a binary search tree T, defined as a list of four components, as described in Section 5.3.

Exercise 5.7 — Let $\mathbf{a}$ be a sorted list (in ascending order) of *different* integers. The goal of this problem is to *efficiently* search for an element that is equal to its position (i.e., index) along the list. In other words, we want to find an element i for which $a_i = i$. For example, if $\mathbf{a} = [-3, -1, 2, 5, 6, 7, 9]$, then the output of the method will be 2, since $a_2 = 2$. Note that the first element of the list is located at position 0. For simplicity, assume that $\mathbf{a}$ will have at most one element that satisfies $a_i = i$. If the list does not contain any element that satisfies that property the function will return the value -1. Finally, specify its asymptotic computational cost.

Exercise 5.8 — Let $\mathbf{a}$ be a list of n integers, arranged so that even numbers appear before odd numbers (the index of any even number will be smaller than the index of any odd number), like, for example, $\mathbf{a} = [2, -4, 10, 8, 0, \mathbf{12}, 9, 3, -15, 3, 1]$. The goal of the following problem is to develop an efficient recursive algorithm for determining the largest index associated with the even numbers. In other words, the index i for which a_i is even, but a_j will be odd for any $j > i$. For the example the method would return $i = 5$, since $a_5 = 12$ is even, but $a_6 = 9$ is odd (and so are a_7, a_8, and so on). If $\mathbf{a}$ does not contain even numbers the result will be -1, while if it is composed entirely of even numbers the output will be $n - 1$. Finally, specify its asymptotic computational cost.

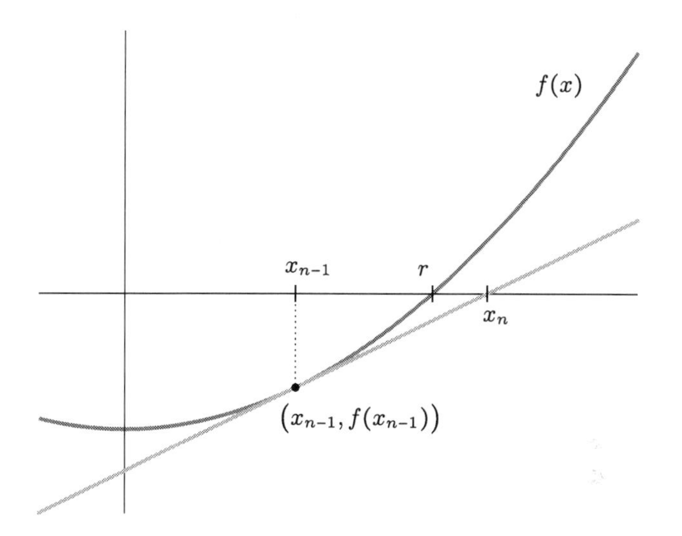

Figure 5.15 Main idea behind Newton's method.

Exercise 5.9 — Newton's method is an approach for finding successively better approximations to the roots of a real-valued function $f(x)$ that is differentiable. It is based on the following recursive rule:

$$x_n = x_{n-1} - \frac{f(x_{n-1})}{f'(x_{n-1})},$$

which is explained geometrically in Figure 5.15. Say we have an initial approximation x_{n-1} to the root r of function $f(x)$. The procedure finds the tangent line to $f(x)$ at x_{n-1}, and computes a new approximation (x_n) to r that is the value of x where the line crosses the X-axis (note that x_n is closer to r). Thus, starting with some initial value x_0, $\hat{z} = x_n$ will be the approximation of the root of $f(x)$ after applying the recursive rule n steps.

This procedure can be used, for example, to obtain very accurate approximations to the square root of a number, in only a few steps. Say we want to calculate $\sqrt{a}$, which is some unknown value x. In that case we have the following identities: $x = \sqrt{a}$, where $x^2 = a$, which implies that $x^2 - a = 0$. Therefore, the root of the function $x^2 - a$ will be the square root of a. In that case, the recursive formula associated with Newton's method is:

$$x_n = x_{n-1} - \frac{x_{n-1}^2 + a}{2x_{n-1}}. \tag{5.8}$$

Implement linear and tail-recursive functions that receive the value a, an initial positive estimate x_0 of $\sqrt{a}$, and a certain number of steps n, and return the final estimate $\hat{z} = x_n$ of $\sqrt{a}$ by applying (5.8) n times. Finally, specify their asymptotic computational cost.

CHAPTER 6

Multiple Recursion I: Divide and Conquer

Διαίρει και βασίλευε (divide and conquer).

— Philip II of Macedon

THE advantages of recursion over iteration, such as code clarity or avoiding managing a stack explicitly (see Section 10.3.5), are mainly due to the possibility of using multiple recursion. Methods based on this type of recursion invoke themselves several times in at least one recursive case. Therefore, these algorithms solve more than one simpler self-similar problem, and must combine, extend, and/or modify their results in order to obtain the solution to the original problem.

The book devotes three chapters to multiple recursion, and contains additional examples throughout it as well. In this chapter we will cover an important class of algorithms based on multiple recursion that decompose problems by dividing their size by some constant. These algorithms are said to follow the "divide and conquer" approach, which is one of the most important algorithm design paradigms. It can be viewed as a general problem solving strategy that consists of breaking up a problem into several self-similar subproblems whose size is a fraction of the original problem's size. The approach is therefore closely related to recursion (iterative algorithms can also follow the divide and conquer strategy), since it relies on recursive problem decomposition (see Figure 1.4). The function in Listing 1.2 and the third method of Listing 1.5 constitute examples of this algorithm design approach.

In some references the term "divide and conquer" is also applied to algorithms where the decomposition is based on a single subproblem

Listing 6.1 Function that determines whether a list is sorted in ascending order.

```
def is_list_sorted(a):
    n = len(a)
    if n <= 1:
        return True
    else:
        return (is_list_sorted(a[0:n // 2])
                and a[n // 2 - 1] <= a[n // 2]
                and is_list_sorted(a[n // 2:n]))
```

of half the size of the original's (the method would invoke itself only once). However, many authors consider that the term should only be used when the solution relies on breaking up a problem into two or more subproblems. This book adopts this last convention. Thus, the methods that invoke themselves once, dividing the size of the problem by two, have been covered in earlier chapters (mostly in Chapter 5). The following sections describe classical recursive divide and conquer algorithms.

6.1 IS A LIST SORTED IN ASCENDING ORDER?

In this problem the input is a list **a** of n items that can be sorted through the $\leq$ operator (or some function that allows us to compare elements by implementing a total order binary relation), and the output is a Boolean value that can be expressed as:

$$f(\mathbf{a}) = \bigwedge_{i=0}^{n-2} (a_i \leq a_{i+1}). \tag{6.1}$$

Thus, the result is `True` when the elements of the list appear in (non-strictly) increasing or nondecreasing order.

The size of this problem is the number of elements in the list. If the list contains one element the result is trivially `True`. In addition, we can consider that an empty list is also ordered.

The problem can be solved by a linear-recursive method that reduces the size of the problem one unit in the decomposition stage. However, we will present a solution that decomposes the problem by dividing the input list in two halves, and which leads to the following partition of

(6.1):

$$f(\mathbf{a}) = \underbrace{(a_0 \le a_1) \wedge \cdots \wedge (a_{\lfloor n/2 \rfloor -2} \le a_{\lfloor n/2 \rfloor -1})}_{\text{Subproblem 1}}$$
$$\wedge \; (a_{\lfloor n/2 \rfloor -1} \le a_{\lfloor n/2 \rfloor}) \wedge \underbrace{(a_{\lfloor n/2 \rfloor} \le a_{\lfloor n/2 \rfloor +1}) \wedge \cdots \wedge (a_{n-2} \le a_{n-1})}_{\text{Subproblem 2}}.$$

Clearly, if a list is sorted in ascending order then both halves must also be sorted in ascending order. Thus, the method can invoke itself twice with the two corresponding sublists, and perform a logical AND operation with their result. Finally, the combination of the results of the subproblems needs an additional step. In particular, the last element of the first sublist ($a_{\lfloor n/2 \rfloor -1}$) must be less than or equal to the first element of the second sublist ($a_{\lfloor n/2 \rfloor}$). This condition also requires another AND operation in the recursive case. Listing 6.1 shows a possible implementation of the method, which runs in $O(n)$ time, since its cost is characterized by:

$$T(n) = \begin{cases} 1 & \text{if } n \le 1, \\ 2T(n/2) + 1 & \text{if } n > 1. \end{cases}$$

6.2 SORTING

Sorting a general list (or array, sequence, etc.) is one of the most studied problems in computer science. It can be solved by numerous algorithms that can be used to introduce essential concepts related to computational complexity and runtime analysis, algorithm design paradigms, or data structures. The sorting algorithms covered in this chapter will assume that the input to the problem is a list $\mathbf{a}$ of n real numbers. The output is a rearrangement (or permutation) of the elements that provides another list $\mathbf{a}'$ where $a'_i \le a'_{i+1}$, for $i = 0, \ldots, n-2$ ($\le$ can be thought of as Boolean function that allows us to determine if an element precedes another, implementing a total order binary relation). The choice of real numbers for the type of the elements of the list implies that the algorithms need to carry out comparisons through $\le$. It can be shown that any algorithm that solves this problem requires $\Omega(n \log n)$ comparisons (i.e., at least on the order of $n \log n$ decisions using $\le$). Lastly, there exist algorithms for sorting lists that do not need to compare elements, which require $\Omega(n)$ operations. However, the elements of the list must satisfy certain conditions in order to apply them. For example, the "counting

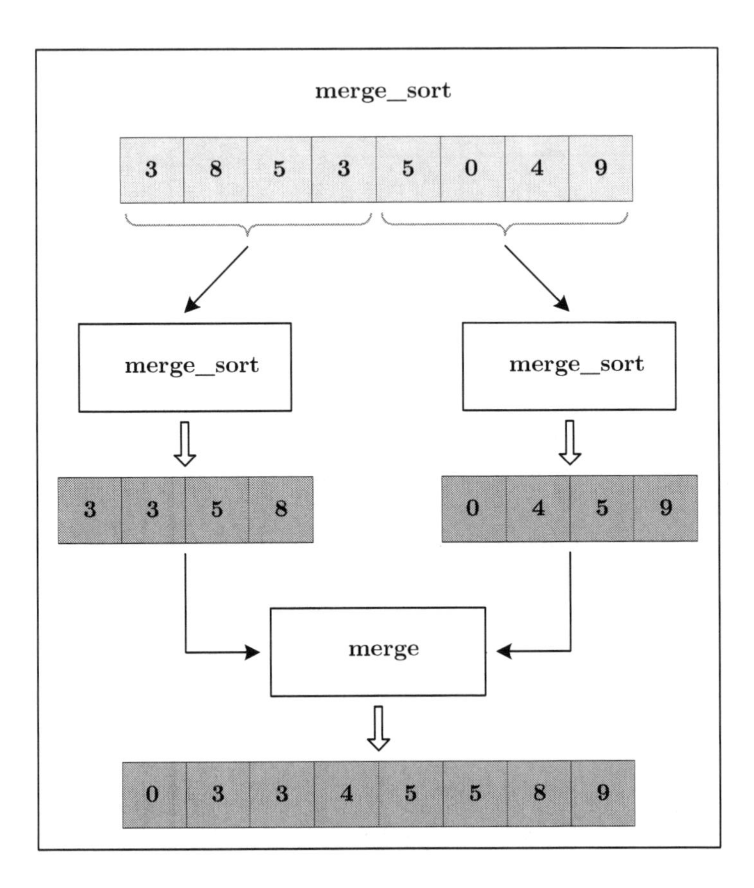

Figure 6.1 Merge sort algorithm.

sort" algorithm sorts integers that belong to a small interval $[0, k]$ (see Exercise 5.3).

6.2.1 The merge sort algorithm

The merge sort algorithm is one of the most extensively used examples to show the potential of the divide and conquer approach. While many sorting algorithms run in $\mathcal{O}(n^2)$ time in the worst case (see Section 4.4.1 and Exercise 4.17), the merge sort algorithm runs in $\Theta(n \log n)$. In other words, its efficiency is optimal from a computational complexity perspective.

The size of the problem is the length of the list (n). The smallest instances occur when $n \leq 1$, where the algorithm can simply return the input list. In the recursive case the algorithm decomposes the problem by

Listing 6.2 Merge sort method.

```
def merge_sort(a):
    n = len(a)
    if n <= 1:
        return a
    else:
        a1 = merge_sort(a[0:n // 2])
        a2 = merge_sort(a[n // 2:n])
        return merge(a1, a2)
```

dividing the input list in two halves, giving rise to two different subproblems of (roughly) half the size of the original's. The following diagram shows a concrete example of the recursive thought process associated with the algorithm:

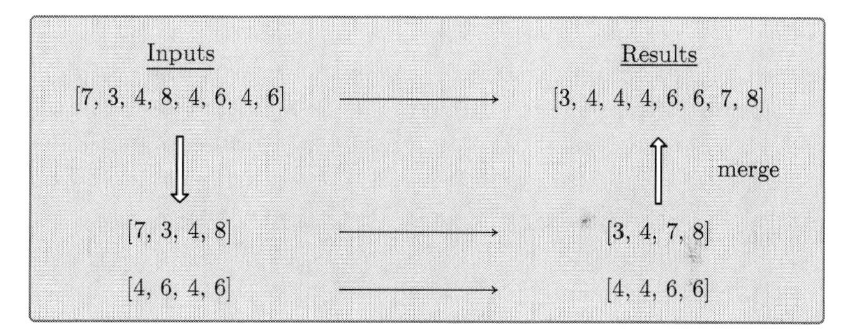

The recursive case is more complicated than in previous examples, since the final sorted list cannot be obtained through a simple operation (e.g., addition, computing the maximum of two numbers, concatenating two lists, etc.). In this case, it is necessary to solve a new problem in order to combine the solutions to the subproblems. In particular, the recursive case needs to "merge" two sorted lists (the outputs of the subproblems) in order to produce the final sorted list. Assuming we have implemented a function `merge` that solves this problem (see below), the succinct code in Listing 6.2 implements the merge sort method. Note the readability of the code and the simplicity of the algorithm, which conforms perfectly to the divide and conquer strategy.

The runtime cost for this algorithm is:

$$T(n) = \begin{cases} 1 & \text{if } n <= 1, \\ 2T(n/2) + f(n) & \text{if } n > 1, \end{cases}$$

where we have also assumed that the partition of the list can be obtained in a constant number of operations. In other words, we have considered that it is possible to obtain `a[0:n//2]` and `a[n//2:n]` in $\Theta(1)$ time. In addition, $f(n)$ measures the number of operations needed by the `merge` method in order to combine two sorted lists of (approximately) $n/2$ elements. In this case, due to the master theorem (see (3.28)), if $f(n)$ were a linear function, the merge sort algorithm would run in the (optimal) order of $\Theta(n \log n)$. We will see shortly that indeed it is possible to solve the merge problem in linear time. In general, when faced with a similar problem, algorithm designers focus their efforts on developing the most efficient combination method, which may reduce the order of growth of the divide and conquer algorithm.

The inputs to the merging problem are two sorted lists **a** and **b**, of lengths n_a and n_b, respectively. The output is another sorted list of length $n = n_a + n_b$. We can interpret that the size of the problem is the number of operations needed by the algorithm until it can return a trivial answer. In this scenario, the size of the problem is $m = \min(n_a, n_b)$, since the solution to the problem if one of the lists is empty is obviously the other list. This constitutes the base cases. For the recursive case we can use the following diagram with the sublists used in the previous example:

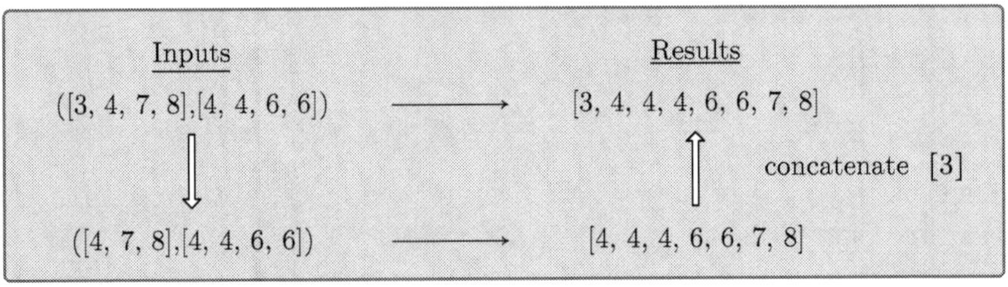

Inputs		Results	
([3, 4, 7, 8],[4, 4, 6, 6])	⟶	[3, 4, 4, 4, 6, 6, 7, 8]	
⇓		⇑	concatenate [3]
([4, 7, 8],[4, 4, 6, 6])	⟶	[4, 4, 4, 6, 6, 7, 8]	

The decomposition reduces the size of the problem by a unit, by discarding the smallest element in either list. Naturally, it must appear at the initial position of the lists, since they are sorted in ascending order. In the example, the smallest element (3) is found in the first input list. Clearly, the recursive case simply needs to concatenate this element with the result of the subproblem. Listing 6.3 shows the corresponding code. Its runtime cost is characterized by the following function:

$$T(m) = \begin{cases} 1 & \text{if } m = 0, \\ T(m-1) + 1 & \text{if } m > 0, \end{cases}$$

assuming that the tail of a list (i.e., `x[1:]`) can be obtained in constant time. Since its nonrecursive expression is $T(m) = m + 1$, the merge func-

Listing 6.3 Method for merging two sorted lists.

```
1  # lists a and b are sorted in ascending order
2  def merge(a, b):
3      if a == []:
4          return b
5      elif b == []:
6          return a
7      else:
8          if a[0] < b[0]:
9              return [a[0]] + merge(a[1:], b)
10         else:
11             return [b[0]] + merge(a, b[1:])
```

tion runs in linear time with respect to m, and also n, which implies that the merge sort algorithm runs in $\Theta(n \log n)$.

6.2.2 The quicksort algorithm

The quicksort algorithm is another method developed by Tony Hoare. It is based on the divide and conquer approach and receives its name due to its remarkable efficiency. Unlike the merge sort algorithm, its running time in the worst case is $O(n^2)$. However, in the best and average cases it runs in $\Theta(n \log n)$ time, and can be several times quicker than the merge sort algorithm in practice.

To understand the difference between both methods, observe that the decomposition in merge sort is easy. In particular, the input list can be divided in two halves by simply using appropriate index ranges (`0:n//2` and `n//2:n`). However, combining the results of the subproblems requires solving yet another problem (merge) that is not immediately straightforward. In contrast, in the quicksort algorithm the decomposition is hard, but the combination stage is not only trivial, but may not even be necessary in some implementations.

Specifically, instead of simply dividing the list in two halves, quicksort's decomposition transforms the input by using a partitioning scheme like the ones presented in Section 5.4. Figure 6.2 illustrates this type of decomposition, where the sublists within the original list are separated by a pivot (the sublist to the right of the pivot contains elements that are less than or equal to the pivot, and the sublist to the left is composed of numbers that are greater than the pivot). A concrete recursive diagram based on the previous decomposition could be:

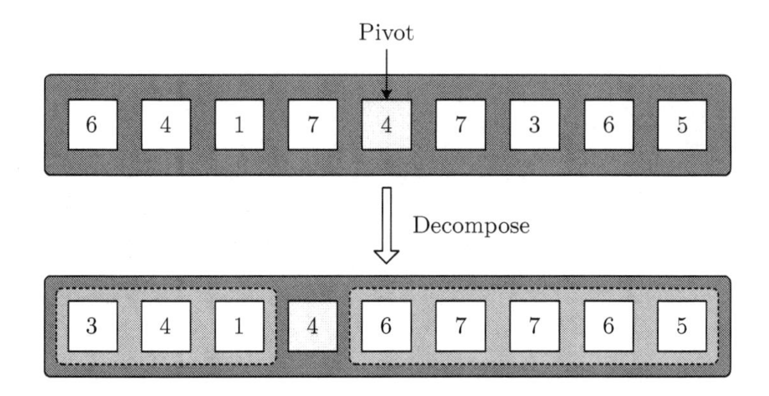

Figure 6.2 Decomposition of the quicksort algorithm.

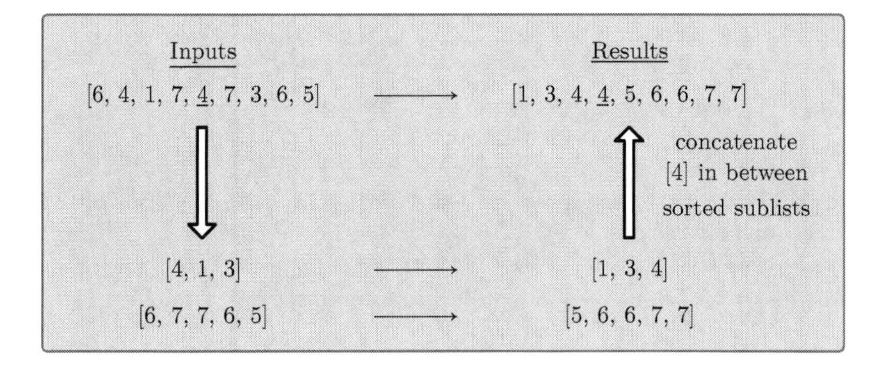

The diagram clearly shows that sorting the original list simply requires solving the two subproblems through recursive calls, and concatenating the sorted sublists while maintaining the pivot in between them. Thus, after solving the subproblems recursively, the combination of the respective solutions is straightforward. Lastly, an important detail regarding this decomposition is that the pivot is removed from the list that contains the elements that are smaller than or equal to it. This is necessary in order to guarantee that the size of the subproblem is indeed smaller than the size of the original problem.

Listing 6.4 implements a slower variant of the method that is based on the basic partitioning schemes described in Section 5.4.1. Firstly, it checks whether the input corresponds to the base case (which is the same as in the merge sort method). In the recursive case a common strategy is to consider that the pivot is the first element of the list. With this choice the worst case occurs when the input list is already sorted. Moreover, the algorithm would also perform poorly if the input

Listing 6.4 Variant of the quicksort algorithm.

```
def quicksort_variant(a):
    n = len(a)
    if n <= 1:
        return a
    else:
        pivot = a[n // 2]
        v1 = get_smaller_than_or_equal_to(a, pivot)
        v1.remove(pivot)
        v2 = get_greater_than(a, pivot)
        return (quicksort_variant(v1) + [pivot]
                + quicksort_variant(v2))
```

Listing 6.5 In-place quicksort algorithm.

```
def quick_sort_inplace(a, lower, upper):
    if lower < upper:
        pivot_index = partition_Hoare_wrapper(a, lower, upper)

        quick_sort_inplace(a, lower, pivot_index - 1)
        quick_sort_inplace(a, pivot_index + 1, upper)
```

is almost sorted. Since these situations may occur frequently in practice, the function uses the middle element of the input list instead as the pivot. Subsequently, the method obtains the two sublists associated with the decomposition (removing the pivot from the sublist that contains it), and finally concatenates the results of the subproblems, leaving the pivot in between them.

The merge sort method in Listing 6.2, and the quicksort function in Listing 6.4, do not alter the input list and return their result in a new list, which requires twice as much storage space. These types of algorithms are denoted as "out-of-place." Instead, it is possible to implement variants that overwrite the input list, which do not need to allocate storage space for a full list of length n. These algorithms are known as "in-place." Listing 6.5 shows an in-place variant of the quicksort algorithm that uses lower and upper limits to indicate the boundaries of a sublist within the original one. The code includes the recursive version of Hoare's partitioning method described in Listing 5.13 (naturally, the iterative version, which is more efficient, can be used as well), which carries out the decomposition of the problem.

The runtime cost analysis of the algorithm is similar to that of the quickselect algorithm. The main difference is that, instead of solving one subproblem, quicksort invokes itself twice. The best case occurs when the pivot is always located at the middle position of the list. In that case, the running time can be characterized by:

$$T(n) = \begin{cases} 1 & \text{if } n \leq 1, \\ 2T(n/2) + cn & \text{if } n > 1, \end{cases}$$

where $T(n) \in \Theta(n \log n)$. Instead, the worst case occurs when the pivot is always located at an extreme of the list. In that case the runtime cost is determined by:

$$T(n) = \begin{cases} 1 & \text{if } n \leq 1, \\ T(n-1) + cn & \text{if } n > 1, \end{cases}$$

which is a quadratic function ($T(n) \in \Theta(n^2)$).

6.3 MAJORITY ELEMENT IN A LIST

A list is said to have a "majority element" if *more* than half of its entries are identical. In this classical problem the goal consists of determining whether a list **a** of length n has a majority element, and, if so, to find that element. In addition, the problem assumes that the elements can only be compared in $\Theta(1)$ (constant) time by using the equality (==) operator. In other words, we are not allowed to compare them through relations such as $<$, $>$, $\leq$, or $\geq$ (the examples will use integers, but not those operators).

There are several divide and conquer variants that solve the problem. We will consider a method that returns a tuple (it could also be a list) containing three values that indicate:

a) Whether the list contains a majority element (Boolean).

b) The majority element if it exists. Otherwise this value is irrelevant (the function could simply return `None`).

c) The number of occurrences of the majority element (0 if it does not exist).

The size of the problem is n. A first base case can correspond to an empty list, where the result would be the tuple (`False`, x, 0). The value

x would be irrelevant. Additionally, if the list contains one element then the function would return (True, a_0, 1).

The recursive case can rely on decomposing the list in two halves, and solving the corresponding subproblems. We can form initial recursive diagrams with concrete examples in order to understand how we can combine the solutions to the subproblems. For example:

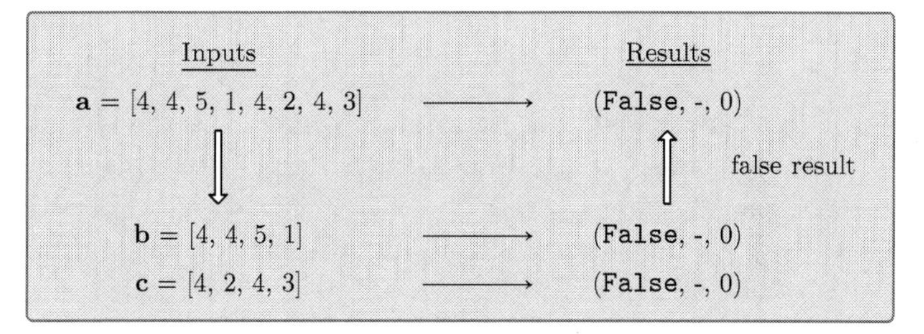

In this case the result of both subproblems is False, which implies that the input list cannot contain a majority element (even though there are exactly $n/2$ occurrences of the element 4 in the example, it is not enough to produce a true result). In general, this can be shown as follows. Firstly, the initial list is divided into a sublist **b** of length $\lfloor n/2 \rfloor$, and another sublist **c** of length $\lceil n/2 \rceil$, regardless of whether n is even or odd, since $n = \lfloor n/2 \rfloor + \lceil n/2 \rceil$. If the result for both subproblems is False then an element can appear at most $\lfloor \lfloor n/2 \rfloor /2 \rfloor$ times in **b**, and $\lfloor \lceil n/2 \rceil /2 \rfloor$ times in **c**. Adding these quantities yields:

$$\left\lfloor \frac{\lfloor n/2 \rfloor}{2} \right\rfloor + \left\lfloor \frac{\lceil n/2 \rceil}{2} \right\rfloor \leq \left\lfloor \frac{\lfloor n/2 \rfloor + \lceil n/2 \rceil}{2} \right\rfloor = \left\lfloor \frac{n}{2} \right\rfloor \leq \frac{n}{2}.$$

Therefore, we can conclude that an element cannot appear more than $n/2$ times in the initial list.

Another concrete diagram could be:

Inputs		Results
a = [4, 4, 5, 4, 1, 2, 4, 3]	⟶	(False, -, 0)
⇓		⇑ $\underline{3} + \underbrace{\#(\mathbf{c},4)}_{=1} > n/2$?
b = [4, 4, 5, 4]	⟶	(True, 4, $\underline{3}$)
c = [1, 2, 4, 3]	⟶	(False, -, 0)

Listing 6.6 Code for counting the number of times an element appears in a list.

```
1 def occurrences_in_list(a, x):
2     if a == []:
3         return 0
4     else:
5         return int(a[0] == x) + occurrences_in_list(a[1:], x)
```

In this case the element 4 appears three times in the first sublist, and is therefore a majority element since $3 > n/2 = 2$. The algorithm must therefore count the number of occurrences of 4 in the second list (denoted through $\#(\mathbf{c}, 4)$) in order to determine whether it is also a majority element of the initial list $\mathbf{a}$. This can be computed through a simple linear-recursive function (see Listing 6.6) that receives an input list $\mathbf{a}$ and an element x. If $\mathbf{a}$ is empty the result is obviously 0. Otherwise, the output can consist of the method applied to the tail of $\mathbf{a}$ ($\mathbf{a}_{1..n-1}$) and x, plus a unit only if $a_0 = x$.

Listing 6.7 shows a possible implementation of the function that solves the majority element problem. Lines 3–6 code the base cases. Lines 8 and 9 decompose the input list into two halves. Line 11 invokes the method on the first sublist, and if there exists a majority element (line 12), then line 13 computes the number of occurrences of the element in the second sublist. If the total number of occurrences of the element (in both sublists) is greater than $n/2$ (line 14), then the method returns a tuple in line 15 with the values: `True`, the majority element, and the number of times it appears in the input list. Lines 17–21 are analogous, but switch the roles of the sublists. Finally, if the function has not returned, then the list does not contain a majority element (line 23).

In the recursive cases the method needs to invoke itself twice with one half of the input list, and also needs to compute the occurrences of an element on two sublists of length $n/2$ (approximately). Since this last auxiliary function runs in linear time, the time complexity of the method can be characterized by:

$$T(n) = \begin{cases} 1 & \text{if } n \leq 1, \\ 2T(n/2) + cn & \text{if } n > 1. \end{cases}$$

Listing 6.7 Code for solving the majority element problem.

```
def majority_element_in_list(a):
    n = len(a)
    if n == 0:
        return (False, None, 0)
    elif n == 1:
        return (True, a[0], 1)
    else:
        b = a[0:n // 2]
        c = a[n // 2:n]

        t = majority_element_in_list(b)
        if t[0]:
            occurrences = occurrences_in_list(c, t[1])
            if t[2] + occurrences > n / 2:
                return (True, t[1], t[2] + occurrences)

        t = majority_element_in_list(c)
        if t[0]:
            occurrences = occurrences_in_list(b, t[1])
            if t[2] + occurrences > n / 2:
                return (True, t[1], t[2] + occurrences)

        return (False, None, 0)
```

Therefore, the order of growth of the algorithm is $\Theta(n \log n)$ (see (3.28)). Lastly, this problem can be solved through the Boyer–Moore majority vote algorithm in linear time.

6.4 FAST INTEGER MULTIPLICATION

The classical algorithm taught in school to multiply two nonnegative n-digit integers requires n^2 digit-times-digit multiplications. In this section we will analyze Karatsuba's algorithm, which is a faster approach. The method can be applied to numbers expressed in any base, but we will focus on multiplying binary numbers. In particular, let x and y be two nonnegative integers represented by b_x and b_y bits, respectively. Applying a divide and conquer approach, each binary number can be partitioned in two as follows:

$$\begin{aligned} x &= a \cdot 2^m + b, \\ y &= c \cdot 2^m + d, \end{aligned} \tag{6.2}$$

where $m = \min(\lfloor b_x/2 \rfloor, \lfloor b_y/2 \rfloor)$. For example, for $x = 594$, and $y = 69$, the decomposition is:

$$x = 1001010010_2 = \underbrace{1001010}_{a=74}\,\underbrace{010}_{b=2} = 74 \cdot 2^3 + 2,$$

$$y = 1000101_2 = \underbrace{1000}_{c=8}\,\underbrace{101}_{d=5} = 8 \cdot 2^3 + 5,$$

where $b_x = 10$, $b_y = 7$, $m = 3$, $a = 74$, $b = 2$, $c = 8$, and $d = 5$. Thus, the smaller number (in this case, y) is partitioned in two parts with (roughly) the same number of bits, and the numbers associated with the lower-significant parts (b and d) are expressed with the same number of bits.

In many programming languages it is possible to compute the values of a, b, c, and d by relying on bit shift operations. In Python they can be carried out through the << and >> operators. On the one hand, $(x << m)$ is equivalent to $x2^m$, which shifts the bits of x m times towards the left, appending m least-significant zeros. On the other hand, $(x >> m)$ performs $\lfloor x/2^m \rfloor$, shifting the bits of x m times towards the right, which discards those m bits. These bitwise operations are therefore useful for multiplying (or performing an integer division) by a power of two, and can be used as follows to decompose x and y according to (6.2):

$$\begin{aligned} a &= x >> m, \\ b &= x - (a << m), \\ c &= y >> m, \\ d &= y - (c << m), \end{aligned}$$

where the parentheses are necessary due to operator precedence rules.

According to the decomposition, the product of x and y can be written as:

$$xy = (a \cdot 2^m + b)(c \cdot 2^m + d) = ac2^{2m} + (ad + bc)2^m + bd. \tag{6.3}$$

This initial (naive) approach can break up the original problem (a multiplication) into four smaller subproblems: ac, ad, bc, and bd, which can be computed through four recursive calls (we can ignore the cost of multiplications times powers of two, since these can be implemented very efficiently as bit shifts). However, it is not more efficient than the

Listing 6.8 Karatsuba's fast algorithm for multiplying two nonnegative integers.

```
def number_of_bits(n):
    if n < 2:
        return 1
    else:
        return 1 + number_of_bits(n >> 1)


def multiply_karatsuba(x, y):
    if x == 0 or y == 0:
        return 0
    elif x == 1:
        return y
    elif y == 1:
        return x
    else:
        n_bits_x = number_of_bits(x)
        n_bits_y = number_of_bits(y)

        m = min(n_bits_x // 2, n_bits_y // 2)

        a = x >> m
        b = x - (a << m)
        c = y >> m
        d = y - (c << m)

        ac = multiply_karatsuba(a, c)
        bd = multiply_karatsuba(b, d)
        t = multiply_karatsuba(a + b, c + d) - ac - bd

        return (ac << (2 * m)) + (t << m) + bd
```

"school" method, since it also requires n^2 bit-times-bit multiplications for two n-bit numbers. The algorithm proposed by Karatsuba is able to reduce such quantity to approximately $n^{1.585}$ by rearranging the terms and including more addition/subtraction operations, which turn out to be negligible regarding asymptotic computational complexity. In particular, the product xy can be reformulated as:

$$xy = ac2^{2m} + \left[(a+b)(c+d) - ac - bd\right]2^m + bd, \tag{6.4}$$

which requires only three simpler products: ac, bd, and $(a+b)(c+d)$, leading to a faster algorithm that only carries out three recursive calls.

Listing 6.8 implements Karatsuba's method, which contains an auxiliary function that computes the number of bits of the binary representation of a nonnegative integer n (which is equal to $\lfloor \log_2 n \rfloor + 1$, for $n \geq 1$). The size of the problem can be $\min(b_x, b_y)$. Thus, the base cases can check if the inputs are equal to zero or one, which lead to trivial results. The recursive case implements (6.4), using three recursive calls.

Regarding its efficiency, consider that both inputs have the same number of bits n, which is a power of two (i.e., $n = 2^k$). We can make this assumption since it is always possible to append 0 leading bits to the input integers until the assumption is satisfied. In that case, the running time of the algorithm is characterized by:

$$T(n) = \begin{cases} 1 & \text{if } n \leq 1, \\ 3T(n/2) + cn + d & \text{if } n > 1. \end{cases}$$

The term $cn + d$ is due to additions, subtractions, shift operations, and calls to `number_of_bits`, which are carried out in linear time with respect to n. Applying the master theorem, $T(n) \in \Theta(n^{log_2 3}) = \Theta(n^{1.585\ldots})$. Thus, the method is more efficient than the approach related to (6.3), whose computational cost is described by:

$$T(n) = \begin{cases} 1 & \text{if } n \leq 1, \\ 4T(n/2) + en + f & \text{if } n > 1, \end{cases}$$

where $T(n) \in \Theta(n^2)$.

Karatsuba's algorithm computes less multiplications (which correspond to recursive calls) than the school method, but performs more additions and subtractions. In practice, the algorithm will be quicker for large values of n. However, if n is small, the extra operations may make it run slower than the traditional approach. In any case, the method is noteworthy since it is able to reduce the number of multiplication operations, which are considerably more costly than additions or subtractions.

6.5 MATRIX MULTIPLICATION

Two matrices can be multiplied by partitioning them into block matrices as described in Section 2.4. We will now examine a straightforward divide and conquer recursive method that requires n^3 elementary (number) multiplications in order to compute the product of two $n \times n$ matrices (defined through the NumPy package). In addition, we will cover Strassen's algorithm, which is able to obtain the result using approximately $n^{2.8}$ basic scalar multiplications.

6.5.1 Divide and conquer matrix multiplication

Let $\mathbf{A}$ and $\mathbf{B}$ be $p \times q$ and $q \times r$-dimensional matrices, respectively. The size of the problem depends on the three dimensions p, q, and r. A trivial base case occurs when $p = q = r = 1$, where the result is a simple scalar number. Additionally, some implementations may require considering situations where a dimension is 0. In those cases the output should be an empty matrix, as will be addressed shortly.

One fairly straightforward way to decompose the problem consists of partitioning each matrix into four block matrices (forming a 2×2 array of block matrices). In that case their product can be defined as follows:

$$\begin{aligned} \mathbf{AB} &= \begin{bmatrix} \mathbf{A}_{1,1} & \mathbf{A}_{1,2} \\ \mathbf{A}_{2,1} & \mathbf{A}_{2,2} \end{bmatrix} \begin{bmatrix} \mathbf{B}_{1,1} & \mathbf{B}_{1,2} \\ \mathbf{B}_{2,1} & \mathbf{B}_{2,2} \end{bmatrix} \\ &= \begin{bmatrix} \mathbf{A}_{1,1}\mathbf{B}_{1,1} + \mathbf{A}_{1,2}\mathbf{B}_{2,1} & \mathbf{A}_{1,1}\mathbf{B}_{1,2} + \mathbf{A}_{1,2}\mathbf{B}_{2,2} \\ \mathbf{A}_{2,1}\mathbf{B}_{1,1} + \mathbf{A}_{2,2}\mathbf{B}_{2,1} & \mathbf{A}_{2,1}\mathbf{B}_{1,2} + \mathbf{A}_{2,2}\mathbf{B}_{2,2} \end{bmatrix}. \end{aligned} \tag{6.5}$$

Notice that the formula is analogous to multiplying two 2×2 matrices. For example, the top-left block of the result ($\mathbf{A}_{1,1}\mathbf{B}_{1,1} + \mathbf{A}_{1,2}\mathbf{B}_{2,1}$) can be viewed as the product between the first (block) row of $\mathbf{A}$ and the first (block) column of $\mathbf{B}$.

The decomposition involves computing eight simpler matrix products. Thus, the method will invoke itself eight times in the recursive case. The results of each product need to be added and stacked appropriately in order to form the output matrix. Listing 6.9 shows a possible implementation. The recursive case first defines each of the smaller block matrices, adds the simpler products, and builds the output matrix through the methods `vstack` and `hstack`. One of the base cases computes a simple product when $p = q = r = 1$. In addition, the code also considers the possibility of receiving empty input matrices, since they appear when partitioning the matrices in the recursive case if one of the dimensions is equal to one (obviously, a vector cannot be partitioned into four vectors as described in (6.5)). Thus, if any of the dimensions is 0 a special base case returns an empty matrix of dimensions $p \times r$, which can be handled appropriately in Python.

The previous method creates 1×1 matrices (in a base case) and progressively stacks them together to form the final $p \times r$ matrix. In addition, note that the dimensions of the input matrices to the methods are not fixed.

Another more efficient alternative consists of passing the entire matrices $\mathbf{A}$ and $\mathbf{B}$ in each call, and specifying the blocks that need to be

Listing 6.9 Divide and conquer matrix multiplication.

```
import numpy as np


def matrix_mult(A, B):
    p = A.shape[0]
    q = A.shape[1]
    r = B.shape[1]

    if p == 0 or q == 0 or r == 0:
        return np.zeros((p, r))
    elif p == 1 and q == 1 and r == 1:
        return np.matrix([[A[0, 0] * B[0, 0]]])
    else:
        A11 = A[0:p // 2, 0:q // 2]
        A21 = A[p // 2:p, 0:q // 2]
        A12 = A[0:p // 2, q // 2:q]
        A22 = A[p // 2:p, q // 2:q]

        B11 = B[0:q // 2, 0:r // 2]
        B21 = B[q // 2:q, 0:r // 2]
        B12 = B[0:q // 2, r // 2:r]
        B22 = B[q // 2:q, r // 2:r]

        C11 = matrix_mult(A11, B11) + matrix_mult(A12, B21)
        C12 = matrix_mult(A11, B12) + matrix_mult(A12, B22)
        C21 = matrix_mult(A21, B11) + matrix_mult(A22, B21)
        C22 = matrix_mult(A21, B12) + matrix_mult(A22, B22)

        return np.vstack([np.hstack([C11, C12]),
                          np.hstack([C21, C22])])


A = np.matrix([[2, 3, 1, -3], [4, -2, 1, 2]])
B = np.matrix([[2, 3, 1], [4, -1, -5], [0, -6, 3], [1, -1, 1]])
print(matrix_mult(A, B))
```

multiplied through appropriate limits passed as parameters, similarly to Listing 1.6. In addition, the result can be stored in a $p \times r$ matrix parameter $\mathbf{C}$ (passed by reference). Listing 6.10 shows a possible implementation of this alternative solution. The method `matrix_mult_limits` always passes the entire matrices $\mathbf{A}$ and $\mathbf{B}$ in each call, storing the result in a $p \times r$ matrix (its third parameter). Additionally, it specifies the submatrices that it will actually multiply through the rest of the param-

Listing 6.10 Alternative divide and conquer matrix multiplication.

```
1  import numpy as np
2
3
4  def add_matrices_limits(A, B, C, lp, up, lr, ur):
5      for i in range(lp, up + 1):
6          for k in range(lr, ur + 1):
7              C[i, k] = A[i, k] + B[i, k]
8
9
10 def matrix_mult_limits(A, B, C, lp, up, lq, uq, lr, ur):
11     mp = (lp + up) // 2
12     mq = (lq + uq) // 2
13     mr = (lr + ur) // 2
14
15     if lp == up and lq == uq and lr == ur:
16         C[mp, mr] = A[mp, mq] * B[mq, mr]
17     elif lp <= up and lq <= uq and lr <= ur:
18
19         M1 = np.zeros((A.shape[0], B.shape[1]))
20         M2 = np.zeros((A.shape[0], B.shape[1]))
21
22         matrix_mult_limits(A, B, M1, lp, mp, lq, mq, lr, mr)
23         matrix_mult_limits(A, B, M2, lp, mp, mq + 1, uq, lr, mr)
24         add_matrices_limits(M1, M2, C, lp, mp, lr, mr)
25
26         matrix_mult_limits(A, B, M1, lp, mp, lq, mq, mr + 1, ur)
27         matrix_mult_limits(
28             A, B, M2, lp, mp, mq + 1, uq, mr + 1, ur)
29         add_matrices_limits(M1, M2, C, lp, mp, mr + 1, ur)
30
31         matrix_mult_limits(A, B, M1, mp + 1, up, lq, mq, lr, mr)
32         matrix_mult_limits(
33             A, B, M2, mp + 1, up, mq + 1, uq, lr, mr)
34         add_matrices_limits(M1, M2, C, mp + 1, up, lr, mr)
35
36         matrix_mult_limits(
37             A, B, M1, mp + 1, up, lq, mq, mr + 1, ur)
38         matrix_mult_limits(
39             A, B, M2, mp + 1, up, mq + 1, uq, mr + 1, ur)
40         add_matrices_limits(M1, M2, C, mp + 1, up, mr + 1, ur)
41
42
43 def matrix_mult_limits_wrapper(A, B):
44     C = np.zeros((A.shape[0], B.shape[1]))
45     matrix_mult_limits(A, B, C, 0, A.shape[0] - 1,
46                        0, A.shape[1] - 1, 0, B.shape[1] - 1)
47     return C
```

eters, which indicate lower and upper limits related to the dimensions p, q, and r. The base case of `matrix_mult_limits` occurs when both of the submatrices correspond to scalar numbers, say $a_{i,j}$ and $b_{j,k}$. In that case the method simply stores their product in row i and column k of $\mathbf{C}$. Lastly, the method is not a function, since it does not *return* a matrix. Instead, it is a procedure that modifies the parameter $\mathbf{C}$, where it stores the result. Finally, the iterative method `add_matrices_limits` adds the elements of submatrices passed as the first two matrix input parameters, and stores the result in its third parameter (the submatrices are specified through parameter limits).

6.5.2 Strassen's matrix multiplication algorithm

The most expensive arithmetic operation carried out by the previous algorithms is the scalar multiplication in the base cases. In particular, both require pqr of these multiplications, similarly to the straightforward iterative version that uses three loops. It is nevertheless interesting to analyze the time complexity assuming that the input matrices are $n \times n$. In that case the runtime can be specified through the following function:

$$T(n) = \begin{cases} 1 & \text{if } n \leq 1, \\ 8T(n/2) + 4\Theta(n^2) & \text{if } n > 1, \end{cases} \tag{6.6}$$

since the methods invoke themselves eight times, and need to perform four matrix additions whose cost is quadratic with respect to n. Therefore, according to the master theorem (see (3.28)), $T(n) \in \Theta(n^{\log_2 8}) = \Theta(n^3)$. We will now describe Strassen's algorithm, which is a well-known method that can reduce the time complexity to $\Theta(n^{\log_2 7}) = \Theta(n^{2.807\ldots})$.

The method also decomposes each of the input matrices into four block matrices as in the standard algorithm. Thus, $\mathbf{AB} = \mathbf{C}$ can be expressed as:

$$\begin{bmatrix} \mathbf{A}_{1,1} & \mathbf{A}_{1,2} \\ \mathbf{A}_{2,1} & \mathbf{A}_{2,2} \end{bmatrix} \begin{bmatrix} \mathbf{B}_{1,1} & \mathbf{B}_{1,2} \\ \mathbf{B}_{2,1} & \mathbf{B}_{2,2} \end{bmatrix} = \begin{bmatrix} \mathbf{C}_{1,1} & \mathbf{C}_{1,2} \\ \mathbf{C}_{2,1} & \mathbf{C}_{2,2} \end{bmatrix}$$

The key to the method is the definition of the following new matrices that involve one matrix multiplication operation:

$$\begin{aligned}
\mathbf{M}_1 &= (\mathbf{A}_{1,1} + \mathbf{A}_{2,2})(\mathbf{B}_{1,1} + \mathbf{B}_{2,2}),\\
\mathbf{M}_2 &= (\mathbf{A}_{2,1} + \mathbf{A}_{2,2})\mathbf{B}_{1,1},\\
\mathbf{M}_3 &= \mathbf{A}_{1,1}(\mathbf{B}_{1,2} - \mathbf{B}_{2,2}),\\
\mathbf{M}_4 &= \mathbf{A}_{2,2}(\mathbf{B}_{2,1} - \mathbf{B}_{1,1}),\\
\mathbf{M}_5 &= (\mathbf{A}_{1,1} + \mathbf{A}_{1,2})\mathbf{B}_{2,2},\\
\mathbf{M}_6 &= (\mathbf{A}_{2,1} - \mathbf{A}_{1,1})(\mathbf{B}_{1,1} + \mathbf{B}_{1,2}),\\
\mathbf{M}_7 &= (\mathbf{A}_{1,2} - \mathbf{A}_{2,2})(\mathbf{B}_{2,1} + \mathbf{B}_{2,2}).
\end{aligned} \tag{6.7}$$

Finally, these matrices can be combined as follows to form the output's block matrices:

$$\begin{aligned}
\mathbf{C}_{1,1} &= \mathbf{M}_1 + \mathbf{M}_4 - \mathbf{M}_5 + \mathbf{M}_7,\\
\mathbf{C}_{1,2} &= \mathbf{M}_3 + \mathbf{M}_5,\\
\mathbf{C}_{2,1} &= \mathbf{M}_2 + \mathbf{M}_4,\\
\mathbf{C}_{2,2} &= \mathbf{M}_1 - \mathbf{M}_2 + \mathbf{M}_3 + \mathbf{M}_6.
\end{aligned} \tag{6.8}$$

The algorithm therefore computes seven products and 18 additions (or subtractions) in every recursive call. Thus, its runtime cost is described through:

$$T(n) = \begin{cases} 1 & \text{if } n \le 1,\\ 7T(n/2) + 18\Theta(n^2) & \text{if } n > 1, \end{cases} \tag{6.9}$$

which implies that $T(n) = \Theta(n^{\log_2 7}) = \Theta(n^{2.807\dots})$. The algorithm can be faster than the standard method that runs in $\Theta(n^3)$ for large values of n. Nevertheless, for small or medium-sized matrices it may be slower due to the larger multiplicative constants that play a role in practice.

Finally, from a theoretical point of view, the inputs for this algorithm need to be square $n \times n$ matrices, where n is a power of two. Nevertheless, in practice efficient implementations split the matrices into numerous square submatrices, and apply the algorithm repeatedly. A simpler, but slower, alternative consists of padding (i.e., extending) the input matrices with zeros, in order for them to be $2^k \times 2^k$ matrices (see Exercise 6.6).

6.6 THE TROMINO TILING PROBLEM

A tromino is a polygon formed by connecting three equal-sized squares by their edges. Without considering rotations and reflections, there are

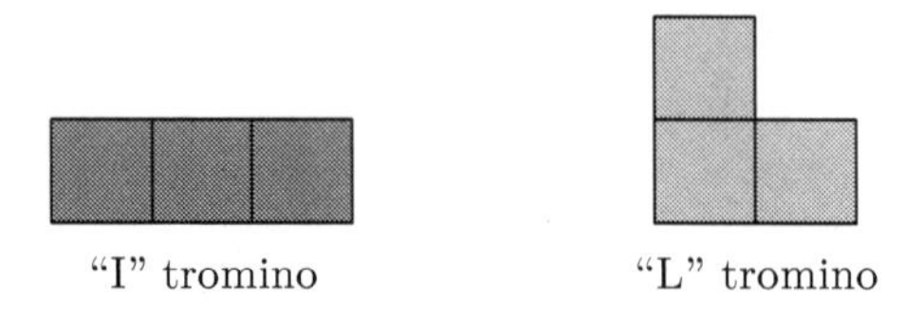

Figure 6.3 Types of trominoes ignoring rotations and reflections.

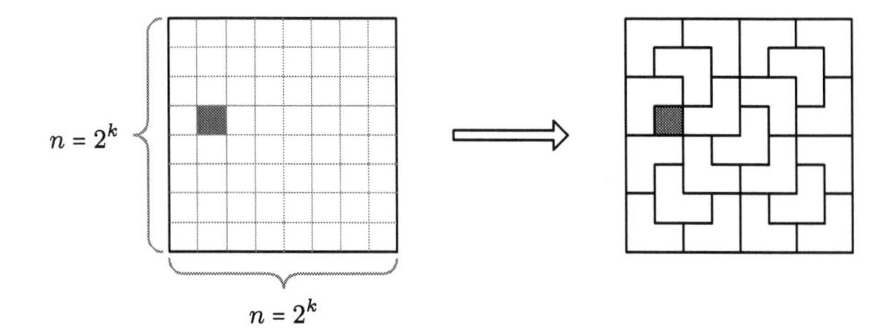

Figure 6.4 Tromino tiling problem.

only two types of trominoes: the "I" and the "L" trominoes, as illustrated in Figure 6.3. The following problem consists of covering a square $n \times n$ board, where $n \geq 2$ is a power of two, which contains a "hole" that cannot be covered, with L trominoes. Figure 6.4 explains the problem graphically with an example.

The size of the problem is clearly n. The smallest instances of the problem correspond to 2×2 boards, whose solutions are trivial. Figure 6.5 illustrates the divide and conquer decomposition used in the recursive case. The initial board in (a) is divided into four smaller square boards of size $n/2$, as shown in (b). However, only one of these smaller boards will contain the initial hole. Therefore, the other three boards do not

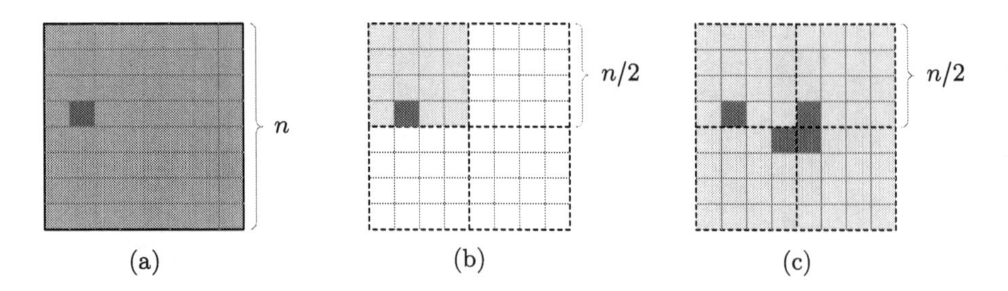

Figure 6.5 Decomposition of the tromino tiling problem.

Listing 6.11 Auxiliary functions for drawing trominoes.

```
def draw_L1(x, y):
    plt.plot([x, x + 2], [y + 2, y + 2], 'k-')
    plt.plot([x, x + 1], [y + 1, y + 1], 'k-')
    plt.plot([x + 1, x + 2], [y, y], 'k-')
    plt.plot([x, x], [y + 1, y + 2], 'k-')
    plt.plot([x + 1, x + 1], [y, y + 1], 'k-')
    plt.plot([x + 2, x + 2], [y, y + 2], 'k-')


def draw_L2(x, y):
    plt.plot([x, x + 2], [y + 2, y + 2], 'k-')
    plt.plot([x, x + 1], [y, y], 'k-')
    plt.plot([x + 1, x + 2], [y + 1, y + 1], 'k-')
    plt.plot([x, x], [y, y + 2], 'k-')
    plt.plot([x + 1, x + 1], [y, y + 1], 'k-')
    plt.plot([x + 2, x + 2], [y + 1, y + 2], 'k-')


def draw_L3(x, y):
    plt.plot([x, x + 2], [y, y], 'k-')
    plt.plot([x, x + 1], [y + 1, y + 1], 'k-')
    plt.plot([x + 1, x + 2], [y + 2, y + 2], 'k-')
    plt.plot([x, x], [y, y + 1], 'k-')
    plt.plot([x + 1, x + 1], [y + 1, y + 2], 'k-')
    plt.plot([x + 2, x + 2], [y, y + 2], 'k-')


def draw_L4(x, y):
    plt.plot([x, x + 2], [y, y], 'k-')
    plt.plot([x, x + 1], [y + 2, y + 2], 'k-')
    plt.plot([x + 1, x + 2], [y + 1, y + 1], 'k-')
    plt.plot([x, x], [y, y + 2], 'k-')
    plt.plot([x + 1, x + 1], [y + 1, y + 2], 'k-')
    plt.plot([x + 2, x + 2], [y, y + 1], 'k-')
```

constitute self-similar subproblems. This can be solved by placing a tromino in the center of the board, where its three squares will constitute holes in the smaller boards, creating valid subproblems, as illustrated in (c).

We will use the package Matplotlib in order to generate images showing the solutions to the problem. In particular, the trominoes can be drawn by plotting six line segments. Since there are four possible L trominoes considering rotations (see Figure 6.6), we can use the four auxil-

Listing 6.12 Recursive method for drawing trominoes.

```
def trominoes(x, y, n, p, q):
    if n == 2:
        if y == q:  # hole in bottom tiles
            if x == p:  # hole in bottom-left tile
                draw_L1(x, y)
            else:  # hole in bottom-right tile
                draw_L2(x, y)
        else:  # hole in top tiles
            if x == p:  # hole in top-left tile
                draw_L3(x, y)
            else:  # hole in top-right tile
                draw_L4(x, y)

    else:
        mid_x = x + n // 2
        mid_y = y + n // 2

        if q < mid_y:  # hole in bottom squares

            if p < mid_x:  # hole in bottom-left square
                draw_L1(mid_x - 1, mid_y - 1)
                trominoes(x, y, n // 2, p, q)
                trominoes(x, mid_y, n // 2, mid_x - 1, mid_y)
                trominoes(mid_x, y, n // 2, mid_x, mid_y - 1)
                trominoes(mid_x, mid_y, n // 2, mid_x, mid_y)
            else:  # hole in bottom-right square
                draw_L2(mid_x - 1, mid_y - 1)
                trominoes(x, y, n // 2, mid_x - 1, mid_y - 1)
                trominoes(x, mid_y, n // 2, mid_x - 1, mid_y)
                trominoes(mid_x, y, n // 2, p, q)
                trominoes(mid_x, mid_y, n // 2, mid_x, mid_y)

        else:  # hole in top squares

            if p < mid_x:  # hole in top-left square
                draw_L3(mid_x - 1, mid_y - 1)
                trominoes(x, y, n // 2, mid_x - 1, mid_y - 1)
                trominoes(x, mid_y, n // 2, p, q)
                trominoes(mid_x, y, n // 2, mid_x, mid_y - 1)
                trominoes(mid_x, mid_y, n // 2, mid_x, mid_y)
            else:  # hole top-right square
                draw_L4(mid_x - 1, mid_y - 1)
                trominoes(x, y, n // 2, mid_x - 1, mid_y - 1)
                trominoes(x, mid_y, n // 2, mid_x - 1, mid_y)
                trominoes(mid_x, y, n // 2, mid_x, mid_y - 1)
                trominoes(mid_x, mid_y, n // 2, p, q)
```

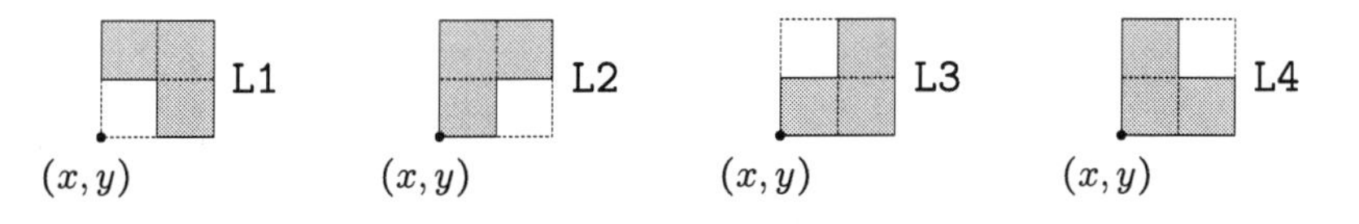

Figure 6.6 L trominoes considering rotations.

Listing 6.13 Code for calling the trominoes method.

```
import random
import matplotlib.pyplot as plt
from matplotlib.patches import Rectangle

# Include tromino methods here

fig = plt.figure()
fig.patch.set_facecolor('white')
ax = plt.gca()
n = 16  # power of 2
p = random.choice([i for i in range(n)])
q = random.choice([i for i in range(n)])
ax.add_patch(Rectangle((p, q), 1, 1, facecolor=(0.5, 0.5, 0.5)))
trominoes(0, 0, n, p, q)
plt.axis('equal')
plt.axis('off')
plt.show()
```

iary functions in Listing 6.11 to draw each one. The functions receive the coordinates (x, y) of the bottom-left corner corresponding to the square surrounding the tromino. The command `plot(`$[x_1, x_2]$`,`$[y_1, y_2]$`,'k-')` draws a black line segment with endpoints (x_1, y_1) and (x_2, y_2).

Listing 6.12 shows a possible implementation of the recursive method. The procedure needs to know which problem/subproblem it should solve. This information is provided by the first three parameters. The first two indicate the bottom-left coordinates (x, y) of the board, while the third is the size of the board (n). The last two indicate the location of the hole (in particular, (p, q) specifies the bottom-left corner of the 1×1 square). In both base and recursive cases the method uses conditions in order to determine the relative position of the hole, and draws the appropriate tromino. Finally, the method invokes itself four times in the recursive case, with different parameters indicating the new subproblems, together with the new holes on three of them.

Finally, Listing 6.13 shows a fragment of code that can be used to call the `trominoes` method. Line 7 creates a figure, line 8 sets its background color to white, and `ax` captures the axes of the figure in line 9. After defining the size of the initial board (line 10), the hole is chosen within it at random, and then drawn in line 13. When using the Matplotlib package a rectangle can be formed by calling the method `Rectangle`. It receives the coordinates of the bottom-left vertex, together with the width and height, and other possible arguments. Line 14 calls the main method, and the last lines are included to avoid scaling factors, to eliminate the axes in the final image, and to draw it.

6.7 THE SKYLINE PROBLEM

The skyline problem consists of finding the outline of a set of rectangular buildings against the sky. Figure 6.7 explains the idea. The input to the problem is a list of $n \geq 1$ rectangles that represent buildings, as shown in (a). The bottom side of the rectangles is always located at level 0. Thus, each building can be specified by using only three parameters. In particular, we will use tuples of the form (x_1, x_2, h), where x_1 marks the location of left side of the building, x_2 indicates the right side, and h contains its height. Thus, $x_1 < x_2$, and $h > 0$. The buildings in the example are: [(1,7,7), (18,20,7), (2,9,5), (17,19,2), (12,24,3), (3,8,8), (11,13,5), (15,21,6)]. Note that it is not necessary to sort them in any way (e.g., according to the value of x_1).

The skyline consists of a curve whose height at any point on the X axis is the maximum height of the buildings at such point, as illustrated through the thick dark segments in (b). Since the buildings are rectangular the skyline can be specified through a set of coordinates on the plane (x, h) that mark the location x when its height changes to h, as illustrated in (c). The output to the problem will therefore be a list of these coordinates, which can be coded as tuples as well, and will appear sorted in ascending order. The output for the example is: [(1,7), (3,8), (8,5), (9,0), (11,5), (13,3), (15,6), (18,7), (20,6), (21,3), (24,0)].

The size of the problem is the number of buildings n. For the base case, the smallest instance occurs when there is only one building. If it is specified through the tuple (x_1, x_2, h) then the output is the list $[(x_1, h), (x_2, 0)]$, as shown in Figure 6.8.

This problem can be decomposed by reducing the size of the problem by a unit. However, this leads to an algorithm whose runtime is $\mathcal{O}(n^2)$ in the worst case. Instead, we will examine a divide and conquer approach

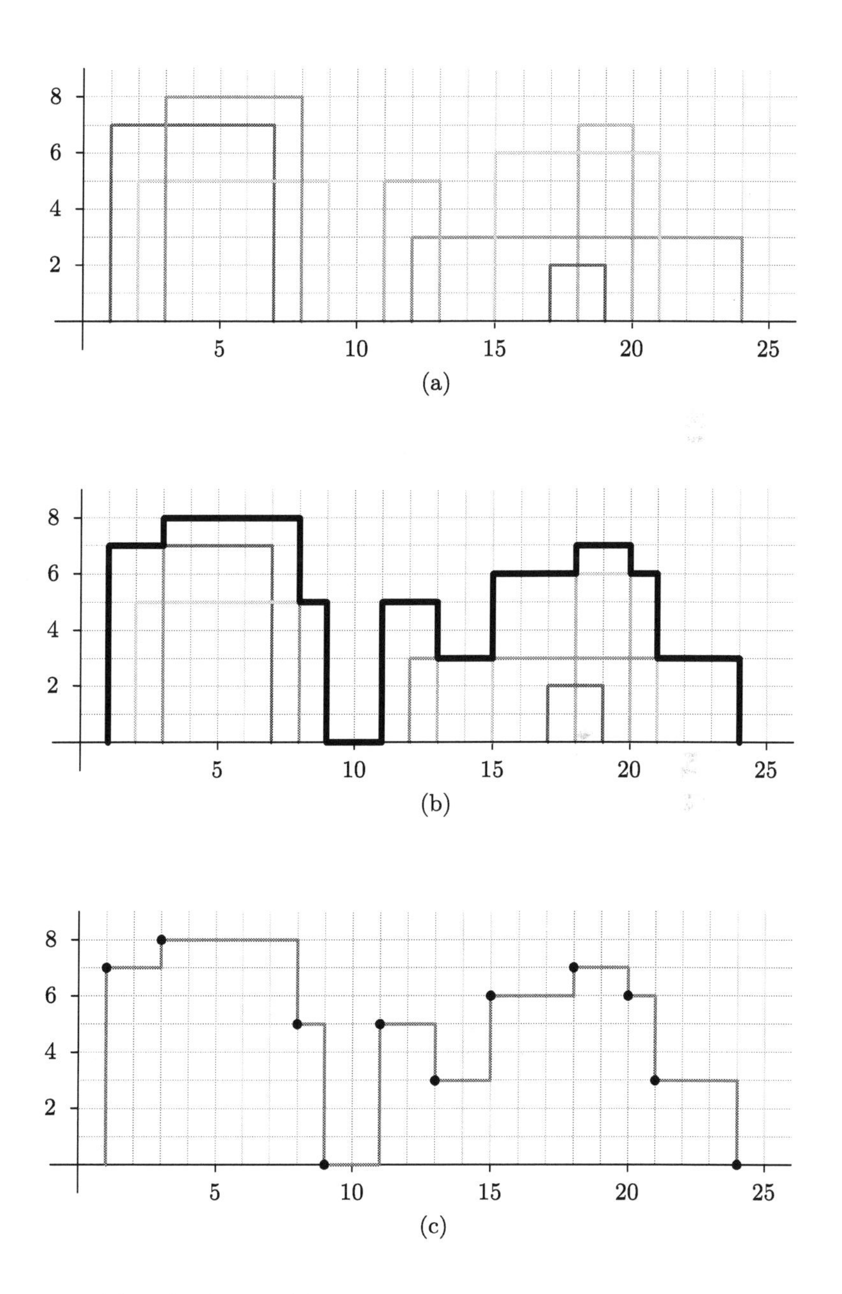

Figure 6.7 The skyline problem.

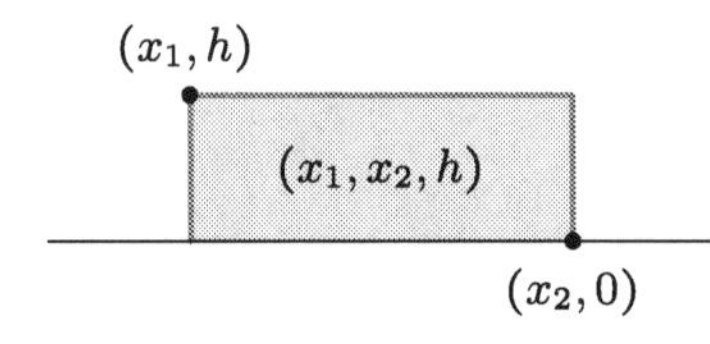

Figure 6.8 Base case for the skyline problem with one building.

Listing 6.14 Main recursive method for computing skylines.

```
1 def compute_skyline(buildings):
2     n = len(buildings)
3     if n == 1:
4         return ([(buildings[0][0], buildings[0][2]),
5                  (buildings[0][1], 0)])
6     else:
7         skyline1 = compute_skyline(buildings[0:n // 2])
8         skyline2 = compute_skyline(buildings[n // 2:n])
9         return merge_skylines(skyline1, skyline2, 0, 0)
```

that is able to solve the problem in $\Theta(n \log n)$ time. The idea, illustrated in Figure 6.9, is similar to the approach used in the merge sort algorithm. The decomposition step consists of dividing the input list in two smaller lists of approximately $n/2$ buildings. The method then carries out two recursive calls on those sublists that return two independent skylines. Assuming that the skylines have been constructed correctly (by applying induction), the final, and challenging, step consists of merging the skylines in order to produce a final one. Listing 6.14 shows the associated divide and conquer method, whose structure is essentially identical to that of Listing 6.2.

The skyline merging problem is a new computational problem in its own right. While the majority of solutions in texts are iterative, we will now examine a linear-recursive method. The inputs are the two input lists of sorted tuples representing skylines. In addition, since a tuple indicates a change in the height of a skyline, the method needs to access the previous height before such change. Moreover, since the proposed algorithm will process the first tuples from the lists (until one list is empty), but progressively discard them as they are analyzed, these previous heights will not be contained in the input lists. Thus, the method will need two additional parameters, say p_1 and p_2, in order to store the previous heights of the skylines. Naturally, both of these parameters will

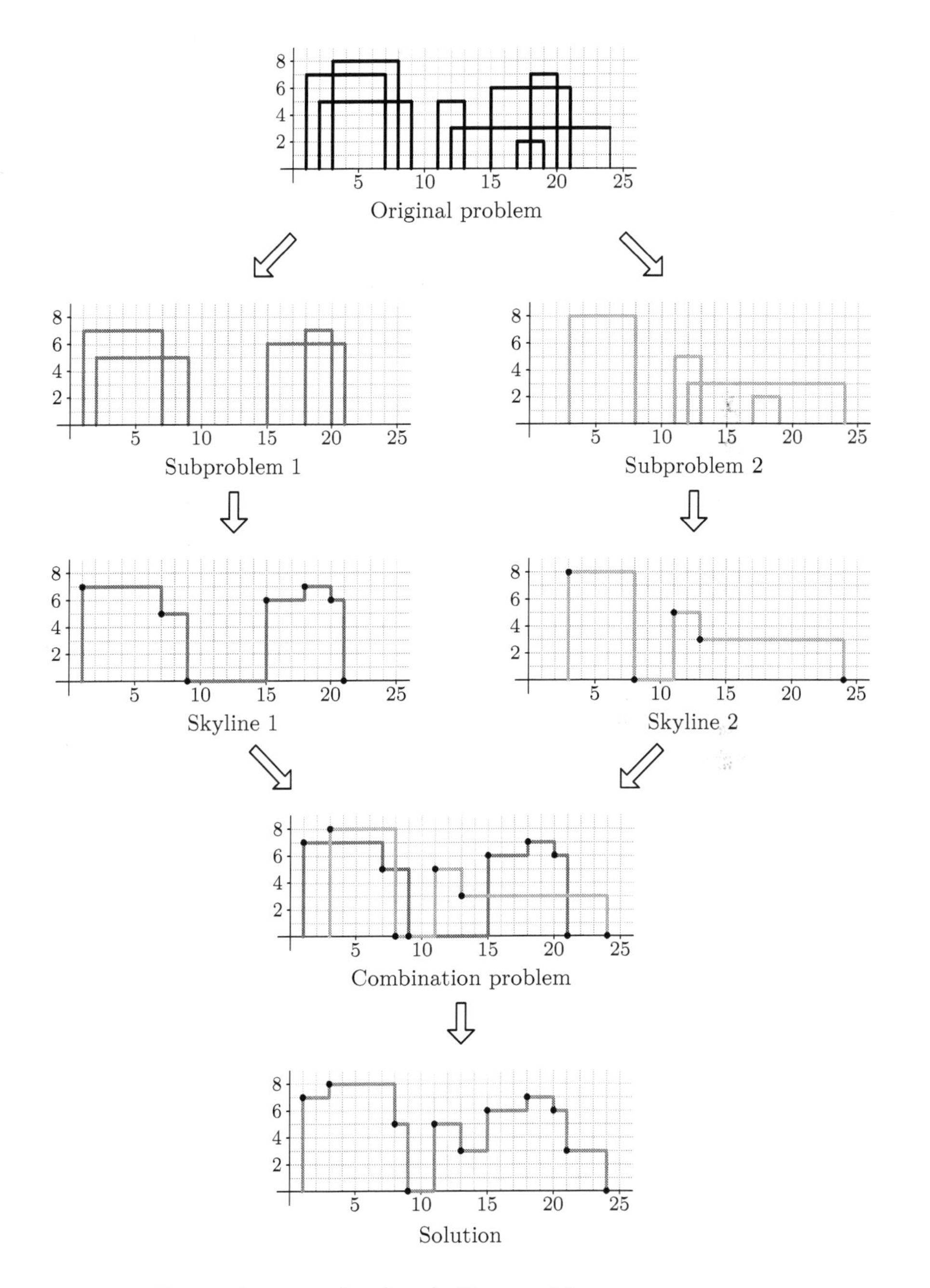

Figure 6.9 Recursive case for the skyline problem.

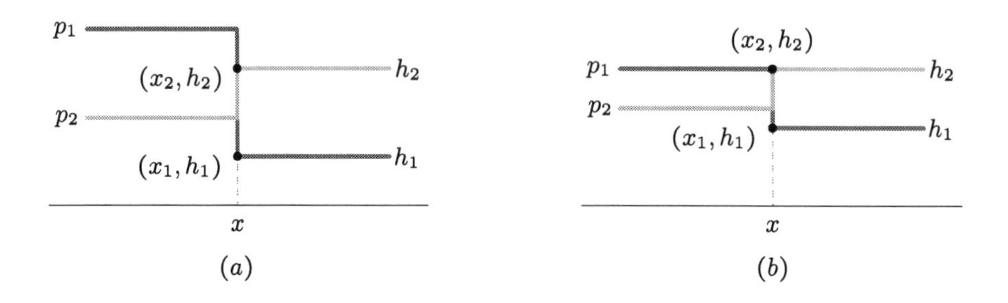

Figure 6.10 Possible situations when merging skylines that change at the same location x.

be initialized to zero when calling the method within the main skyline function (see line 9 of Listing 6.14).

The size of the problem depends on the lengths of the input skyline lists, say n_1 and n_2. We can consider that the base case occurs when one of the lists is empty, where the method must trivially return the other list. Listing 6.15 codes the method, where the base cases are described in lines 2–5.

A key observation for determining an appropriate decomposition is that the output of the merging function produces a new skyline whose tuples are sorted in ascending order according to their x value. Thus, the algorithm will analyze the first tuples of each list, and process the one with a smaller x value (or both if their x values are the same). Thus, in the recursive case we need the first tuples of the skylines (x_1, h_1) and (x_2, h_2), together with their previous heights p_1 and p_2.

Firstly, let us consider the situation when $x_1 = x_2 = x$. Since the tuples mark changes in the skyline, we may need to include in the solution the one with larger height. For example, in Figure 6.10(a) the point (x, h_2) would be included in the final skyline. Furthermore, the recursive call will use the tails of both input lists, discarding (x, h_1) and (x, h_2), since the possible changes at x will have been processed correctly. This is accomplished in lines 19–21. Lastly, there is a situation where a new tuple is not included in the solution. This occurs when the largest new height is equal to the largest previous height of the skyline (i.e., when $\max(h_1, h_2) = \max(p_1, p_2)$), since there would be no change of heights at x. Figure 6.10(b) illustrates this case, where the point (x, h_2) would not be included in the final skyline (see lines 16 and 17).

We will now analyze possible scenarios when the x values of the first tuples of the skylines are not equal. Without loss of generality, assume

Listing 6.15 Recursive method for merging skylines.

```
def merge_skylines(sky1, sky2, p1, p2):
    if sky1 == []:
        return sky2
    elif sky2 == []:
        return sky1
    else:
        x1 = sky1[0][0]
        x2 = sky2[0][0]
        h1 = sky1[0][1]
        h2 = sky2[0][1]

        if x1 == x2:
            h = max(p1, p2)
            new_h = max(h1, h2)
            if h == new_h:
                return merge_skylines(sky1[1:], sky2[1:],
                                      h1, h2)
            else:
                return ([(x1, new_h)]
                        + merge_skylines(sky1[1:], sky2[1:],
                                         h1, h2))

        elif x1 < x2:
            if h1 > p2:
                return ([(x1, h1)]
                        + merge_skylines(sky1[1:], sky2,
                                         h1, p2))
            elif p1 > p2:
                return ([(x1, p2)]
                        + merge_skylines(sky1[1:], sky2,
                                         h1, p2))
            else:
                return merge_skylines(sky1[1:], sky2,
                                      h1, p2)

        else:
            if h2 > p1:
                return ([(x2, h2)]
                        + merge_skylines(sky1, sky2[1:],
                                         p1, h2))
            elif p2 > p1:
                return ([(x2, p1)]
                        + merge_skylines(sky1, sky2[1:],
                                         p1, h2))
            else:
                return merge_skylines(sky1, sky2[1:], p1, h2)
```

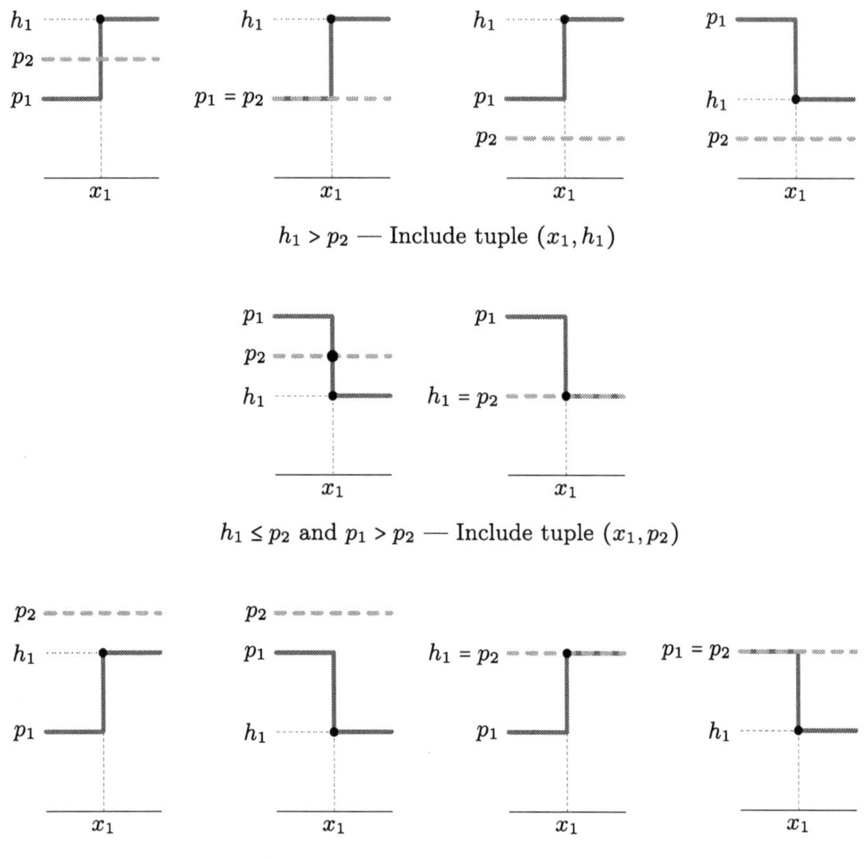

Figure 6.11 Possible situations when merging skylines and $x_1 < x_2$.

$x_1 < x_2$. In that case, the algorithm must decide whether to include the tuple (x_1, h_1), or (x_1, p_2), or none at all, as illustrated in Figure 6.11. If $h_1 > p_2$ then the first skyline is above the second one at location x_1, and therefore must include the tuple (x_1, h_1) as part of the merged skyline (see lines 25–27). If $h_1 \leq p_2$ then the algorithm must check if $p_1 > p_2$. If the result is `True` then the method includes the tuple (x_1, p_2) (see lines 29–31). Notice that when $h_1 < p_2$ this produces a new tuple that does not appear in the input skyline lists. Lastly, in other situations, $p_2 \geq h_1$ and $p_2 \geq p_1$, which implies that the merged skyline will not change at x_1. Finally, having processed the first tuple from the first skyline (x_1, h_1), the method discards it when invoking itself in the corresponding recursive cases. In addition, the arguments that specify the previous heights of the skylines will be h_1 and p_2 (see lines 27, 31, and 34). The rest of the code at lines 36–46 is analogous to the code at lines 23–34, and handles the case when $x_2 < x_1$.

6.8 EXERCISES

Exercise 6.1 — Implement a divide and conquer algorithm that determines whether a list **a** contains an element x.

Exercise 6.2 — Let **a** be a list of n nonnegative integers. Write a function based on the divide and conquer technique that returns the set of digits shared amongst all of the elements in **a**, and specify its asymptotic computational cost. For example, for $\mathbf{a} = [2348, 1349, 7523, 3215]$, the solution is $\{3\}$. The function should call another one that provides the set of digits in a nonnegative integer. Code this function as well.

Exercise 6.3 — The maximum sublist problem consists of finding the sublist of contiguous elements within a list of numbers that has the largest sum of its components. For example, given the list $[-1, -4, 5, 2, -3, 4, 2, -5]$, the optimal sublist is $[5, 2, -3, 4, 2]$, whose elements sum up to 10. Given a nonempty input list of numbers **a**, implement a divide and conquer function that returns the sum of the elements of its maximum sublist.

Exercise 6.4 — Design a fast recursive polynomial multiplication algorithm based on a divide and conquer decomposition analogous to the one used in Karatsuba's algorithm (see Section 6.4). Code polynomials through lists, as described in Exercise 5.2. The method will need to call functions that add and subtract polynomials. Code these functions as well.

Exercise 6.5 — Implement a recursive function that receives an $n \times m$ matrix ($\mathbf{A}$) and returns its $m \times n$ transpose ($\mathbf{A}^\mathsf{T}$). Use the NumPy package, and the following divide and conquer decomposition that breaks up the input matrix $\mathbf{A}$ into four block matrices by dividing each dimension by two:

$$\mathbf{A} = \left[\begin{array}{c|c} \mathbf{A}_{1,1} & \mathbf{A}_{1,2} \\ \hline \mathbf{A}_{2,1} & \mathbf{A}_{2,2} \end{array}\right].$$

In that case, the transpose of **A** can be defined as:

$$\mathbf{A}^{\mathsf{T}} = \left[\begin{array}{c|c} \mathbf{A}_{1,1}^{\mathsf{T}} & \mathbf{A}_{2,1}^{\mathsf{T}} \\ \hline \mathbf{A}_{1,2}^{\mathsf{T}} & \mathbf{A}_{2,2}^{\mathsf{T}} \end{array}\right].$$

Exercise 6.6 — Implement Strassen's matrix multiplication algorithm. The code should include a wrapper function that will allow multiplying a $p \times q$ matrix times a $q \times r$ matrix, by padding the input matrices with zeros.

Exercise 6.7 — Implement a matrix multiplication algorithm by decomposing the input matrices as follows:

$$\mathbf{A} \cdot \mathbf{B} = \left[\begin{array}{c|c} \mathbf{A}_1 & \mathbf{A}_2 \end{array}\right] \cdot \left[\begin{array}{c} \mathbf{B}_1 \\ \hline \mathbf{B}_2 \end{array}\right] = \left[\begin{array}{c} \mathbf{A}_1\mathbf{B}_1 + \mathbf{A}_2\mathbf{B}_2 \end{array}\right].$$

Exercise 6.8 — Implement a matrix multiplication algorithm by decomposing the input matrices as follows:

$$\mathbf{A} \cdot \mathbf{B} = \left[\begin{array}{c} \mathbf{A}_1 \\ \hline \mathbf{A}_2 \end{array}\right] \cdot \left[\begin{array}{c|c} \mathbf{B}_1 & \mathbf{B}_2 \end{array}\right] = \left[\begin{array}{cc} \mathbf{A}_1\mathbf{B}_1 & \mathbf{A}_1\mathbf{B}_2 \\ \mathbf{A}_2\mathbf{B}_1 & \mathbf{A}_2\mathbf{B}_2 \end{array}\right].$$

CHAPTER 7

Multiple Recursion II: Puzzles, Fractals, and More...

Calculus, the electrical battery, the telephone, the steam engine, the radio — all these groundbreaking innovations were hit upon by multiple inventors working in parallel with no knowledge of one another.

— Steven Johnson

THE previous chapter introduced divide and conquer algorithms that used multiple recursion and decomposed problems by dividing their size by some constant. Alternatively, this chapter presents solutions to challenging problems that also decompose them into several subproblems, but reduce their size by one or two units. Previous examples of these algorithms are the function in (1.2) that defines Fibonacci numbers, or in (3.2) that can be used to compute binomial coefficients (Exercise 7.1 consists of implementing the function). The chapter includes the classical towers of Hanoi problem, which is one of the most commonly used examples for illustrating multiple recursion. In addition, it analyzes problems related to fractal images, where we will use the popular Matplotlib package in order to generate figures.

7.1 SWAMP TRAVERSAL

In this problem (which is also known as the "moor traversal problem") we are given an $n \times m$ matrix $\mathbf{A}$ that represents a rectangular swamp

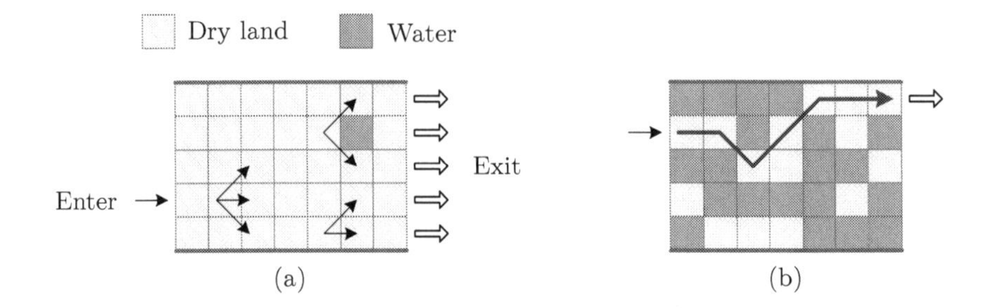

Figure 7.1 Swamp traversal problem.

composed of square patches, and an initial row r. The entries of the matrix $(a_{i,j})$ can only take two values that indicate if a patch in the swamp corresponds to dry land or water. The goal is to determine if it is possible to find a path through a swamp, from left to right, by starting at a particular given patch on its left-hand side $(a_{r,0})$, according to the following rules (see Figure 7.1(a)):

- A path through the swamp can only traverse patches of dry land.
- When advancing it is only possible to move towards the right, and to a new patch that is adjacent to the old one. In other words, when standing at patch $a_{i,j}$, the next patch in the path can only be either $a_{i-1,j+1}$, $a_{i,j+1}$, or $a_{i+1,j+1}$. Note that it is not possible to move vertically.
- The path cannot cross the bottom and top boundaries of the swamp.
- The path can finish at any dry patch of land on the right extreme of the swamp.

Figure 7.1(b) shows a path through a swamp where patches of dry land and water are represented through light and dark squares, respectively.

We can assume that the size of the problem is the width of the swamp (m). In a first base case the method can return `False` trivially if r is not a valid row, according to the boundaries of the swamp ($r < 0$ or $r \geq n$), or if the initial patch $a_{r,0}$ is water (note that this last condition would not be necessary if we impose a precondition on the inputs that forces that first patch to be dry land). Otherwise, the function can return `True` if the width of the swamp is only 1 (it will have already checked that the patch corresponds to dry land in the previous base case).

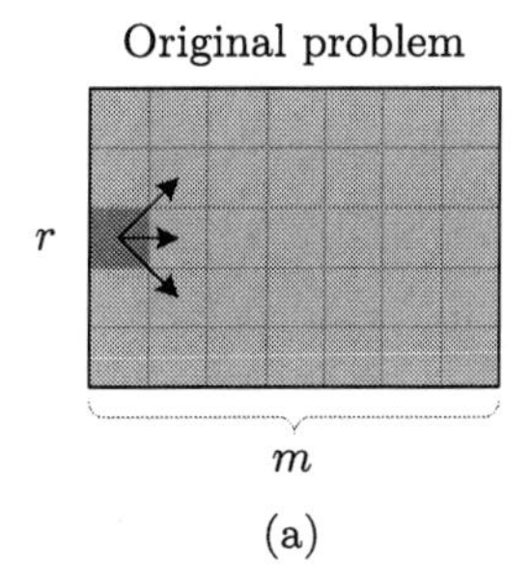

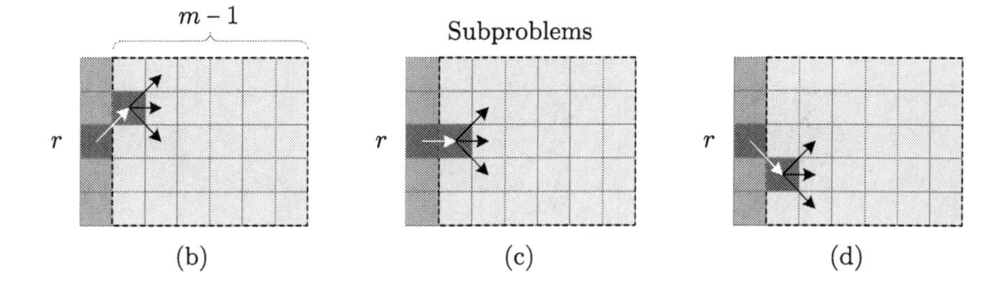

Figure 7.2 Decomposition of the swamp traversal problem.

Listing 7.1 Function that determines whether there exists a path through a swamp.

```
def exists_path_swamp(A, r):
    if r < 0 or r >= A.shape[0] or A[r, 0] == 'W':
        return False
    elif A.shape[1] == 1:
        return True
    else:
        return (exists_path_swamp(A[:, 1:], r - 1)
                or exists_path_swamp(A[:, 1:], r)
                or exists_path_swamp(A[:, 1:], r + 1))
```

For the recursive case we can decompose the problem by using the three subproblems of size $m-1$ illustrated in Figure 7.2. In particular, we can assume that we know whether there is a valid path through the swamp starting at patches $a_{r-1,1}$, $a_{r,1}$, or $a_{r+1,1}$. In other words, we can assume that we know the solutions to the three smaller subproblems that arise from discarding the first column of the swamp, and starting at patches in rows $r-1$, r, and $r+1$. Thus, the recursive case can be based on the three corresponding recursive calls.

Listing 7.2 Alternative function that determines whether there exists a path through a swamp.

```
def exists_path_swamp_alt(A, r):
    if A.shape[1] == 1:
        return A[r, 0] != 'W'
    else:
        if r == 0 or A[r - 1, 1] == 'W':
            diag_up = False
        else:
            diag_up = exists_path_swamp_alt(A[:, 1:], r - 1)

        if not diag_up:
            if r == A.shape[0] - 1 or A[r + 1, 1] == 'W':
                diag_down = False
            else:
                diag_down = exists_path_swamp_alt(
                    A[:, 1:], r + 1)

            if not diag_down:
                if A[r, 1] == 'W':
                    horizontal = False
                else:
                    horizontal = exists_path_swamp_alt(
                        A[:, 1:], r)

        return diag_up or diag_down or horizontal
```

Listing 7.1 shows one of many possible implementations of the function. The matrix **A** (from the NumPy package) contains characters, where 'W' corresponds to a water patch. Lines 2–5 define the base cases. In lines 7, 8, and 9 the method determines whether it is possible to find a path by advancing to $a_{r-1,1}$ (diagonally upwards), $a_{r,1}$ (horizontally rightwards), or $a_{r+1,1}$ (diagonally downwards), respectively.

It is interesting to analyze how the preconditions can affect an algorithm. Listing 7.2 is built by using the precondition on the inputs that forces $a_{r,0} \neq$ 'W'. Thus, it does not need to check whether $a_{r,0} =$ 'W' (see line 2 in Listing 7.1). In addition, any call to the method must make sure that $a_{r,0} \neq$ 'W', which introduces additional conditional statements. Lastly, the efficiency of the code is enhanced by incorporating the `if` statements in lines 10 and 17, which avoid unnecessary recursive calls when a path through the swamp has been found. For example, if there

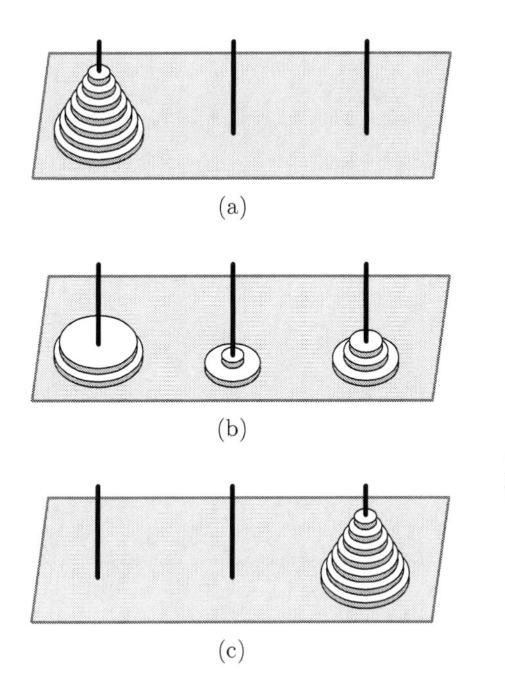

Figure 7.3 The towers of Hanoi puzzle.

exists a path moving diagonally upwards there is no need to check if there are paths moving horizontally, nor diagonally downwards.

Finally, in Listings 7.1 and 7.2 the size of the matrix is reduced explicitly by calling the methods with `A[:,1:]`. Another possibility is to always pass the entire $n \times m$ matrix, and control the size of the problem with an input parameter that indicates the column c where the method should start searching for a path. In this scenario the paths would stem from patch $a_{r,c}$. Thus, the resulting code would be more general, allowing us to determine if there is a valid path that permits exiting the swamp not only from its left margin, but by starting at any patch $a_{r,c}$. Exercise 7.2 proposes implementing this variant.

7.2 TOWERS OF HANOI

The towers of Hanoi puzzle is one of the most popular problems used to illustrate recursion. Although it can be solved using relatively simple iterative algorithms, it also allows a concise and elegant recursive solution that emphasizes the role of problem decomposition and induction in recursion, as well as showing its potential for problem solving.

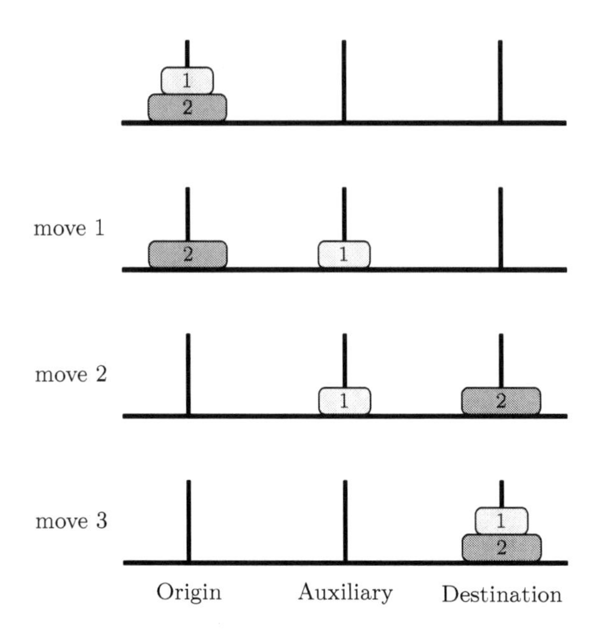

Figure 7.4 Solution to the towers of Hanoi puzzle for $n = 2$ disks.

The game consists of three vertical rods and n disks that can slide onto any rod, as shown in Figure 7.3 for $n = 7$ disks. The disks have different radii, where we can assume that the radius of disk i is precisely i. In the initial configuration, the n disks are stacked on one of the rods in decreasing order (of index or radius), as shown in (a). The goal is to move the entire tower of stacked disks from the initial (origin) rod onto a final (destination) one, with the help of a third (auxiliary) rod, by obeying the following rules:

- Only one disk can be moved at a time.
- Each move takes the upper disk staked on one of the rods, and places it on top of a stack in another rod.
- A larger disk cannot be placed on top of a smaller one on the same rod.

Figure 7.3(b) shows a configuration of disks in an intermediate step after performing a few moves. Finally, the problem is solved when all of the disks appear on the destination rod, as shown in (c).

The size of the problem is determined by the number of disks. The base case occurs when $n = 1$, where the solution is naturally trivial, involving a single basic move of the disk from the origin rod to the destination rod. For two disks the solution is also simple (see Figure 7.4). The first move places the smaller disk on the auxiliary rod, which allows us to move the larger disk onto the destination rod. The last move simply places the smaller disk on the destination rod.

It can be shown that the minimum number of moves required to solve the problem for n disks is $2^n - 1$. For $n = 4$, the problem can be solved through the 15 moves illustrated in Figure 7.5. Although the number of moves increases exponentially as n grows, the difficulty of the problem is essentially the same as the case for $n = 2$ if we apply recursion. The most interesting configurations of disks in the solution for $n = 4$ are the ones after moves 7 and 8. Observe that after the seventh move there are $n - 1$ disks on the auxiliary rod. This situation is necessary in order to move the largest disk onto the destination rod. Thus, the first seven steps consist of solving the puzzle for $n - 1 = 3$ disks, where the goal consists of moving them from the origin to the auxiliary rod, with the help of the destination rod. In other words, the roles of the destination and auxiliary rods are switched. The eighth move is just a basic operation that involves moving the largest disk. After it is placed on the destination rod, there is a stack of $n - 1$ disks on the auxiliary rod that needs to be moved onto the destination rod. Therefore, the seven remaining steps consist of solving another problem of $n - 1$ disks, where in this case the roles of the auxiliary and origin rods are switched. This reasoning based on problem decomposition and induction is illustrated in Figure 7.6. In particular, the decomposition considers two problems of size $n-1$, where the solution consists of three operations. The first and third steps solve one of these subproblems, while the second operation simply moves the largest disk onto the destination rod. A recursive solution can therefore rely on induction in order to assume that we can move entire stacks of $n - 1$ disks in a single step (which corresponds to a recursive call). In that case note that the solution is conceptually very similar to the one for $n = 2$. In particular, observe the similarity between Figures 7.4 and 7.6. Both solutions consist of three steps, where disk 1 is replaced by a full stack of $n - 1$ disks in Figure 7.6 (and disk 2 plays the same role as disk n).

Listing 7.3 shows an implementation of the procedure, which prints one line in the output console per move. The last line in the code simply calls the method, which produces the output shown in Figure 7.7 for $n = 4$. The parameters related to the rods are necessary in order to specify the particular subproblems (and original problem) correctly. The

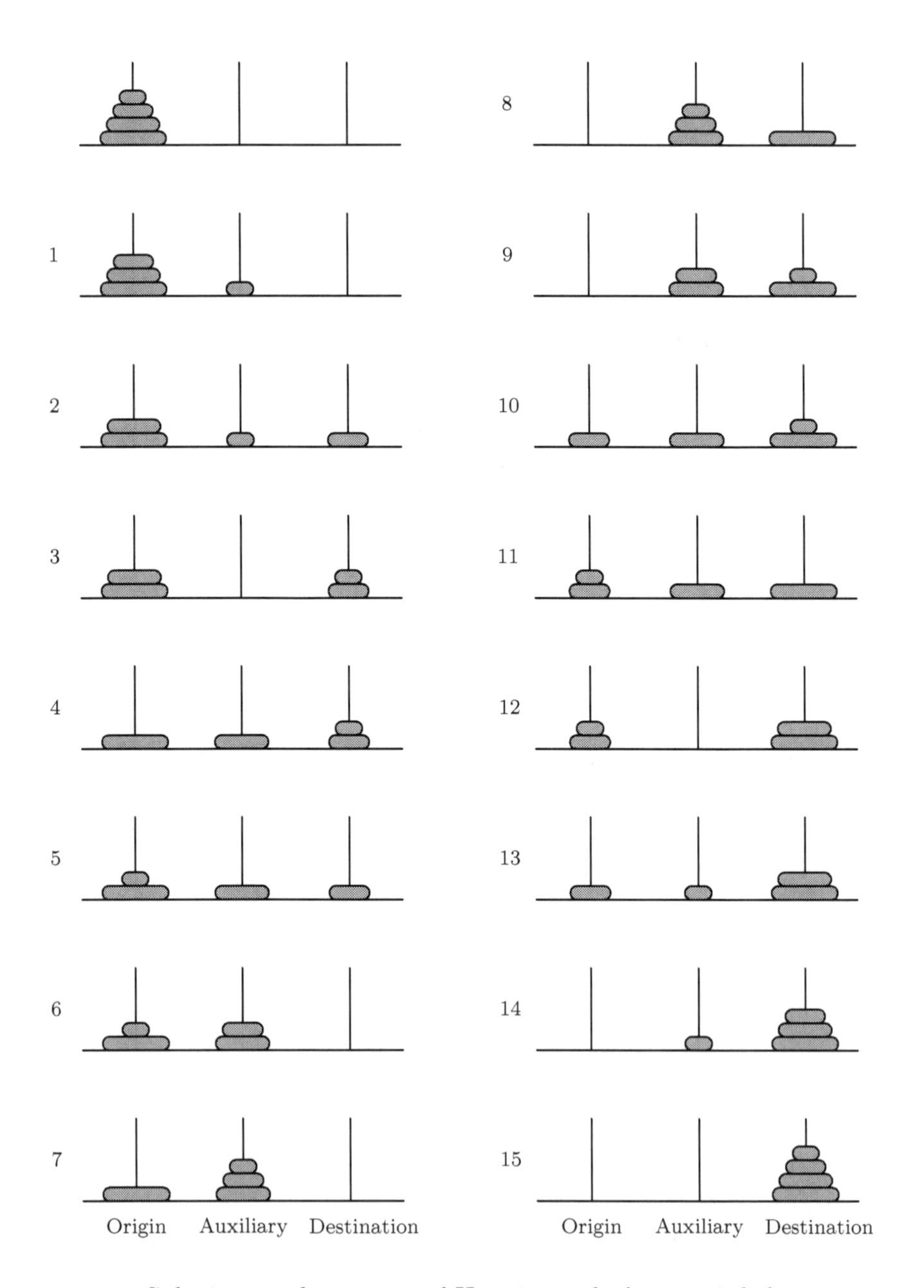

Figure 7.5 Solution to the towers of Hanoi puzzle for $n = 4$ disks.

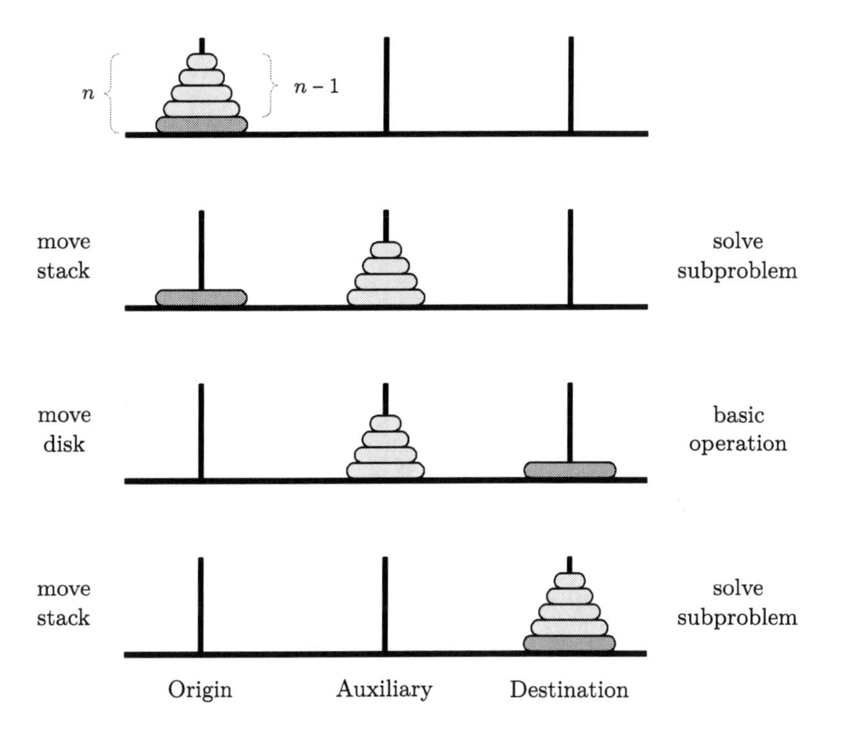

Figure 7.6 Decomposition of the towers of Hanoi problem.

code also shows an alternative procedure that considers a base case for $n = 0$, where it simply does not need to carry out any action.

7.3 TREE TRAVERSALS

A tree traversal is the process of visiting each of the nodes of a tree (understood as an undirected graph) only once in some particular order. In this section we will examine three ways of traversing a tree in some depth-first order. This means that we will examine the entire subtree of a node before analyzing its siblings and their corresponding subtrees. In addition, we will only consider left-to-right traversals, which process sibling nodes from left to right. Recursive tree traversals are appealing for their simplicity, and do not require stack or queue data structures as do iterative versions.

Listing 7.3 Towers of Hanoi procedure.

```
def towers_of_Hanoi(n, o, d, a):
    if n == 1:
        print('Move disk', n, 'from rod', o, 'to rod', d)
    else:
        towers_of_Hanoi(n - 1, o, a, d)
        print('Move disk', n, 'from rod', o, 'to rod', d)
        towers_of_Hanoi(n - 1, a, d, o)


def towers_of_Hanoi_alt(n, o, d, a):
    if n > 0:
        towers_of_Hanoi_alt(n - 1, o, a, d)
        print('Move disk', n, 'from rod', o, 'to rod', d)
        towers_of_Hanoi_alt(n - 1, a, d, o)


towers_of_Hanoi(4, 'O', 'D', 'A')
```

```
Move disk 1 from rod O to rod A
Move disk 2 from rod O to rod D
Move disk 1 from rod A to rod D
Move disk 3 from rod O to rod A
Move disk 1 from rod D to rod O
Move disk 2 from rod D to rod A
Move disk 1 from rod O to rod A
Move disk 4 from rod O to rod D
Move disk 1 from rod A to rod D
Move disk 2 from rod A to rod O
Move disk 1 from rod D to rod O
Move disk 3 from rod A to rod D
Move disk 1 from rod O to rod A
Move disk 2 from rod O to rod D
Move disk 1 from rod A to rod D
```

Figure 7.7 Output of Listing 7.3, which represents the solution to the towers of Hanoi puzzle for $n = 4$ disks.

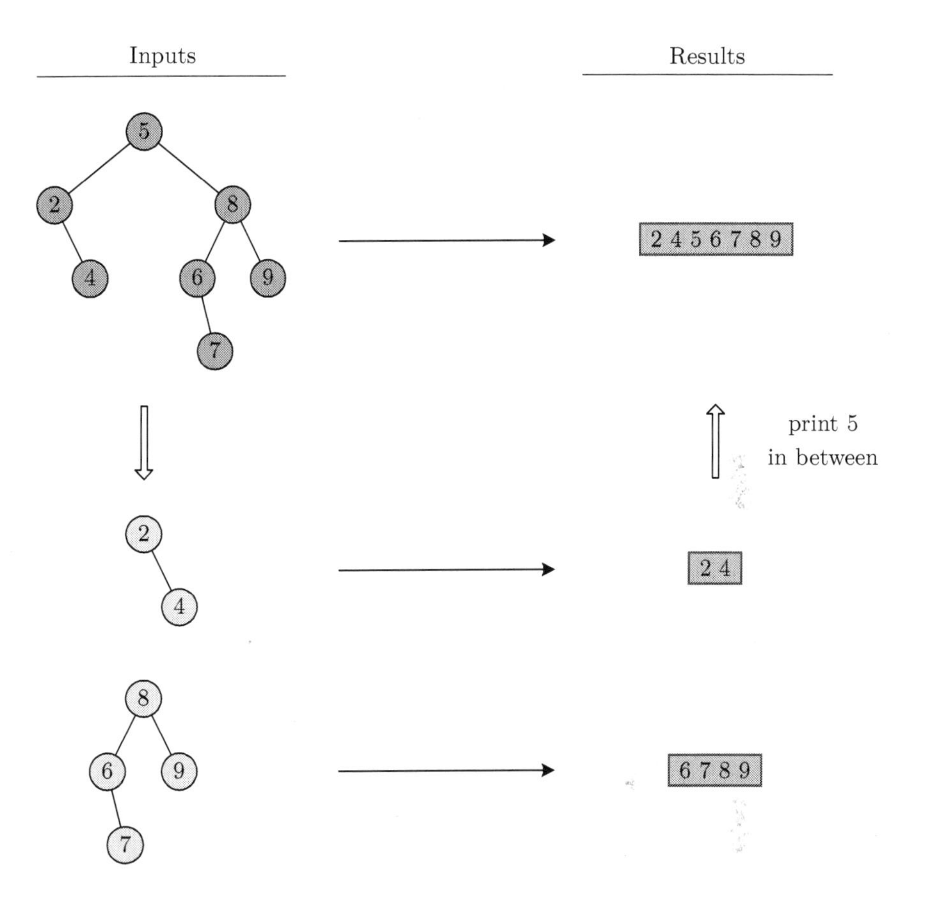

Figure 7.8 Concrete example of the decomposition of the inorder traversal problem.

7.3.1 Inorder traversal

Consider the problem of printing the items in a binary search tree (see Section 5.3) according to the order of their keys. We can develop a recursive algorithm that solves the problem as follows. Naturally, if the tree is empty the algorithm does not perform any action. This would be the base case. For the recursive case consider the diagram in Figure 7.8, which discards the root note and decomposes the tree into the root's left and right subtrees. Firstly, we can be sure that all of the keys of the root's left subtree are less than the key of root node. In addition, due to the binary search tree property, the left subtree is also a binary search tree. Therefore, we can assume by induction that a recursive call on the left subtree will print its keys properly ordered. Similarly, a recursive call

Listing 7.4 Inorder traversal of a binary tree.

```
1 def inorder_traversal(T):
2     if T != []:
3         inorder_traversal(T[2])  # process left subtree
4         print(T[0], ': ', T[1], sep='')  # print key and item
5         inorder_traversal(T[3])  # process right subtree
```

Listing 7.5 Preorder and postorder traversals of a binary tree.

```
1 def preorder_traversal(T):
2     if T != []:
3         print(T[0], ': ', T[1], sep='')  # print key and item
4         preorder_traversal(T[2])  # process left subtree
5         preorder_traversal(T[3])  # process right subtree
6
7
8 def postorder_traversal(T):
9     if T != []:
10        postorder_traversal(T[2])  # process left subtree
11        postorder_traversal(T[3])  # process right subtree
12        print(T[0], ': ', T[1], sep='')  # print key and item
```

on the right subtree will print a sorted sequence of the keys in the right subtree. Clearly, the method simply has to solve both subproblems, and print the key associated with the root node in between the outputs of the subproblems.

The procedure can be coded as in Listing 7.4, where we have assumed that the input binary search tree is a list of four elements, as described in Section 5.3. For example, running the code on the birthday calendar tree specified in (5.1) prints the elements of the tree in alphabetical order according to the names of the persons (keys). Thus, we say that the method performs an **inorder** traversal of the binary search tree.

7.3.2 Preorder and postorder traversals

Instead of printing the key (and the item) in between recursive calls, **preorder** and **postorder** traversals print (or process) the key before and after the recursive calls, respectively. Listing 7.5 shows methods that implement these traversals. Observe that the preorder traversal prints the key of the root node of the entire tree, or any of its subtrees, as soon as the program executes a method call with a nonempty tree. Therefore, a preorder traversal indicates the order in which the recursive calls are carried out (one per node of the binary tree). In particular, the method `preorder_traversal` produces the following output when executed on the list in (5.1):

```
Emma: 2002/08/23
Anna: 1999/12/03
Paul: 2000/01/13
Lara: 1987/08/23
John: 2006/05/08
Luke: 1976/07/31
Sara: 1995/03/14
```

Alternatively, the sequence of keys related to a postorder traversal reflects the order in which the associated recursive calls finish. In particular, a postorder traversal of the birthday calendar binary search tree produces the following sequence of keys: 'Anna', 'John', 'Luke', 'Lara', 'Sara', 'Paul', 'Emma'. Observe that the key of the initial root node ('Emma') appears in last place, since the method call with the entire tree is the last one to terminate.

7.4 LONGEST PALINDROME SUBSTRING

In this problem, given an initial input string $\mathbf{s} = s_0 s_1 \cdots s_{n-2} s_{n-1}$ of length n, where the s_i represent characters, the goal consists of finding the longest palindrome substring contained within it. If there are several palindrome substrings of the same maximum length the algorithm may return any of them. It is important to understand that a **substring** is a sequence of contiguos elements of $\mathbf{s}$. Thus, it should not be confused with the notion of a **subsequence** (for which the elements do not necessarily have to appear contiguously). Lastly, this is an **optimization problem**, where the goal consists of obtaining a maximum value of some function involving the string. This particular problem can be tackled by using the "dynamic programming" algorithm design technique (see Section 10.4), which can be applied on certain types of optimization problems.

The size of the problem is the length of the string n. The smallest instances correspond to strings of length 0 (empty) or 1 (a single character), where the algorithm can trivially return the input string at a base case. The problem can be simplified by discarding a character at the extremes of the string. Thus, the decomposition can rely on the two subproblems illustrated in Figures 7.9(b) and (c), where the original problem is shown in (a). Clearly, if the longest palindrome substring

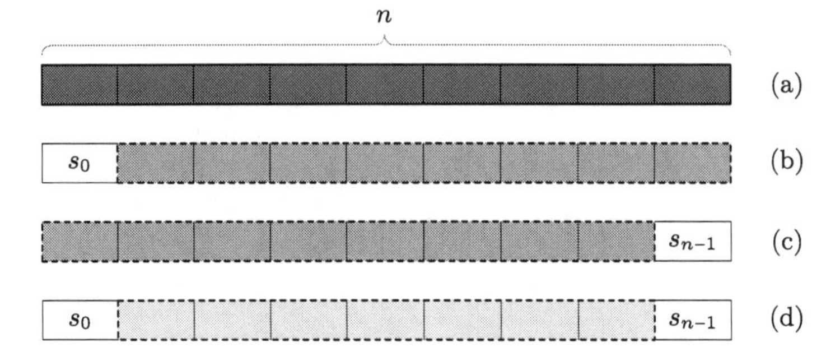

Figure 7.9 Decomposition of the problem of finding the longest palindrome substring.

Listing 7.6 Code for finding the longest palindrome substring (or sublist).

```
def longest_palindrome_substring(s):
    n = len(s)
    if is_palindrome(s):
        return s
    else:
        s_aux_1 = longest_palindrome_substring(s[1:n])
        s_aux_2 = longest_palindrome_substring(s[0:n - 1])
        if len(s_aux_1) > len(s_aux_2):
            return s_aux_1
        else:
            return s_aux_2
```

does not contain the character (s_0) at the left-most location of the input string, the result will be the output of the method applied to the remaining substring ($\mathbf{s}_{1..n-1}$), regardless of whether s_{n-1} belongs to the solution. Similarly, if the longest palindrome substring does not contain s_{n-1}, the algorithm should return the longest palindrome substring of $\mathbf{s}_{0..n-2}$. Lastly, the entire input string could be a palindrome. This can be checked through the function in Listing 4.12, which also returns a true value for the base cases.

Listing 7.6 shows an implementation of the recursive method. Since the problem is analogous to finding the longest palindrome sublist, the code also works if the input is a list. If the entire input $\mathbf{s}$ is a palindrome (see Section 4.3.2) the result is simply $\mathbf{s}$, which naturally covers empty inputs and those that contain a single element. Otherwise, the algorithm searches for the longest palindrome in the two subproblems of size $n-1$, returning the longest solution. Its runtime cost can be specified through:

$$T(n) = \begin{cases} 1 & \text{if } 0 \le n \le 1, \\ 2T(n-1) + n/2 + 1 & \text{if } n > 1, \end{cases}$$

where the $n/2$ term is associated with the call to `is_palindrome`. The nonrecursive expression is $T(n) = (7/4)2^n - n/2 - 2$. Thus, $T(n) \in \Theta(2^n)$ has an exponential growth.

The recursive algorithm is clearly inefficient since the problem can be solved in $\mathcal{O}(n^3)$ through a brute force approach. Note that there are on the order of n^2 substrings (which can be specified by two limiting indices controlled by two loops), and it takes on the order of n operations to determine if a string is a palindrome. The inefficiency of Listing 7.6 stems from solving numerous identical **overlapping subproblems**. For example, both recursive calls solve the subproblem of length $n-2$ in Figure 7.9(d), which entails repeating identical, and therefore redundant, operations. Moreover, smaller identical subproblems are solved an exponential number of times. These recursive solutions can nevertheless be useful in practice, since developing them can be the first step towards designing more efficient algorithms. Section 10.4 discusses how how to avoid overlapping subproblems in recursive solutions through an approach known as memoization, or by applying dynamic programming, which is a prominent algorithm design technique. Solutions based on dynamic programming for this problem can run in $\mathcal{O}(n^2)$ time. Lastly, several algorithms have been developed that are capable of solving it in linear time.

Finally, Listing 7.6 relies on the function `is_palindrome` in order to determine if a string is a palindrome. The following example replaces that function by another recursive call. Obviously, a string $\mathbf{s}$ of length n is a palindrome if its longest palindrome substring has length n. Thus, we can check if $\mathbf{s}$ is a palindrome by evaluating whether the longest palindrome substring of $\mathbf{s}_{1..n-2}$ has length $n-2$, with $s_0 = s_{n-1}$. Listing 7.7 shows this alternative method, which only invokes itself, employing the solution to the subproblem in Figure 7.9(d). Its running time can be characterized by:

$$T(n) = \begin{cases} 1 & \text{if } 0 \le n \le 1, \\ 2T(n-1) + T(n-2) + 1 & \text{if } n > 1, \end{cases}$$

whose order of growth is $\Theta((1+\sqrt{2})^n) = \Theta((2.41\ldots)^n)$. The method is therefore more inefficient than the one in Listing 7.6, since it computes the solutions to even more identical overlapping subproblems.

Listing 7.7 Alternative code for finding the longest palindrome substring (or sublist).

```
def longest_palindrome_substring_alt(s):
    n = len(s)
    if n <= 1:
        return s
    else:
        s_aux_1 = longest_palindrome_substring_alt(s[1:n - 1])
        if len(s_aux_1) == n - 2 and s[0] == s[n - 1]:
            return s
        else:
            s_aux_2 = longest_palindrome_substring_alt(s[1:n])
            s_aux_3 = longest_palindrome_substring_alt(
                s[0:n - 1])
            if len(s_aux_2) > len(s_aux_3):
                return s_aux_2
            else:
                return s_aux_3
```

7.5 FRACTALS

A fractal can be understood as a geometrical objet whose structure is repeated at different scales. In this section we will learn how to generate the Koch snowflake and Sierpiński's carpet, which are two of the first developed fractal images.

7.5.1 Koch snowflake

The Koch snowflake is a fractal image formed by multiple line segments associated with "Koch curves." These curves are formed by iteratively modifying the segments of an image in the following way. Consider an initial line segment, as shown in Figure 7.10, of length L. The idea consists of transforming it into the curve shown at iteration 1 that has four smaller line segments of length $L/3$. The process simply divides the initial line segment into three equal segments, but replaces the middle one by two sides of an equilateral triangle whose length is $L/3$. Thus, the new curve has length $4L/3$. A Koch curve fractal is formed by repeating the process with every line segment of an image. For example, in Figure 7.10 the curve at iteration 2 is formed by applying the transformation to the four segments of the image at iteration 1. The new curve contains 16 segments of length $L/9$. Thus, its total length is $(4/3)^2L$. Observe that

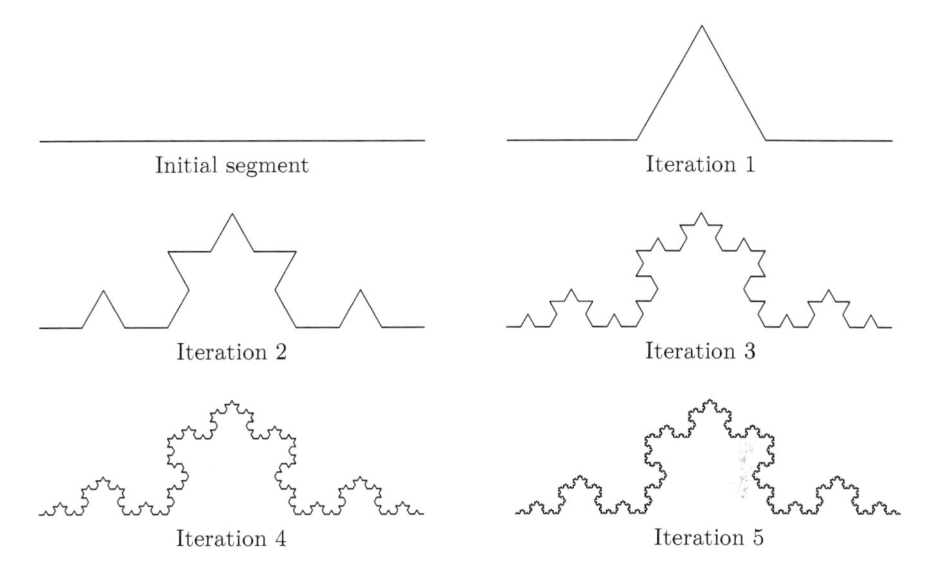

Figure 7.10 Koch curve fractal.

after n iterations the length of the curve is $(4/3)^n$, which approaches infinity as n approaches infinity.

The Koch snowflake is constructed by applying the same process starting with the three segments of an equilateral triangle (i.e., by generating three Koch curves). Figure 7.11 shows the first iterations of the process that generates a Koch snowflake.

From a computational point of view, the problem of generating a Koch curve requires three input parameters: the two endpoints, $\mathbf{p}$ and $\mathbf{q}$ (which are two-dimensional points on a plane) of an initial line segment, and a finite number of iterations n. The size of the problem is clearly n, since it determines the number of steps needed to generate the fractal image. The simplest case occurs when $n = 0$, where the algorithm simply draws a line from $\mathbf{p}$ to $\mathbf{q}$.

Figure 7.12 shows the decomposition of the problem that we will use for the recursive case. Note that a Koch curve after n iterations consists of four simpler Koch curves after $n-1$ iterations. The main mathematical challenge is to define the endpoints of each simpler Koch curve. Figure 7.13 explains one way to obtain them by using vectors. Firstly, the points $\mathbf{p}$ and $\mathbf{q}$ can be understood as vectors in a plane, as shown in (a). In addition, let $\mathbf{v} = \mathbf{q} - \mathbf{p}$. The endpoints on the original segment at one and two thirds of the distance from $\mathbf{p}$ to $\mathbf{q}$ are $\mathbf{p} + \mathbf{v}/3$, and $\mathbf{p} + 2\mathbf{v}/3$, respectively, as shown in (b) and (c). The last endpoint

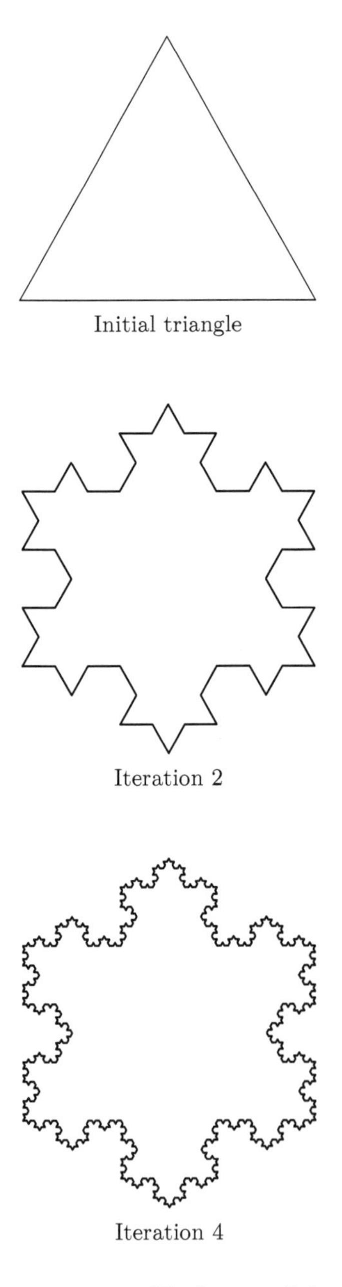

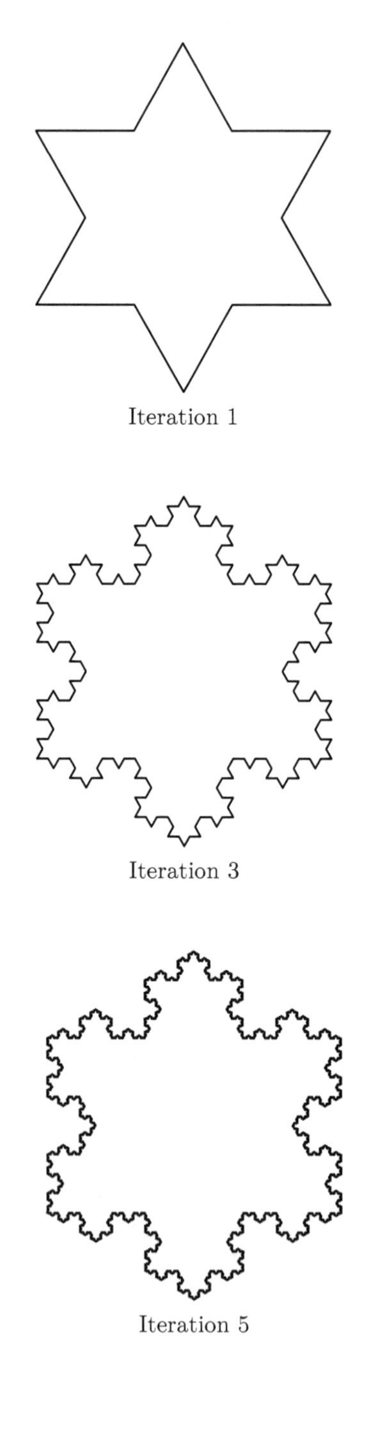

Figure 7.11 Koch snowflake fractal.

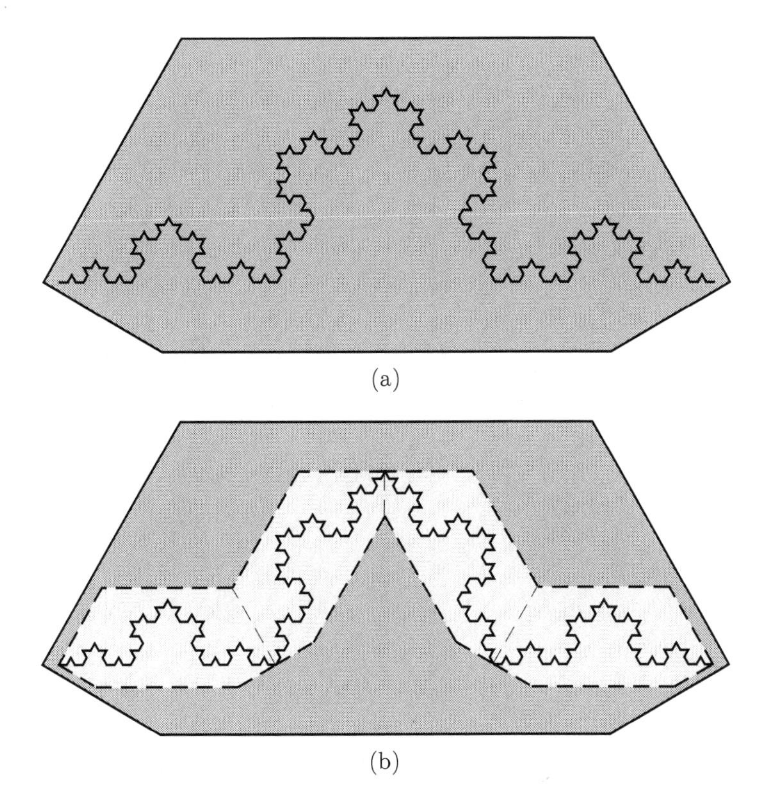

Figure 7.12 Koch curve decomposition.

($\mathbf{x}$), which does not lie in the segment between $\mathbf{p}$ and $\mathbf{q}$, can be obtained by adding $\mathbf{p} + \mathbf{v}/3 + \mathbf{R}_{60^\circ}\mathbf{v}/3$, where:

$$\mathbf{R}_{60^\circ} = \begin{bmatrix} \cos(\pi/3) & -\sin(\pi/3) \\ \sin(\pi/3) & \cos(\pi/3) \end{bmatrix} = \begin{bmatrix} 1/2 & -\sqrt{3}/2 \\ \sqrt{3}/2 & 1/2 \end{bmatrix}$$

is a rotation matrix that rotates vectors 60 degrees counterclockwise, as illustrated in (d).

Listing 7.8 relies on the Numpy and Matplotlib packages in order to generate Koch curves. The base case simply plots a line segment from $\mathbf{p}$ to $\mathbf{q}$. The recursive case invokes the method four times with appropriate endpoints, reducing the number of iterations by a unit. The Koch snowflake can be obtained by generating Koch curves for three initial segments that form an equilateral triangle. Finally, note that the code defines column vectors, in accordance with most texts in which vectors correspond to column vectors.

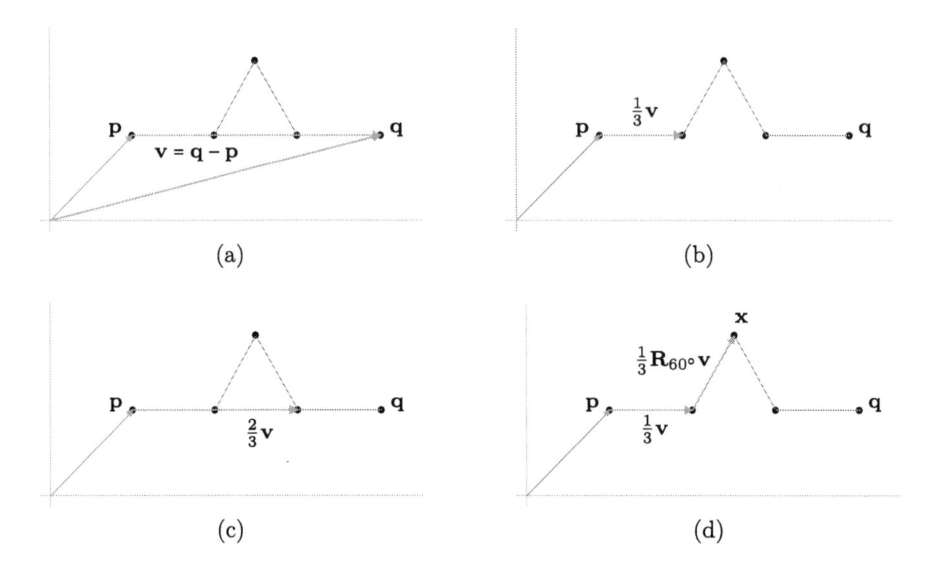

Figure 7.13 New endpoints of shorter segments when applying an iteration related to the Koch curve.

7.5.2 Sierpiński's carpet

Sierpiński's carpet is another classical fractal that is formed in a top-down manner as follows. Given an empty square whose side measures a certain length s, divide it into 9 squares of height and width $s/3$, and draw (i.e., fill with some color) the middle square, as illustrated in iteration 1 of Figure 7.14. Subsequently, the process can be repeated with the eight smaller empty squares surrounding the middle square. This would lead to the image in iteration 2 of Figure 7.14. The process can be further repeated up to some n-th iteration. Lastly, Figure 7.14 draws a square frame bounding the images for reference, but it does not form part of the fractal.

The inputs to the problem can be an initial 2-dimensional point $\mathbf{p}$ that marks the center of the initial empty square (without edges), its length s, and the number of desired iterations n. Clearly, the size of the problem is determined by n, and the base case occurs when $n = 0$, which does not require any action.

Figure 7.15 shows the decomposition of the problem for a general recursive case, where the square is divided into nine subsquares whose sides measure $s/3$. Listing 7.9 shows a possible implementation of the method. When $n > 0$ it first draws a square centered at $\mathbf{p}$, where its bottom-left corner is located at point $\mathbf{p} - [s/6, s/6]$, while its width and

Listing 7.8 Code for generating Koch curves and the Koch snowflake.

```
import math
import numpy as np
import matplotlib.pyplot as plt


def koch_curve(p, q, n):
    if n == 0:    # The base case is just a line segment
        plt.plot([p[0, 0], q[0, 0]], [p[1, 0], q[1, 0]], 'k-')
    else:

        v = q - p
        koch_curve(p, p + v / 3, n - 1)

        R_60 = np.matrix([[math.cos(math.pi / 3),
                           -math.sin(math.pi / 3)],
                          [math.sin(math.pi / 3),
                           math.cos(math.pi / 3)]])

        x = p + v / 3 + R_60 * v / 3
        koch_curve(p + v / 3, x, n - 1)

        koch_curve(x, p + 2 * v / 3, n - 1)

        koch_curve(p + 2 * v / 3, q, n - 1)


def koch_snowflake(n):
    p = np.array([[0], [0]])
    q = np.array([[1], [0]])
    r = np.array([[0.5], [math.sqrt(3) / 2]])
    koch_curve(p, r, n)
    koch_curve(r, q, n)
    koch_curve(q, p, n)


fig = plt.figure()
fig.patch.set_facecolor('white')
koch_snowflake(3)
plt.axis('equal')
plt.axis('off')
plt.show()
```

height is $s/3$ (these values are used by the `Rectangle` method in Matplotlib). Subsequently, it recursively draws eight Sierpiński carpets of

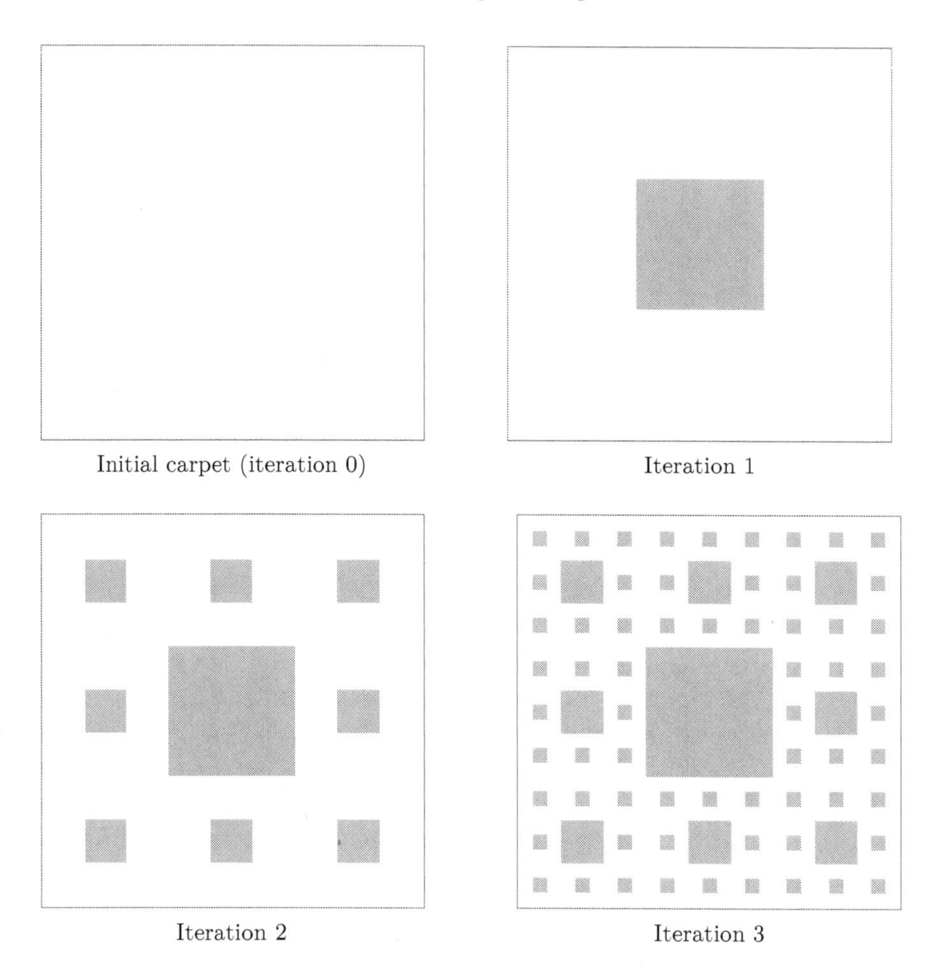

Figure 7.14 Sierpiński's carpet after 0, 1, 2, and 3 iterations.

$n - 1$ iterations, and size $s/3$, on the remaining eight smaller squares. The coordinates of the centers of those squares are simply $\mathbf{p}$ plus $s/3$, 0, or $-s/3$ on each dimension. Lastly, the code in lines 39–41 simply draws the edges of the initial empty square.

7.6 EXERCISES

Exercise 7.1 — Implement the function in (3.2) that computes binomial coefficients.

Exercise 7.2 — In this exercise the goal is to develop a solution for a variant of the swamp traversal problem described in Section 7.1. In

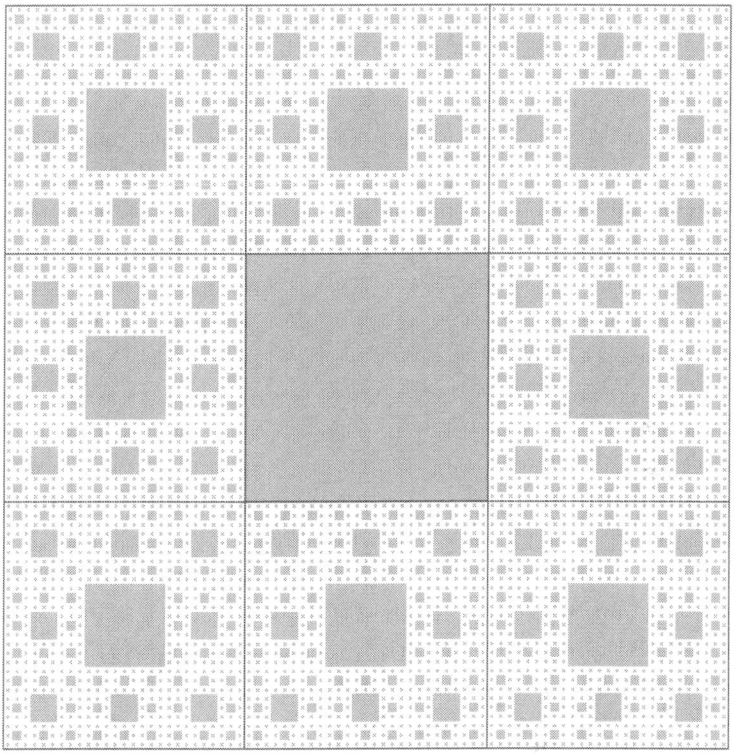

Figure 7.15 Sierpiński's carpet decomposition.

Listings 7.1 and 7.2 the input matrix is reduced at each new function call until it is simply a column vector. In some programming languages it is simpler (and in some cases necessary) to pass the entire input matrix in each recursive call. In these cases the size of the problem can be determined with additional parameters. Implement a variant of the mentioned algorithms that determines if there exists a path through the swamp (advancing towards the right), which begins at patch $a_{r,c}$. The method therefore needs the extra parameter c, and can pass the entire input matrix in each recursive call.

Exercise 7.3 — Implement a recursive method that simulates printing the tick marks of an English ruler that indicates inches. Given a certain nonnegative integer n, the program should print the tick marks on the console as shown in Figure 7.16, from 0 to n inches. In addition, it will print subdivisions of an inch according to a certain precision specified by a parameter k. In particular, each inch will be divided into 2^k parts (i.e.,

Listing 7.9 Code for generating Sierpiński's carpet.

```
import numpy as np
import matplotlib.pyplot as plt
from matplotlib.patches import Rectangle


def sierpinski_carpet(ax, p, n, size):
    if n > 0:
        ax.add_patch(Rectangle((p[0, 0] - size / 6,
                                p[1, 0] - size / 6),
                               size / 3, size / 3,
                               facecolor=(0.5, 0.5, 0.5),
                               linewidth=0))

        q = np.array([[-size / 3], [-size / 3]])
        sierpinski_carpet(ax, p + q, n - 1, size / 3)
        q = np.array([[-size / 3], [0]])
        sierpinski_carpet(ax, p + q, n - 1, size / 3)
        q = np.array([[-size / 3], [size / 3]])
        sierpinski_carpet(ax, p + q, n - 1, size / 3)

        q = np.array([[0], [-size / 3]])
        sierpinski_carpet(ax, p + q, n - 1, size / 3)
        q = np.array([[0], [size / 3]])
        sierpinski_carpet(ax, p + q, n - 1, size / 3)

        q = np.array([[size / 3], [-size / 3]])
        sierpinski_carpet(ax, p + q, n - 1, size / 3)
        q = np.array([[size / 3], [0]])
        sierpinski_carpet(ax, p + q, n - 1, size / 3)
        q = np.array([[size / 3], [size / 3]])
        sierpinski_carpet(ax, p + q, n - 1, size / 3)


fig = plt.figure()
fig.patch.set_facecolor('white')
ax = plt.gca()
p = np.array([[0], [0]])
sierpinski_carpet(ax, p, 4, 1)
ax.add_patch(Rectangle((-1 / 2, -1 / 2), 1, 1,
                       fill=False, edgecolor=(0, 0, 0),
                       linewidth=0.5))
plt.axis('equal')
plt.axis('off')
plt.show()
```

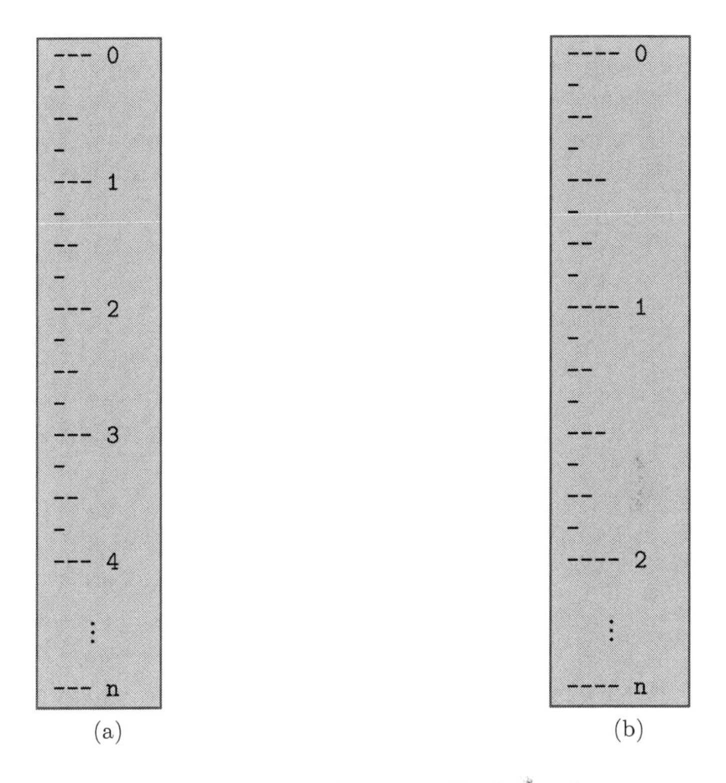

Figure 7.16 Simulation of tick marks on an English ruler.

it will print subdivisions of $1/2^k$ inches). In Figure 7.16(a) $k = 2$, while in (b) $k = 3$. Finally, tick marks indicating subdivisions of $1/2^j$ inches (for the smallest possible value of j) will be represented by $k + 1 - j$ hyphen characters. For example, in (b) tick marks that indicate 1/4 (and 3/4, 5/4, 7/4, etc.) inches appear with $3 + 1 - 2 = 2$ hyphen characters.

Exercise 7.4 — In this exercise the goal is to solve a variant of the towers of Hanoi puzzle that we will call the "sideways" towers of Hanoi problem. Assume that the three rods are arranged from left to right. The rules are the same as in the original problem, but it is not allowed to move individual disks directly between the left and right rods, as illustrated in Figure 7.17. Design a procedure that uses multiple recursion in order to move a stack of n disks from the left rod to the right one, or vice versa (Exercise 9.7 covers methods to move disks between the middle rod and the left or right rods). In addition, determine the number of individual disk moves involved in moving a stack of n disks.

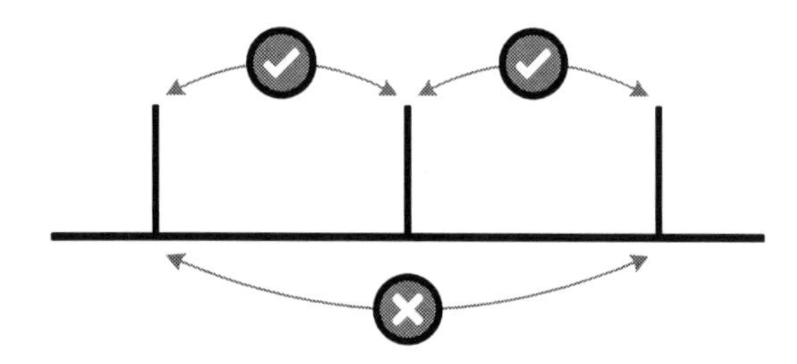

Figure 7.17 Rules for the "sideways" variant of the towers of Hanoi puzzle.

Exercise 7.5 — Implement a function that computes the number of nodes in a binary search tree. Assume it is coded as a list of four components, as described in Section 5.3.

Exercise 7.6 — Implement a function in Python that, given an input list **a** of n elements, determines the longest sublist within it composed of contiguous identical elements. For example, for **a** = $[1, 3, 5, 5, 4, 4, 4, 5, 5, 6]$ the result is $[4, 4, 4]$. Break up the problem similarly to Listing 7.6, and use an auxiliary recursive function that determines if all of the elements of a list are identical. In addition, implement a recursive function that does not need to call the auxiliary method, similarly to Listing 7.7.

Exercise 7.7 — Design and code a recursive method that computes the longest palindrome subsequence in a given list **a** of n elements. This is not the same problem as the one described in Section 7.4, since a subsequence of a list does not necessarily contain contiguous elements of it. For example, the longest palindrome subsequence of $[1, 3, 4, 4, 6, 3, 1, 5, 1, 3]$ is $[1, 3, 4, 4, 3, 1]$, while the longest palindrome (contiguous) sublist is $[3, 1, 5, 1, 3]$.

Exercise 7.8 — Implement a recursive method that draws Sierpiński's triangle (see Figure 1.1). From a top-down perspective Sierpiński's triangle can be generated by starting with an equilateral triangle of the form $\triangle$, which is divided into four smaller equilateral triangles of half the size of the original's. One of them will have the form $\triangledown$, but three will be $\triangle$ triangles. The process is repeated for these three triangles iteratively (see Figure 7.18). The method should receive the coordinates of a two-dimensional point for reference (for example, the bottom-left vertex of the triangle), the length of the triangle's base, and the number of iterations n to carry out. Finally, use packages Numpy and Matplotlib.

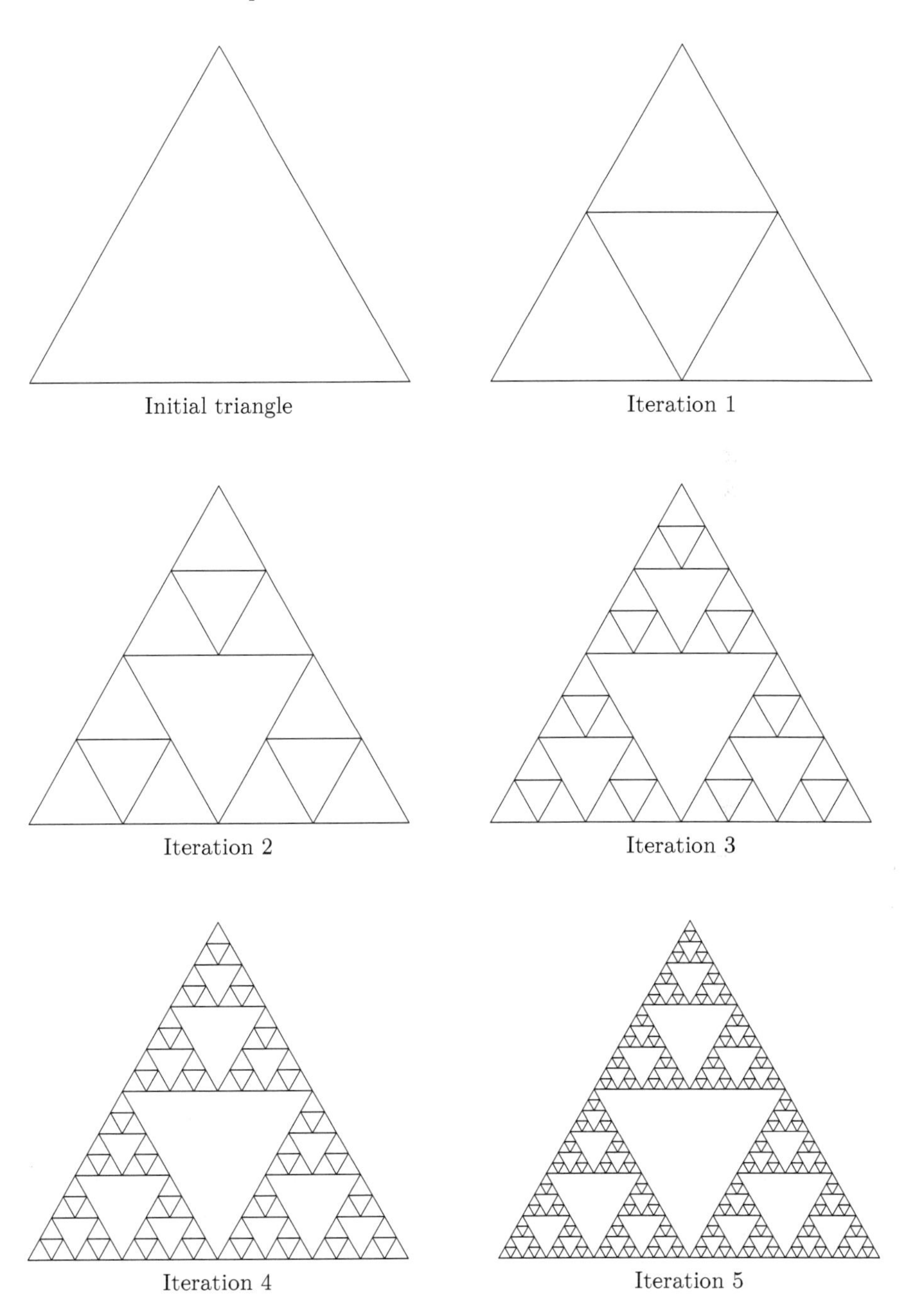

Figure 7.18 Sierpiński's triangle after 0, 1, 2, and 3 iterations.

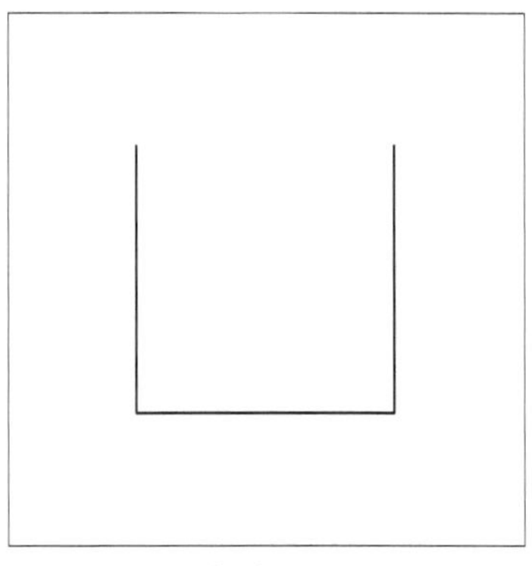

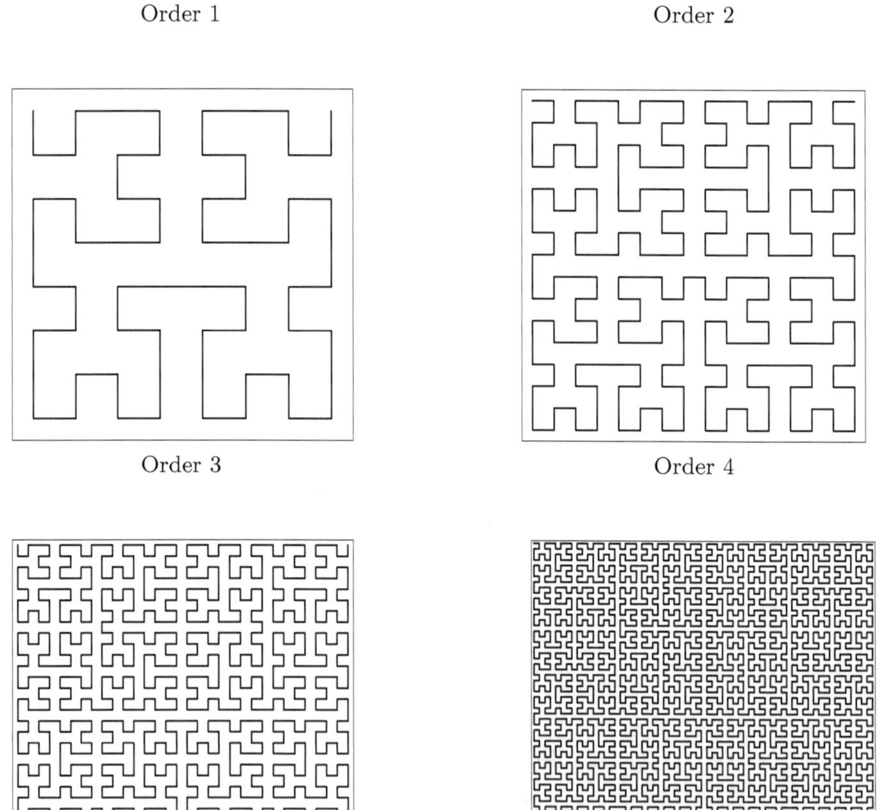

Figure 7.19 Hilbert curves of orders 1–6.

Exercise 7.9 — A Hilbert curve is a continuous space-filling curve fractal. Figure 7.19 shows the first four Hilbert curves. As the order increases, the length of the curve increases exponentially, and progressively fills the bounding square. A recursive decomposition of the problem involves breaking up the fractal into four curves of smaller order, which need to be oriented appropriately. In addition, since the total curve is continuous, the four smaller curves need to be connected, and it is actually these connections that will end up forming the entire curve. In this challenging exercise the goal consists of implementing a recursive method that draws a Hilbert curve of order n.

CHAPTER 8

Counting Problems

Music is the pleasure the human mind experiences from counting without being aware that it is counting.

— Gottfried Leibniz

RECURSION is used extensively in combinatorics, which is a subfield of mathematics that studies counting, and plays an important role in advanced algorithm analysis. This chapter focuses on recursive solutions to computational counting problems, whose goal consists of adding up a certain number of elements, entities, choices, concepts, etc. A common strategy consists of grouping the items to be counted into several disjoint subsets, and adding the number of elements in each one. In terms of recursion, an original problem will be decomposed into several subproblems, and the result will be the sum of the outputs of those smaller problems.

We have seen several of these problems in previous chapters. For example, computing the sum of the first positive integers can be interpreted as counting the number of square blocks of triangular structures like the ones in Figure 1.5(a). Another example is the bit counting problem associated with Exercise 3.8 (and 4.11). In this chapter we will examine two alternative recursive strategies for counting the bits, which will rely on two different decompositions.

We will see that several of the most widely used functions to illustrate recursion are associated with fundamental combinatorial concepts. For instance, the factorial, power, and binomial coefficients are related to permutations, variations (with repetition), and combinations, respectively. In addition, Fibonacci numbers arise in numerous combinatorial problems as well. In this chapter we will derive these functions by decomposing related counting problems, instead of breaking up their mathe-

[1,2,3,4]	[2,1,3,4]	[3,1,2,4]	[4,1,2,3]
[1,2,4,3]	[2,1,4,3]	[3,1,4,2]	[4,1,3,2]
[1,3,2,4]	[2,3,1,4]	[3,2,1,4]	[4,2,1,3]
[1,3,4,2]	[2,3,4,1]	[3,2,4,1]	[4,2,3,1]
[1,4,2,3]	[2,4,1,3]	[3,4,1,2]	[4,3,1,2]
[1,4,3,2]	[2,4,3,1]	[3,4,2,1]	[4,3,2,1]

Figure 8.1 Possible permutations (lists) of the first four positive integers.

matical formulas and definitions. Although the problems are inherently mathematical, we will tackle them by applying the same concepts, skills, and methodology as in the previous chapters.

Finally, since the solutions to the problems are either simple to program, or have appeared earlier throughout the book, the chapter will not contain code (some exercises propose implementing functions). In this regard, its main purpose is to incite recursive thinking, by analyzing a different family of computational problems.

8.1 PERMUTATIONS

A permutation of a collection of elements can be understood as a particular sequence of such elements (it can also signify the act of rearranging the order of the elements in a given sequence). In this section we will analyze permutations "without repetition," where all of the elements are distinct. Specifically, consider the counting problem whose goal is to determine the number of different permutations that it is possible to generate with n distinct items. For simplicity, assume that the n elements are the integers $1, 2, \ldots, n$, and a permutation is a list of these elements. For $n = 4$ there are 24 different permutations, which are illustrated in Figure 8.1 through lists (Section 12.2.2 will describe an algorithm for generating all of these permutations).

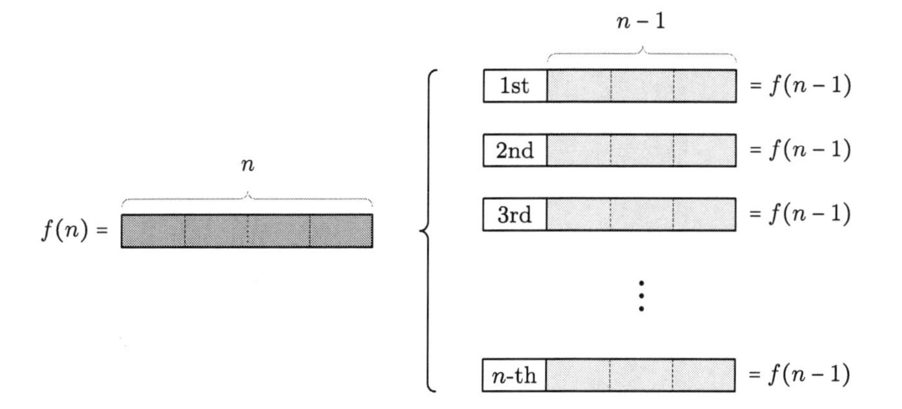

Figure 8.2 Decomposition of the problem that counts the number of possible permutations of n different elements, denoted as $f(n)$.

In order to define a mathematical recursive function that provides the result for a given n, we can proceed as usual by using the template in Figure 2.1. For counting problems the size of the problem is related to the total number of items to be added up. In this problem the size is simply n, since the number of permutations grows as a function of n. A trivial base case occurs when $n = 1$, where there is only one possible permutation that we can form with a single element. Additionally, we can consider the case when $n = 0$. Although it may seem that the result would be 0 permutations, for convenience mathematicians interpret that there is one way to build a permutation. For instance, if a permutation corresponds to a list, then we can think that when $n = 0$ it is possible to build the empty list.

For the recursive case we can break up the result by considering n subproblems of size $n - 1$, as illustrated in Figure 8.2, where each of the subproblems counts a different subset of the possible permutations. Let $f(n)$ represent the function that computes the number of possible permutations of n distinct elements. If we fix an element at the first position of the permutation, then we need to count the number of ways it is possible to arrange the remaining $n - 1$ elements in the rest of the locations. This number, by definition, is $f(n-1)$, which we can assume that we know by induction. Lastly, since any of the n elements can appear in the first position of the permutation, then $f(n)$ is simply defined as $f(n-1)$ added n times, which can obviously be expressed as $n \cdot f(n-1)$.

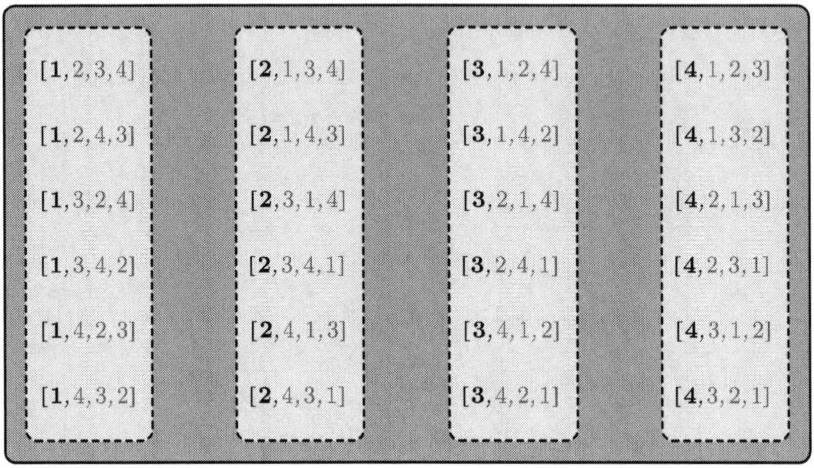

[**1**,2,3,4]	[**2**,1,3,4]	[**3**,1,2,4]	[**4**,1,2,3]
[**1**,2,4,3]	[**2**,1,4,3]	[**3**,1,4,2]	[**4**,1,3,2]
[**1**,3,2,4]	[**2**,3,1,4]	[**3**,2,1,4]	[**4**,2,1,3]
[**1**,3,4,2]	[**2**,3,4,1]	[**3**,2,4,1]	[**4**,2,3,1]
[**1**,4,2,3]	[**2**,4,1,3]	[**3**,4,1,2]	[**4**,3,1,2]
[**1**,4,3,2]	[**2**,4,3,1]	[**3**,4,2,1]	[**4**,3,2,1]

Figure 8.3 Decomposition of the possible permutations of the first four positive integers.

Thus, together with the base cases, the function is defined as:

$$f(n) = \begin{cases} 1 & \text{if } n = 0, \\ n \cdot f(n-1) & \text{if } n > 0, \end{cases}$$

which is the factorial function (the base case for $n = 1$ is redundant). Figure 8.3 shows the reasoning behind the decomposition for $n = 4$, where the permutations to be counted are divided into 4 subsets (corresponding to subproblems of size $n-1$), where each one contains permutations that share the same first element.

Finally, it is not hard to see that the factorial function provides the desired result, since there are n ways to choose a first element in the permutation, $n-1$ ways to select the second (having fixed the first one), $n-2$ ways to choose the third one, and so on. This naturally leads to $n! = n \times (n-1) \times (n-2) \times \cdots \times 1$ different permutations. However, this reasoning is more closely related to iterative thinking. Instead, the recursive solution presented above relies on induction and problem decomposition.

8.2 VARIATIONS WITH REPETITION

Consider some set of n elements. In combinatorics, k-element variations of n items are sequences of $k \leq n$ items from the set, where the order of the elements in the sequence matters. Using lists to denote a variation,

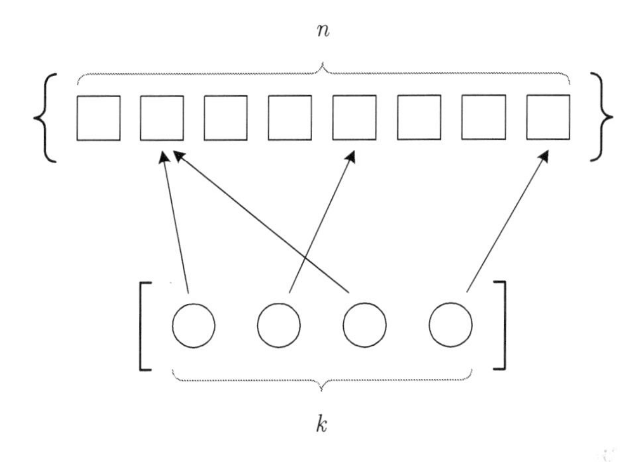

Figure 8.4 Example of a k-element variation with repetition of n items.

$[a_1, a_5, a_8]$ and $[a_8, a_5, a_1]$ would be two different 3-element variations of the elements in the set $\{a_1, a_2, \ldots, a_{10}\}$. In addition, if an element can appear several times within the sequence, the variations are called "with repetition" (e.g., $[a_1, a_5, a_1]$ would be a valid variation). In this section we will study these last variations. Figure 8.4 illustrates a concrete example where the squares represent elements of the set, the circles depict the locations of the items in the sequence, and the arrows indicate the chosen element for a particular location along the sequence (Exercise 8.1 uses an alternative representation). It is not hard to see through the "multiplication principle" that the number of k-element variations with repetition of n items is n^k, since there are n ways to choose an element in each of the k locations along the sequence. Nevertheless, we will deduce this by using recursion.

The problem is defined through the parameters k and n. Although the problem is simplified by decreasing k and/or n, in this case the decomposition is easier if we consider that the size of the problem is k. A trivial base case occurs when $k = 1$, where there are n different ways to select an element from the set. In addition, similarly to the problem of finding permutations, we can also consider a scenario when $k = 0$. In that case we can interpret that we could build a single "empty" variation.

For the recursive case we can use the diagrams in Figure 8.5 in order to develop the recursive rule. The original problem, denoted by $f(n, k)$, is shown in (a), where the thick arrow encompasses all of the possible assignments between the set and the sequence. The diagram in (b) shows the subproblem that stems from decrementing k by a unit, which counts

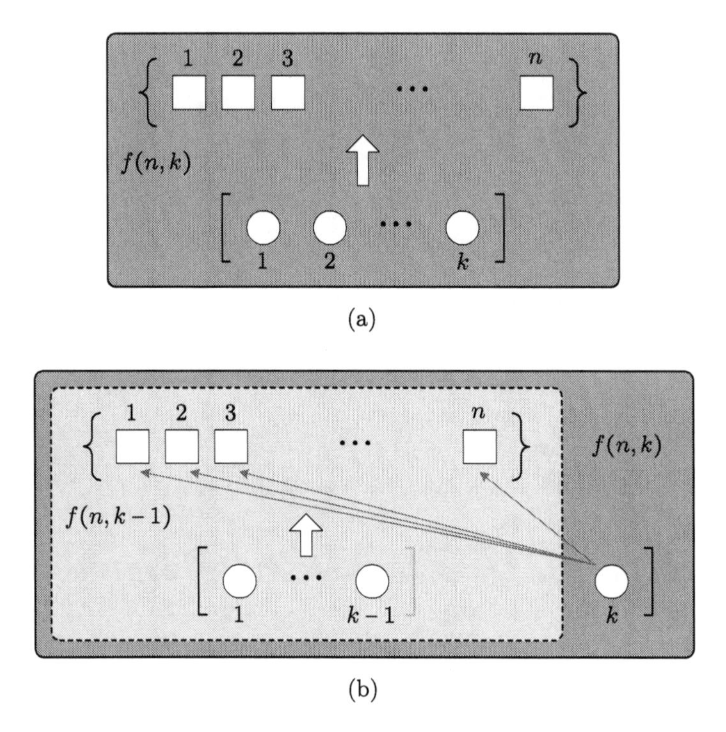

Figure 8.5 Decomposition of the problem that counts the number of k-element variations with repetition of n items.

all of the ways to choose the n elements for $k-1$ locations. Finally, since there are n ways to select the last k-th element, the total number of variations is $f(n, k-1)$ added n times. The recursive function is therefore defined as:

$$f(n,k) = \begin{cases} 1 & \text{if } k = 0, \\ f(n,k-1)\cdot n & \text{if } k > 0, \end{cases}$$

which is the power function n^k (see Listing 4.1).

8.3 COMBINATIONS

A k-element combination, without repetition, of n distinct items is simply a subset containing k of those elements. These combinations are similar to variations, but the order in which the elements appear is irrelevant. Therefore, a combination corresponds to a set instead of a sequence. For example, $\{a_1, a_2, a_3\}$ is a combination of the elements in

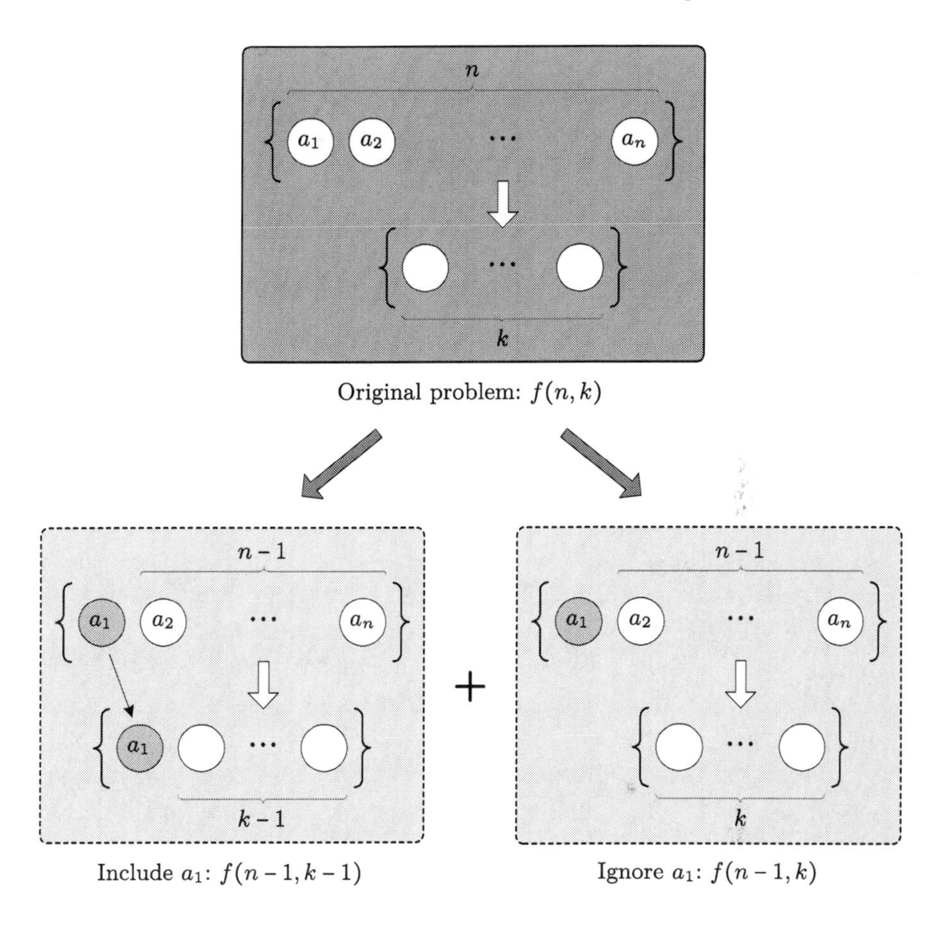

Figure 8.6 Decomposition of the problem that counts the number of k-element combinations of n items.

$\{a_1, a_2, \ldots, a_{10}\}$, where $\{a_1, a_2, a_3\}$ and $\{a_3, a_1, a_2\}$ are identical. In this section we will study the number of different k-element combinations (without repetition) of n distinct items.

In this problem the size depends on both k and n (where $k \leq n$). As we will see shortly, we will have to decrease both of them in order to develop a recursive solution. Nonetheless, the next step consists of determining the base cases. A trivial one occurs when $k = n$, where there is only one valid combination, which is the entire set of n elements. Another simple case occurs when $k = 1$, where the result is clearly n. Moreover, when $k = 0$ we can consider that there is one valid combination, which would be the empty set. As in previous sections, there will be no need to use the condition $k = 1$ if we incorporate the base case for $k = 0$.

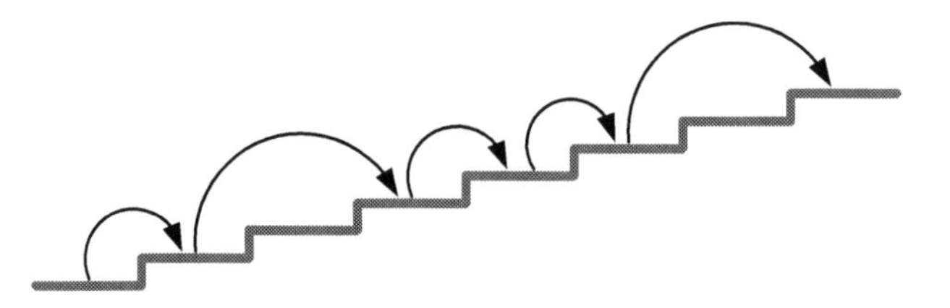

Figure 8.7 A possible way to climb a staircase by taking leaps of one or two steps.

For the recursive case we will decompose the total number of combinations into two sets. Consider an initial collection of n items: $\{a_1, a_2, \ldots, a_n\}$, and a particular element, say a_1. Notice that there are two types of combinations: (a) those that contain a_1, and (b) those that do not. Thus, the original problem, represented by $f(n,k)$, can be decomposed by distinguishing both types of combinations, as shown in Figure 8.6. If the combinations contain a_1, then the remaining $k-1$ elements constitute a $(k-1)$-element combination whose items belong to the subset $\{a_2, \ldots, a_n\}$, of size $n-1$. Therefore, this collection of combinations has size $f(n-1, k-1)$. Instead, if a_1 is not present in the combinations, these will contain k elements, also taken from the subset $\{a_2, \ldots, a_n\}$. The number of combinations of this type is therefore $f(n-1,k)$. Finally, since the result is the sum of both types of combinations, the recursive formula is $f(n,k) = f(n-1,k-1) + f(n-1,k)$. Finally, together with the base cases, the function is defined as follows:

$$f(n,k) = \begin{cases} 1 & \text{if } k = 0 \text{ or } k = n, \\ f(n-1,k-1) + f(n-1,k) & \text{otherwise,} \end{cases}$$

which is the definition of a binomial coefficient (see (3.2)).

8.4 STAIRCASE CLIMBING

The goal of the next problem is to count the number of ways it is possible to climb a staircase of n steps, if we advance by taking individual steps, or leaps of two steps. Figure 8.7 shows a possible way to climb a staircase of seven steps.

The size of the problem is clearly n. We can consider several base cases. If $n = 1$ then there is only one way to climb the staircase. If $n = 2$ then we can either take two individual steps, or a leap of two steps. In addition, if we code the ways to climb the staircase through a list, we

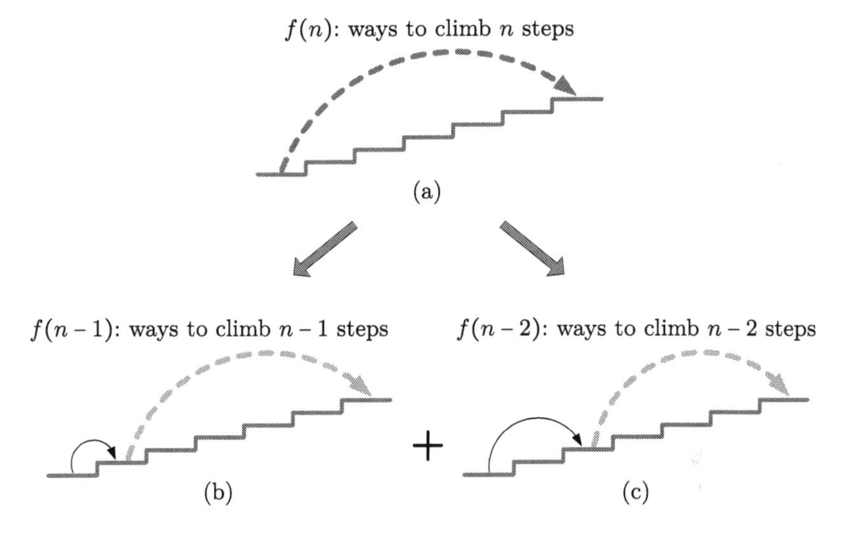

Figure 8.8 Decomposition of the problem that counts the number of ways to climb a staircase by taking leaps of one or two steps.

can also interpret that there is one way to proceed if $n = 0$, where the result would be an empty list.

For the recursive case we can decompose the problem as shown in Figure 8.8. The original problem is shown in (a), where the dark arrow represents the total number of ways to climb a staircase of n steps, denoted by $f(n)$. The ways to climb the staircase can be divided into two groups. One of them will contain the sequences of steps where the first leap is necessarily an individual step. Having reached the first step, there will be $f(n-1)$ ways to get to the top of the staircase (by taking leaps of one or two steps), as depicted in (b). The thicker arrow therefore represents a subproblem of size $n-1$. In the second group of sequences the first step is a leap of two steps. In that case there will be $f(n-2)$ ways to climb the staircase, as shown in (c), where the thicker arrow indicates a subproblem of size $n-2$. Finally, since both groups of sequences represent the total number of ways to climb the stairs, the solution is $f(n-1) + f(n-2)$. Together with the base cases, the function can be defined as follows:

$$f(n) = \begin{cases} 1 & \text{if } n = 1, \\ 2 & \text{if } n = 2, \\ f(n-1) + f(n-2) & \text{if } n \geq 3, \end{cases}$$

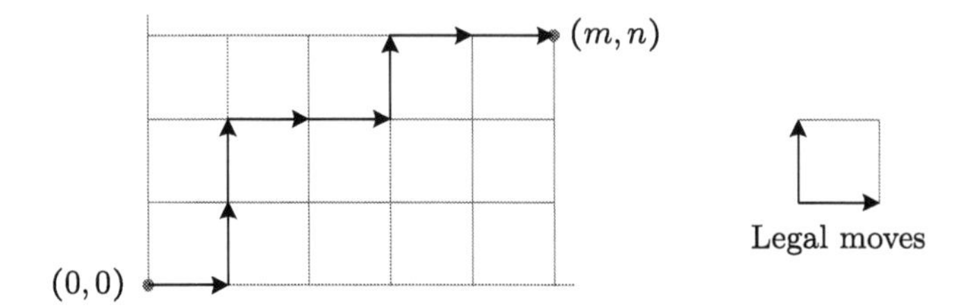

Figure 8.9 Manhattan paths problem.

which is very similar to the Fibonacci function $F(n)$ (see (1.2)). Specifically, $f(n) = F(n+1)$.

8.5 MANHATTAN PATHS

In many cities streets and avenues are perpendicular to each other. Assume that they are arranged in a regular grid in the coordinate system shown in Figure 8.9. In this problem the goal is to determine the number of ways one can walk to the crossing of avenue $m \geq 0$ and street $n \geq 0$ (where m and n are integers) by starting at avenue 0 and street 0. In order to do so, assume you are only allowed to advance one step at a time to a higher-numbered street or avenue, i.e., upwards, or towards the right in the figure.

We can consider that the size of the problem is $\min(m, n)$. In that case the smallest problems arise when either parameter is 0, since there would only be one valid path to the point (m, n), which would form a straight line segment. In addition, if both parameters are zero we can also interpret that there is one "empty" path.

The problem can be decomposed into two separate subproblems as shown in Figure 8.10. Let $f(m, n)$ denote the total number of paths from $(0, 0)$ to (m, n). Notice that we can divide the paths into two disjoint sets depending on how we take the first step. If we start by going towards the right we land on $(1, 0)$. From that point there will be $f(m-1, n)$ ways to reach the (m, n). Alternatively, if the first step is upwards, then there will be $f(m, n-1)$ paths from $(0, 1)$ to (m, n). Adding these quantities provides the recursive rule. The function can be coded as follows:

$$f(m,n) = \begin{cases} 1 & \text{if } m = 0 \text{ or } n = 0, \\ f(m-1,n) + f(m,n-1) & \text{otherwise.} \end{cases}$$

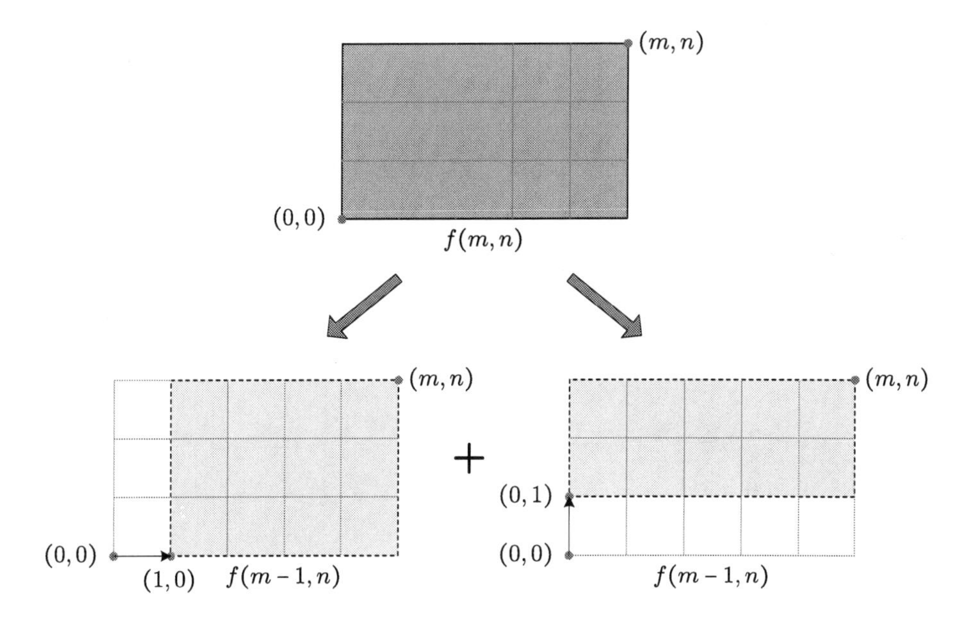

Figure 8.10 Decomposition of the Manhattan paths problem.

Finally, a path to (m, n) consists of a sequence of $m + n$ steps, where m go rightwards, and n go upwards. In other words, a path can be seen as a binary sequence of length $m + n$. Therefore, it can be specified by describing where the moves to the right appear in the sequence (an m-element combination of $n + m$ elements), or equivalently, where the upward moves appear (an n-element combination of $n + m$ elements). Thus, the function is a binomial coefficient:

$$f(m,n) = \binom{m+n}{n} = \binom{m+n}{m} = \frac{(m+n)!}{m! \cdot n!}.$$

Lastly, the formula can also be interpreted as the number of permutations with repetition of $m + n$ elements, where m of them (rightward moves) are identical, and n of them (upward moves) are also the same.

8.6 CONVEX POLYGON TRIANGULATIONS

A triangulation of a polygon can be thought of as a collection of triangles that do not overlap, cover the entire area of the polygon, and are formed with vertices of the polygon. Figure 8.11 shows two triangulations of a convex polygon containing seven vertices (a convex polygon is one such that a segment connecting any two points inside it will also lie in the

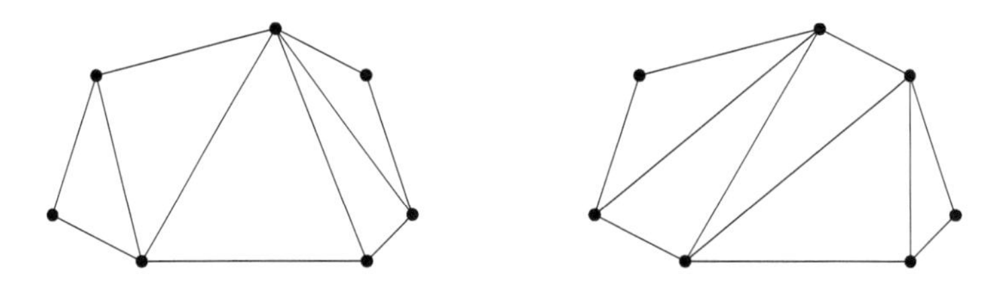

Figure 8.11 Two possible triangulations of the same convex polygon containing seven vertices.

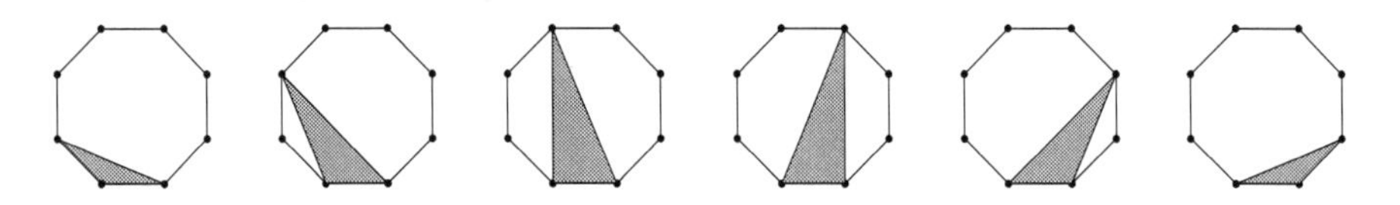

Figure 8.12 Six $(n-2)$ possible triangles associated with an edge of an octagon $(n=8)$.

polygon). The next problem consists of defining a function that determines the different ways it is possible to triangulate a convex polygon containing n vertices.

The size of the problem is n. A trivial base case occurs when $n=3$, where the result is obviously one. In addition, similarly to the previous problems discussed in the chapter, if $n=2$ we can also interpret that the result is one (this will be justified below).

The recursive case is more complex than in previous examples, since the solution will require considering several subproblems of smaller size. To understand the decomposition we will use a concrete example. Consider, for instance, the bottom edge of the octagons in Figure 8.12. Naturally, it must belong to one of the triangles in the triangulation, and there are 6 $(n-2)$ different triangles that contain the edge, together with one of the 6 polygon vertices that are not endpoints of the edge. The triangulations can therefore be divided into 6 disjoint sets, according to the particular triangle associated with the bottom edge. In other words, a shaded triangle defines a particular set of triangulations, and the total number of triangulations will be the sum of the triangulations in each of the six sets.

Once a particular shaded triangle has been fixed, there are two adjacent convex polygons (one possibly empty) at each side of the triangle that need to be triangulated. Note that both polygons will share one of the edges of the shaded triangle. Due to the multiplication principle, the

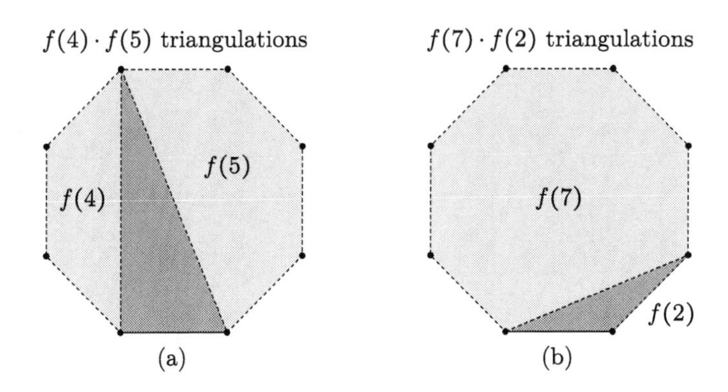

Figure 8.13 Decomposition of the convex polygon triangulation problem for fixed triangles.

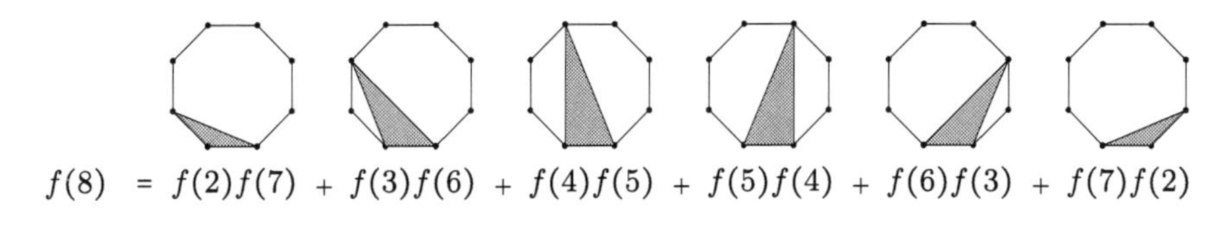

Figure 8.14 Total number of triangulations related to an octagon.

number of ways it is possible to triangulate one of the adjacent polygons, times the number of triangulations that can be constructed with the other, will be the number of triangulations associated with the shaded triangle. This is illustrated in Figure 8.13. Let $f(n)$ denote the function that solves the problem. In (a) the shaded triangle leaves a subproblem of size 4 on the left, and another of size 5 on the right. The number of triangulations that contain the shaded triangle will be $f(4)\cdot f(5)$. In (b), one of the subproblems has size 7, while another has size 2. Thus, there are $f(7)\cdot f(2)$ triangulations associated with that shaded triangle. Note that although one polygon (related to $f(2)$) is empty, the number of possible triangulations that contain the shaded triangle is provided by $f(7)$. This justifies using $f(2) = 1$ as a base case.

Lastly, we must add all of the triangulations related to the six shaded triangles. In the example we have:

$$f(8) = f(2)f(7)+f(3)f(6)+f(4)f(5)+f(5)f(4)+f(6)f(3)+f(7)f(2),$$

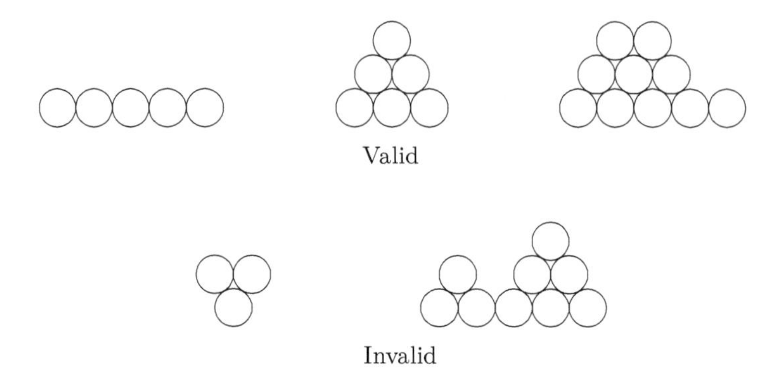

Figure 8.15 Valid and invalid circle pyramids.

as shown in Figure 8.14. In general, the recursive function can be defined as:

$$f(n) = \begin{cases} 1 & \text{if n=2,} \\ \sum_{i=2}^{n-1} f(i) \cdot f(n+1-i) & \text{if n>2.} \end{cases}$$

Note that an additional base case for $n = 3$ is not necessary.

Finally, $f(n)$ is related to Catalan numbers, which appear frequently in combinatorics and can be defined as:

$$C(n) = \begin{cases} 1 & \text{if n=0,} \\ \sum_{i=0}^{n-1} C(i) \cdot C(n-1-i) & \text{if n>0.} \end{cases} \tag{8.1}$$

In particular, $f(n) = C(n-2)$.

8.7 CIRCLE PYRAMIDS

The following problem consists of determining the number of ways it is possible to arrange several circles along rows in a pyramidal structure. Starting with a row of contiguous circles at the bottom, it is possible to place circles on upper rows by obeying the following rules:

a) All of the circles on a row must be placed contiguously. In other words, there cannot be gaps between the circles in a row.

b) Each circle that is not on the bottom row has to be placed on top of two circles on the row immediately below it, forming an equilateral triangular pyramid.

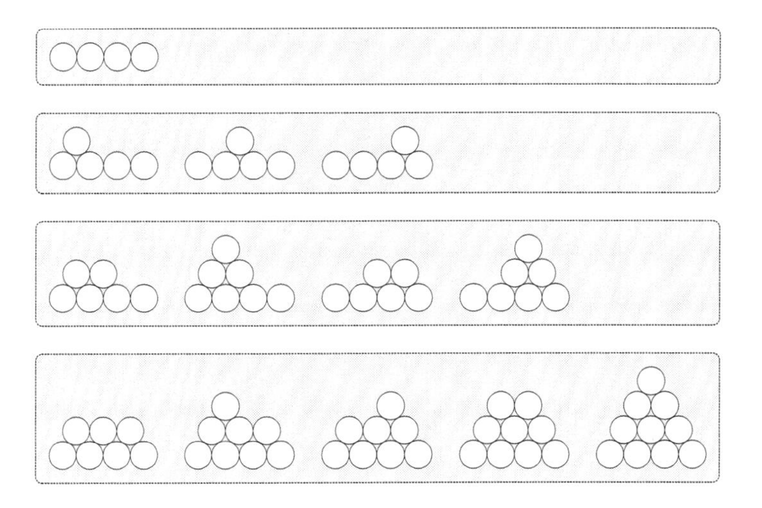

Figure 8.16 Pyramids of $n = 4$ circles on the bottom row, grouped according to how many circles appear in the row immediately above it.

Figure 8.15 shows examples of valid and invalid configurations of circles. The concrete problem is to define a function that counts the number of valid circle pyramids that can be formed by starting with n circles in the bottom row.

The size of the problem is n. A trivial case occurs when $n = 1$, where the output is obviously 1. For the recursive case we will decompose the problem by dividing the possible pyramidal configurations into several disjoint groups, depending on how many circles are placed on the row immediately above the bottom one. Figure 8.16 illustrates the idea for $n = 4$. The key observation is that smaller pyramids (subproblems) with the same number of circles in their bottom row can be placed in several positions on top of the bottom circles of the original problem, as shown in Figure 8.17. In particular, given the initial input parameter n, there are $n - i$ ways to place an identical pyramid of size i. In the figure $i = 3$ and $n = 7$, which implies that there are $n - i = 4$ ways to place identical pyramids of size 3 on top of the initial $n = 7$ circles.

Lastly, since the size of the smaller subproblems ranges from 1 to $n-1$, the function has to add all of those possible configurations of circles. In addition, a valid case is to not place any circle on top of the n initial ones. Therefore, the function must also add a unit for this particular case. The

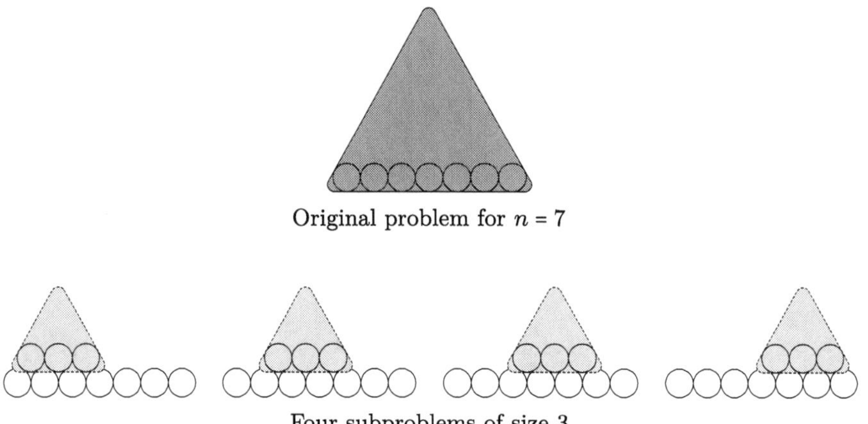

Figure 8.17 Decomposition of the circle pyramids problem for subproblems of a fixed size.

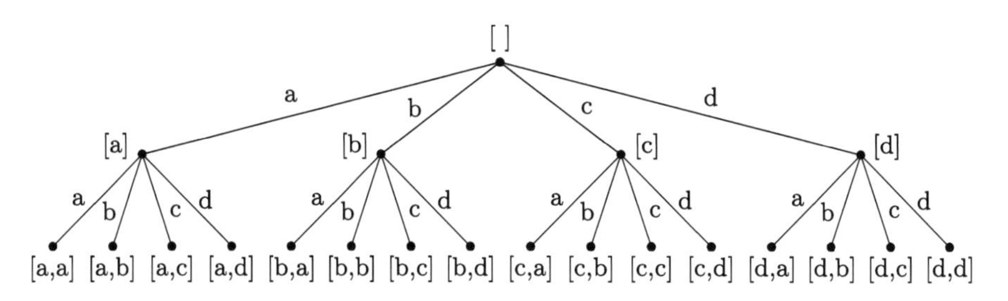

Figure 8.18 Two-element variations with repetition of the four items in {a,b,c,d}.

following definition specifies the function that solves the problem:

$$f(n) = \begin{cases} 1 & \text{if n=1,} \\ 1 + \sum_{i=1}^{n-1} (n-i) \cdot f(i) & \text{if n>1.} \end{cases} \tag{8.2}$$

Finally, the result is a Fibonacci number. In particular, it can be shown that $f(n) = F(2n-1)$, where F is the Fibonacci function.

8.8 EXERCISES

Exercise 8.1 — Let $f(n,k)$ represent the number of k-element variations with repetition of n items. There are n possibilities to choose

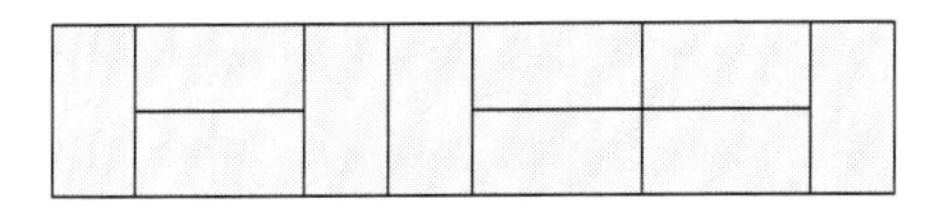

Figure 8.19 Tiling of a 2×10 rectangle with 1×2 or 2×1 domino tiles.

the first element of the variation, another n ways to select the second one, and so on. We can illustrate all of these possibilities through a full tree of height k whose internal nodes always have n children, and whose leaves represent the variations. Figure 8.18 shows an example for all of the 2-element variations with repetition of the 4 items in {a,b,c,d}. The lists depicted at the nodes indicate partial sequences as the variations are progressively built, while the labeled edges indicate the particular chosen item that will be appended to the partial sequence. Thus, a path from the root node to a leaf specifies a particular variation.

Draw diagrams based on these full trees that allow us to derive the recursive rule $f(n,k) = f(n,k-1) \cdot n$. The trees should represent the original problem and the subproblems. However, for simplicity, omit the labels next to the edges and nodes.

Exercise 8.2 — Assuming that we are given n identical rectangular "domino" tiles of size 1×2 (or 2×1 when rotated 90°), define a function that determines the number of ways to create a $2 \times n$ rectangle. Figure 8.19 illustrates a concrete example for $n = 10$, where the tiles are obviously not allowed to overlap.

Exercise 8.3 — In a game of basketball a team can score points in three different ways. A "free throw" scores one point, a "field goal" is worth two points, and a successful shot beyond the "three-point" line scores three points. Define a function that determines the number of ways a team can reach n points. For example, there are four ways to score three points: $1+1+1$, $1+2$, $2+1$, and 3. Thus, assume that the order in which the points are scored matters.

Exercise 8.4 — Solve the problem in Exercise 3.8 (and 4.11) by using a recursive approach, instead of relying on sums. Draw the original problem as in Figure 3.11 and find subproblems within it.

Exercise 8.5 — Implement function $C(n)$ in (8.1).

Exercise 8.6 — In a binary tree the nodes can have zero, one, or two

Figure 8.20 Five different binary trees that contain three nodes.

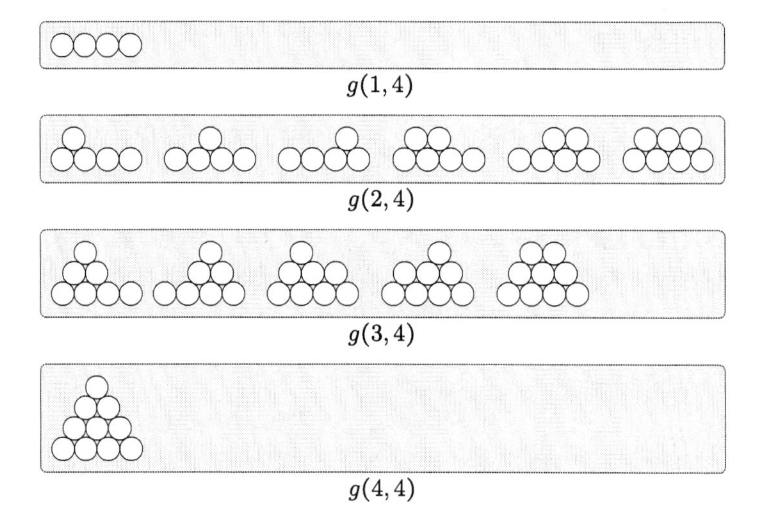

Figure 8.21 Pyramids of $n = 4$ circles on the bottom row, grouped according to their height.

children. Specify a recursive function $f(n)$ that determines the number of different binary trees that contain n nodes. Consider distinguishing between left and right children. For example, Figure 8.20 shows the five possible binary trees that can be formed with $n = 3$ nodes.

Exercise 8.7 — Consider the circle pyramids problem in Section 8.7. This exercise proposes using a different decomposition to solve the problem. In particular, it will categorize the different pyramids according to their height, as shown in Figure 8.21, where $g(h,n)$ denotes the number of pyramids of height h that can be formed when there are n on the bottom row. Thus, the goal of this exercise consists of defining $g(h,n)$ recursively. Finally, check that the function is correct by coding and computing the total number of pyramids that contain n circles on their bottom row:

$$\sum_{i=1}^{n} g(h,n) = f(n) = F(2n-1), \tag{8.3}$$

where f is the function in (8.2), and F is the Fibonacci function.

CHAPTER 9

Mutual Recursion

Alone we can do so little, together we can do so much.
— Helen Keller

A method that invokes itself directly within its body is recursive, by definition. However, a recursive method does not necessarily have to call itself. It may call another one that in turn calls it back. Thus, invoking the method can provoke multiple calls to it. This type of recursion is called mutual or indirect.

In general, several methods are mutually recursive when they invoke themselves in a cyclical order. For instance, consider a set of methods $\{f_1, f_2, \ldots, f_n\}$. If f_1 calls f_2, f_2 calls f_3, and so on, and finally f_n calls back f_1, then the methods are said to be mutually recursive. In addition, they do not necessarily have to call each other in a strict cyclical sequence. They could call several methods, including themselves, as illustrated in Figure 9.1, where the arrows indicate method calls (e.g., f_8 calls f_1 and f_4). In the example, only f_3 invokes itself directly, but all of the methods are mutually recursive.

When applying mutual recursion the first step consists of breaking up a problem into different problems (not simply subproblems). Afterwards, the recursive solutions to these problems are designed by relying on the solutions to their subproblems. The key observation when using mutual recursion is that when solving some problem P we will use recursive solutions to subproblems that are not necessarily instances of P. Naturally, decomposing the original problem into several different problems requires more work. However, once these problems are defined, the recursive design process is analogous to working with a single method. In other words, we will apply problem decomposition and induction as

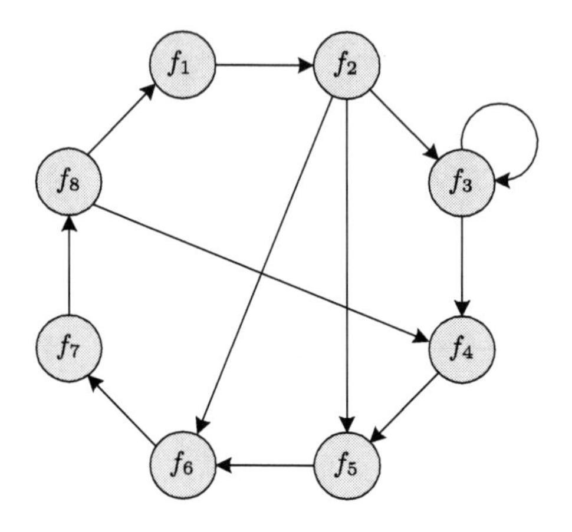

Figure 9.1 Calls of mutually recursive methods.

usual. Moreover, the greater complexity of the problems and the nature of the decompositions will accentuate the importance of these concepts.

Lastly, this chapter contains several challenging problems, some of which are variants of others covered throughout the book. If readers understand the solutions, and are able to develop the algorithms on their own, they will have acquired a solid foundation on recursive thinking skills.

9.1 PARITY OF A NUMBER

One of the simplest and most popular problems used to illustrate mutual recursion consists of determining the parity of a nonnegative integer n. We have already seen a solution that determines whether n is even in Listing 2.6, by reducing the problem by two units. In addition, Exercise 4.1 proposes solving the problem through a single linear-recursive function. In this case we will build two functions that will depend on each other mutually: one that checks if n is even, and another one that determines if n is odd. Thus, besides solving the original problem (whether n is even), we will also develop a solution to the problem of finding whether n is odd.

Let $f(n)$ indicate whether n is even, and $g(n)$ whether n is odd. Assuming that n is the size of the problem, for $f(n)$ the base case is reached when $n = 0$, where the result is obviously `True`. Alternatively, $g(0)$ is `False`. For the recursive cases we can reduce the size of the problem by a unit. Since we are going to build both $f(n)$ and $g(n)$, we

Listing 9.1 Mutually recursive functions for determining the parity of a nonnegative integer n.

```
def is_even(n):
    if n == 0:
        return True
    else:
        return is_odd(n - 1)


def is_odd(n):
    if n == 0:
        return False
    else:
        return is_even(n - 1)
```

can assume by induction that we can apply them to $n-1$. It is crucial to note that we can use $g(n-1)$ to build $f(n)$, and $f(n-1)$ to implement $g(n)$. In particular, we can rely on the trivial properties $f(n) = g(n-1)$, and $g(n) = f(n-1)$, in order to code the two mutually recursive methods as in Listing 9.1.

9.2 MULTIPLAYER GAMES

When coding subroutines, a good general programming practice is to keep them as brief and simple as possible, which often enhances the readability of the code. In this regard, mutual recursion can be useful when implementing different actions or behaviors. For instance, in multiplayer games separate methods can implement different roles. The following example simulates two different strategies when playing a simple game. Starting with n pebbles on a board, two players take turns in order to remove them from the board. They can either take one or two pebbles in each turn, and the player who removes the last pebble wins the game. Imagine that the first player (Bob) decides to remove a single pebble in each turn, while the second player (Alice) withdraws one pebble if the number of pebbles left on the board is odd, and two if it is even.

The game can be coded easily by using a single method. However, Listing 9.2 shows a solution based on two mutually recursive methods, where each procedure implements a different strategy. In particular, the base cases check if a player will win the game, while in the recursive cases

Listing 9.2 Mutually recursive procedures implementing Alice and Bob's strategies when playing a game.

```
def play_Alice(n):
    if n <= 2:
        print('Alice wins')
    elif n & 1:
        # Alice removes one pebble
        play_Bob(n - 1)  # Turn switches to Bob
    else:
        # Alice removes two pebbles
        play_Bob(n - 2)  # Turn switches to Bob


def play_Bob(n):
    if n == 1:
        print('Bob wins')
    else:
        # Bob removes one pebble
        play_Alice(n - 1)  # Turn switches to Alice
```

the turn switches to the other player. The argument passed to the function that implements the other player's strategy is decreased depending on the number of pebbles removed from the board. It is important to note that each behavior is coded in a separate procedure, which makes the code easier to understand and modify. Finally, Alice's more complex strategy is clearly better than Bob's. In particular, Bob only wins if there is just a single pebble when the game begins.

9.3 RABBIT POPULATION GROWTH

The following problem is related to a thought experiment concerning an artificial rabbit population, which grows according to the following rules:

a) Initially, a newly born pair of rabbits, one male and one female, are placed in a field.

b) The rabbits take one month to mature.

c) Mature rabbit pairs mate at the beginning of every month, and give birth to another pair of newly born rabbits at the beginning of the next month.

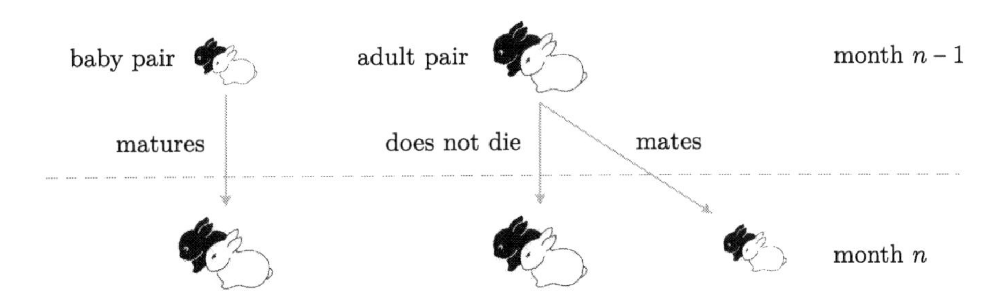

Figure 9.2 Illustration of the rabbit population growth rules.

d) The female rabbit always gives birth to one male and one female rabbit, who will only mate with themselves.

e) The rabbits never die.

The goal of the (counting) problem is to calculate the size of the population of rabbits during the n-th month, denoted as $R(n)$. The next subsections describe two solutions that rely on mutual recursion.

9.3.1 Adult and baby rabbit pairs

Instead of computing the size of the entire population directly, we can proceed by counting the number of adult and baby pairs during each month separately. This leads to two different problems of size n. In particular, let $A(n)$ and $B(n)$ denote the number of adult and baby pairs during month n, respectively. Regarding the base cases, $A(1) = 0$ and $B(1) = 1$, since initially there is only one pair of baby rabbits. In the second month the rabbits will have matured, but they will not have produced any offspring yet. Thus, $B(2) = 0$ and $A(2) = 1$.

For the recursive cases, the rules that determine the population growth are illustrated in Figure 9.2, where the smaller and larger rabbits depict baby and adult pairs, respectively. On the one hand, every pair of baby rabbits in a given month matures into an adult pair in the following month. On the other hand, each adult pair also appears in the following month (since the rabbits never die), and the rabbits also mate, generating a new pair of offspring.

Assuming by induction that we know the population size of adult and baby pairs during month $n-1$ (i.e., $A(n-1)$ and $B(n-1)$), it is fairly straightforward to derive recursive rules for $A(n)$ and $B(n)$. Firstly, the number of baby pairs during the n-th month is equal to the number of adult pairs during the previous month, since the adults always mate

Listing 9.3 Mutually recursive functions for counting the population of baby and adult rabbits after n months.

```
1  def adults(n):
2      if n == 1:
3          return 0
4      else:
5          return adults(n - 1) + babies(n - 1)
6
7
8  def babies(n):
9      if n == 1:
10         return 1
11     else:
12         return adults(n - 1)
```

and take one month to generate a new pair of baby rabbits. Therefore, $B(n) = A(n-1)$. Regarding $A(n)$, all of the adult pairs at month $n-1$ will be alive at month n, since the rabbits do not die. In addition, all of the baby pairs during month $n-1$ will have matured and turned into adults. Thus, the recursive rule is $A(n) = A(n-1) + B(n-1)$. Both functions can be defined as follows:

$$A(n) = \begin{cases} 0 & \text{if } n = 1, \\ A(n-1) + B(n-1) & \text{if } n > 1, \end{cases} \tag{9.1}$$

$$B(n) = \begin{cases} 1 & \text{if } n = 1, \\ A(n-1) & \text{if } n > 1, \end{cases} \tag{9.2}$$

where it is not necessary to explicitly use the base cases for $n = 2$. These are precisely the functions introduced in Section 1.6.4. Clearly, A is recursive since it invokes itself. Method B invokes itself indirectly, by first calling A, which in turn can call B. Listing 9.3 shows the corresponding code for these functions.

It is possible to express the previous functions exclusively in terms of themselves, as shown in (3.38) and (3.39). Finally, the total number of pairs of rabbits is simply $A(n) + B(n)$, which happens to be the n-th Fibonacci number (see Exercise 9.2).

9.3.2 Rabbit family tree

A completely different way to tackle this problem consists of analyzing the rabbit family (i.e., genealogical) tree. Figure 9.3 shows this tree after

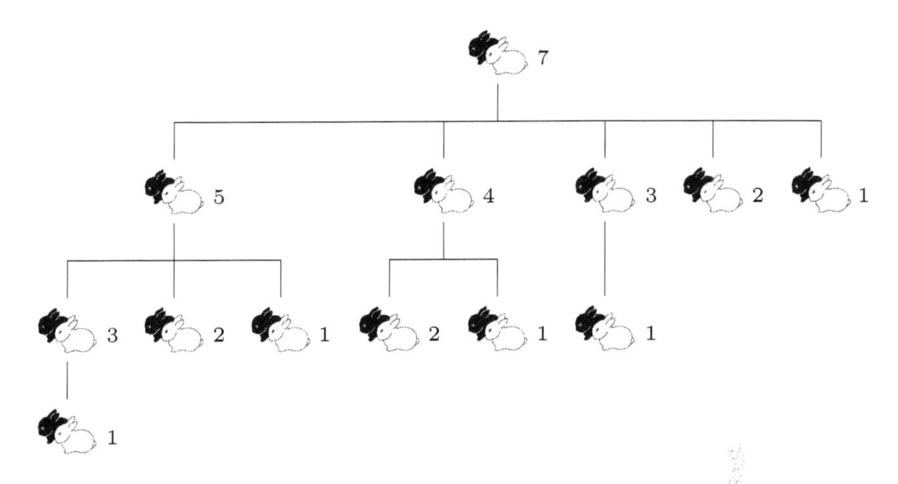

Figure 9.3 Rabbit family tree after seven months.

seven months of the thought experiment, where the labels next to the rabbits indicate their age (in months). The top rabbit pair is the initial pair that was placed in the field at the first month. Observe that it has five pairs of offspring, the first of which was born at the beginning of the third month (therefore, its age is 5), and the last at the beginning of the seventh month (thus, at the end of the seventh month the rabbits are one month old). The following subsections examine solutions based on multiple and mutual recursion.

9.3.2.1 Solution using multiple recursion

The first decomposition that comes to mind when tackling this problem consists of breaking it up as shown in Figure 9.4. In this specific example, a problem of size 7 is broken up into five smaller problems, of sizes from 1 to 5. Observe that the initial pair of rabbits generates a new pair of offspring from the third month on, where each of them will have their own descendants. A particular children pair, together with its descendants, constitutes a smaller self-similar problem to the original. Thus, in the recursive case the function $f(n)$ that represents the solution is a unit corresponding to a rabbit pair that has lived n months, plus the total number of its descendant pairs. Formally, the decomposition leads to the

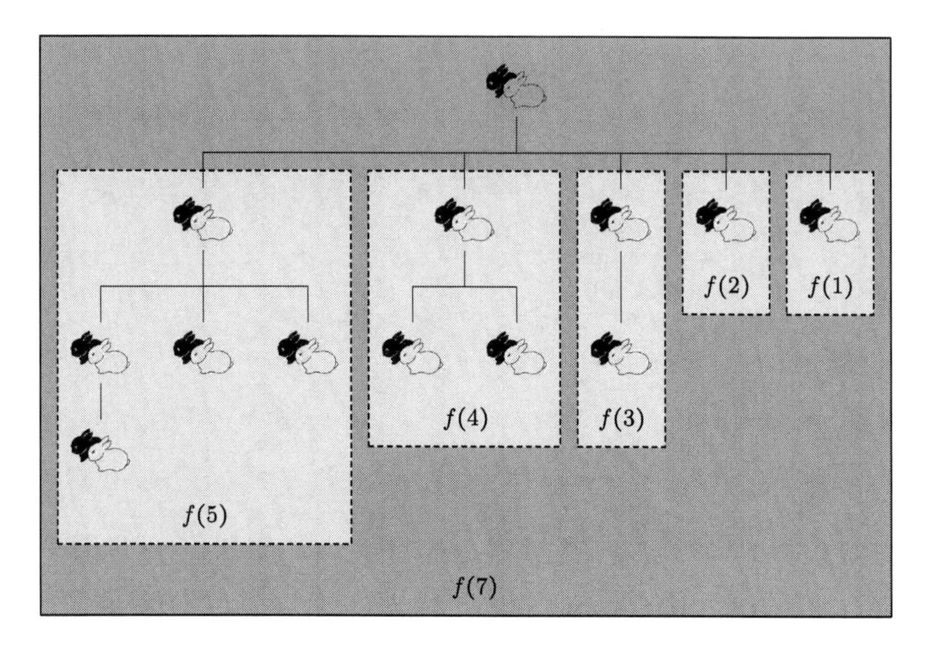

Figure 9.4 Concrete example of the decomposition of the rabbit population growth problem into self-similar subproblems.

following definition:

$$f(n) = \begin{cases} 1 & \text{if } n = 1 \text{ or } n = 2, \\ 1 + \sum_{i=1}^{n-2} f(i) & \text{if } n > 2, \end{cases} \tag{9.3}$$

which uses multiple recursion and is analogous to (1.7). Furthermore, the diagram in Figure 2.4 shows the decomposition of the problem for a general n (recall that $f(n) = F(n)$, where F is the Fibonacci function).

9.3.2.2 Solution using mutual recursion

The recursive expression in (9.3) can be implemented with a basic loop, as shown in Listing 1.4. The solution that we are about to see replaces the sum by a recursive function (this would be necessary if the programming language does not support loops).

Firstly, mutual recursion can only be applied if we identify several problems. In this case, besides providing a function ($f(n)$) that solves the original problem, we will also consider solving a similar but different problem. In particular, let $g(n)$ represent the number of rabbit pairs

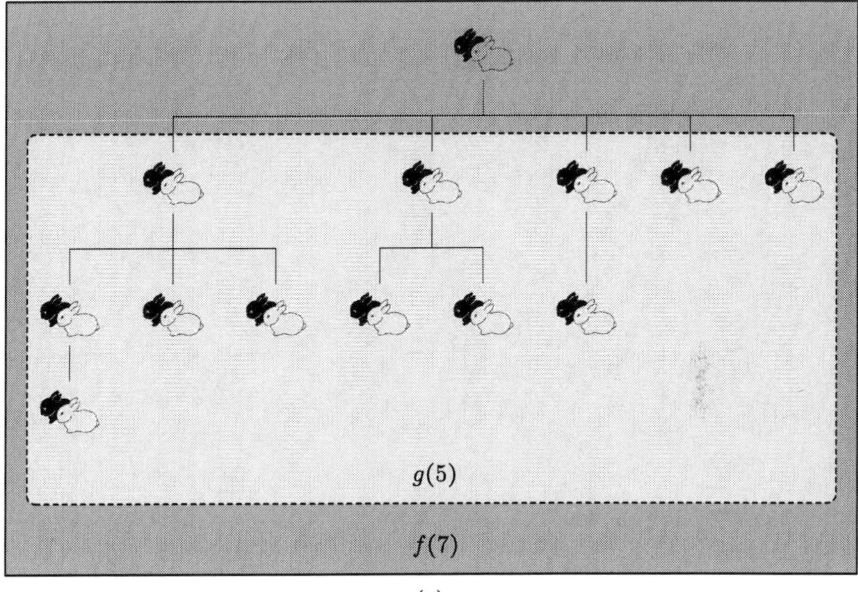

(a)

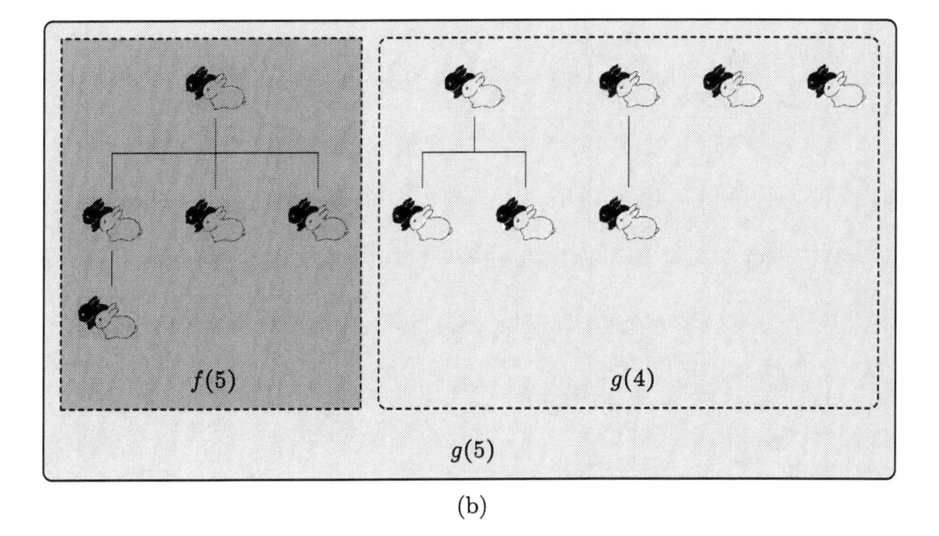

(b)

Figure 9.5 Concrete decompositions of the problems that lead to two mutually recursive methods for solving the rabbit population growth problem.

Listing 9.4 Alternative mutually recursive functions for counting the population of rabbits after n months.

```
def population_rabbits(n):
    if n <= 2:
        return 1
    else:
        return 1 + children_descendants(n - 2)


def children_descendants(n):
    if n == 0:
        return 0
    else:
        return (population_rabbits(n)
                + children_descendants(n - 1))
```

corresponding to a set of n (pair) siblings and all of their descendants, where the oldest is n months old, the second oldest is $n-1$ months old, and so on. With this auxiliary function it is straightforward to define $f(n)$. Specifically, for a given input n, it will count a unit corresponding to the initial pair of rabbits, plus the function g evaluated at $n-2$, since the initial pair will always be two months older than the eldest of its children. Thus, $f(n)$ can be expressed as follows:

$$f(n) = \begin{cases} 1 & \text{if } n = 1 \text{ or } n = 2, \\ 1 + g(n-2) & \text{if } n > 2, \end{cases} \tag{9.4}$$

which corresponds to the decomposition illustrated in Figure 9.5(a) for $n = 7$.

For function g the size of the problem is n. A trivial base case occurs when $n = 1$, where the result is simply 1. However, it is also possible to use the base case $g(0) = 0$. For the recursive case we can rely on the decomposition shown in Figure 9.5(b) for $n = 5$, which is based on reducing the size of the problem by a unit. Observe that the total number of rabbit pairs is equal to the original problem for some initial pair after n months of the experiment (i.e., $f(n)$), plus the family of siblings and their descendants, where the oldest one is $n-1$ months old, which is $g(n-1)$. Therefore, $g(n)$ can be described as:

$$g(n) = \begin{cases} 0 & \text{if } n = 0, \\ f(n) + g(n-1) & \text{if } n > 0. \end{cases} \tag{9.5}$$

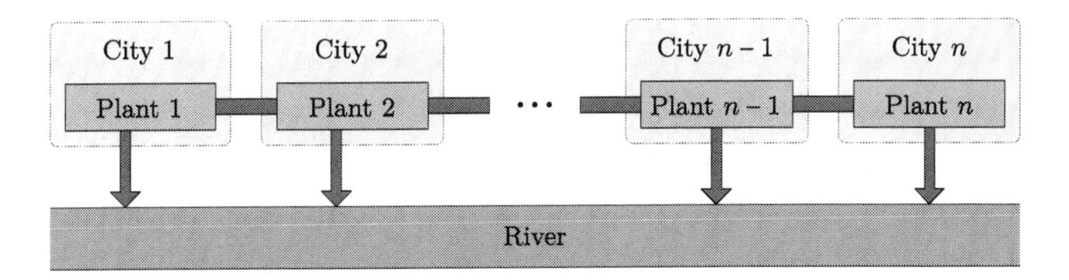

Figure 9.6 Water treatment plants puzzle.

The result is a pair of mutually recursive functions that call each other. In addition, note that $g(n)$ simply implements $\sum_{i=1}^{n} f(i)$. Thus, (9.4) and (9.5) are equivalent to (9.3). Lastly, it is straightforward to code the functions, as shown in Listing 9.4.

9.4 WATER TREATMENT PLANTS PUZZLE

In the "water treatment plants puzzle" there are n cities, each with a water treatment plant, located along a side of a river, as shown in Figure 9.6. We will assume that the cities are arranged from left to right. Each city generates sewage water that must be cleaned in a plant (not necessarily its own) and discharged into the river through pipes, where additional pipes connect neighboring cities. If a plant is working it will clean the sewage water from its city, plus any water coming from a pipe connecting a neighboring city, and discharge it to the river. However, a plant might not be working. In that case the water from its city, plus any water coming from a neighboring city, must be sent to another city. Given that water can flow in only one direction inside a pipe, the problem consists of determining the number of different ways it is possible to discharge the cleaned water into the river, for n cities. Finally, we can assume that at least one of the plants will work.

We will see two recursive solutions that solve this counting problem. The first one models the water flow between cities and uses multiple recursion. A second solution considers how the water is discharged at each city, and is based on three mutually recursive functions.

9.4.1 Water flow between cities

The water can flow in each of the $n-1$ pipes that connect neighboring cities in three ways that we will denote by characters: 'N' if there is no flow, 'R' if water flows to the right, and 'L' if it flows to the left.

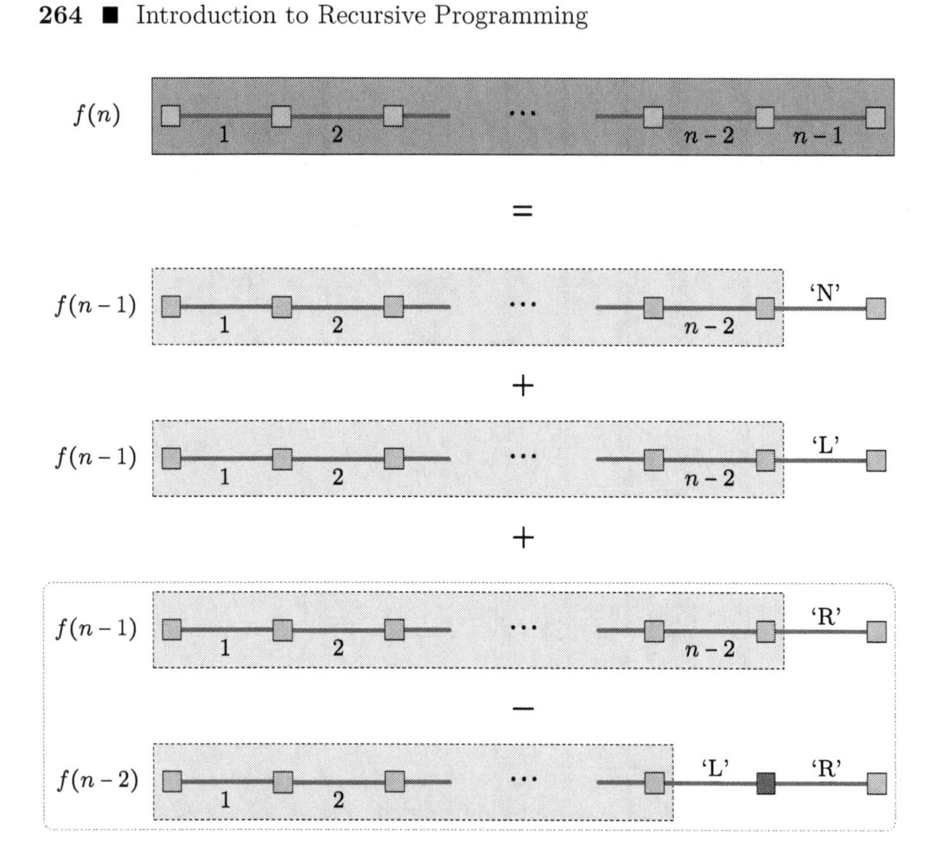

Figure 9.7 Decomposition of the water treatment plants puzzle when modeling the water flow between cities.

Therefore, we can model a way to discharge the water into the river as a string of length $n-1$ containing only these characters. The only restriction that affects the water flow is that a city cannot send its water to both of its neighbors. Thus, the goal of the problem is to find all of the strings in which substring "LR" does not appear.

The size of the problem is clearly n. Let $f(n)$ represent the solution. If there is one city then its plant must discharge its water. Thus, $f(1) = 1$. If there are two cities there are three possibilities associated with a string of size 1 that contains any of the three characters. Therefore, $f(2) = 3$. Observe that if both plants work they will discharge the water of their cities. However, if either of the two plants does not work, then it will send the water to its neighboring city's plant.

Figure 9.7 illustrates a decomposition of the problem that leads to an algorithm based on multiple recursion. At first, the total number of

Listing 9.5 Function based on multiple recursion for solving the water treatment plants puzzle.

```
1 def water_multiple(n):
2     if n == 1:
3         return 1
4     elif n == 2:
5         return 3
6     else:
7         return 3 * water_multiple(n - 1) - water_multiple(n - 2)
```

strings of length $n-1$ that we could form is three times the number of valid strings of length $n-2$, since we could append any of the three characters to their end. This provides a term $3f(n-1)$. However, appending an 'R' to a string that finishes with an 'L' is not allowed (plant $n-1$ would send water in both directions, which is not permitted). Therefore, it is necessary to subtract all of the possible strings of length $n-1$ that end in "LR". This quantity is $f(n-2)$. Therefore, the recursive rule is $f(n) = 3f(n-1) - f(n-2)$. Together with the base cases, the function can be defined as:

$$f(n) = \begin{cases} 1 & \text{if } n = 1, \\ 3 & \text{if } n = 2, \\ 3f(n-1) - f(n-2) & \text{if } n > 2, \end{cases} \tag{9.6}$$

and can be easily implemented as shown in Listing 9.5. Lastly, $f(n) = F(2n)$, where F is the Fibonacci function (see Exercise 9.4).

9.4.2 Water discharge at each city

Another approach for solving the problem consists of modeling the direction in which each plant sends its water. There are three possibilities that we will also denote with characters. In particular: 'v' signifies that the water is discharged into the river, '>' means that it is sent to the city to the right, and '<' indicates that it will be discharged towards the left. In this case, the elements that we need to count are strings of length n that do not begin with '<', nor end with '>', and where the substring "><" is not allowed.

We will use three related but alternative problems in order to construct the algorithm based on mutual recursion. Consider valid strings of the three characters that begin (from left to right) with a non-empty

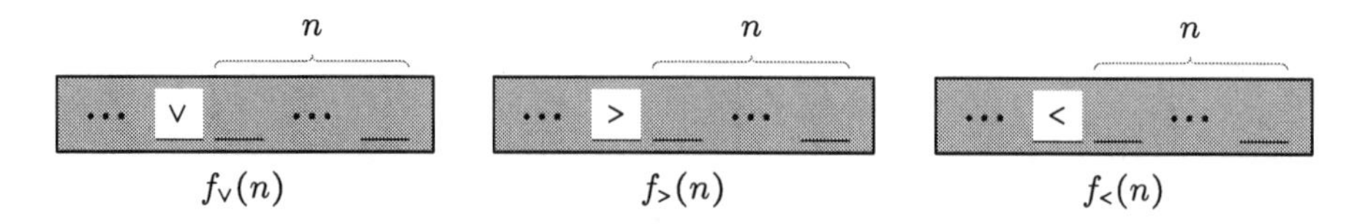

Figure 9.8 Three problems for the mutually recursive solution of the water treatment plants puzzle.

substring, and end with a substring of n characters. There are three scenarios depending on the character that precedes the substring of length n (the substring to the left of the character will be irrelevant). The three problems will consist of determining the number of valid substrings in those three cases. The diagrams in Figure 9.8 represent these problems. Their solutions are provided by the functions $f_\vee(n)$, $f_>(n)$, and $f_<(n)$, where the subscript indicates the character that precedes the substring of length n.

The size of the problems is naturally n, and their base cases occur when $n = 1$. In particular, $f_\vee(1) = f_<(1) = 2$, since there are two options ('v' and '<') for the rightmost character. Instead, $f_>(1) = 1$. Since the rightmost city's plant would receive water from its neighboring city, it must discharge its water into the river.

The decompositions of the problems are illustrated in Figure 9.9. For $f_\vee(n)$ and $f_<(n)$ there is no restriction regarding the first character of the substring of length n. Therefore, the total number of valid strings of length n is equal to the number of valid strings of length $n-1$ for the three possibilities regarding the first character. Formally, the right-hand side of the recursive rule is $f_\vee(n-1) + f_>(n-1) + f_<(n-1)$ for both functions. In addition, since their base cases are also the same, the functions are identical. In particular, we have:

$$f_\vee(n) = f_<(n) = \begin{cases} 2 & \text{if } n = 1, \\ f_\vee(n-1) + f_>(n-1) + f_<(n-1) & \text{if } n > 1. \end{cases} \tag{9.7}$$

Instead, for $f_>(n)$ the first character of the string of length n cannot be '<'. Therefore, the right-hand side of the recursive rule is $f_\vee(n-1) + f_>(n-1)$. Together with its base case, the function is:

$$f_>(n) = \begin{cases} 1 & \text{if } n = 1, \\ f_\vee(n-1) + f_>(n-1) & \text{if } n > 1. \end{cases} \tag{9.8}$$

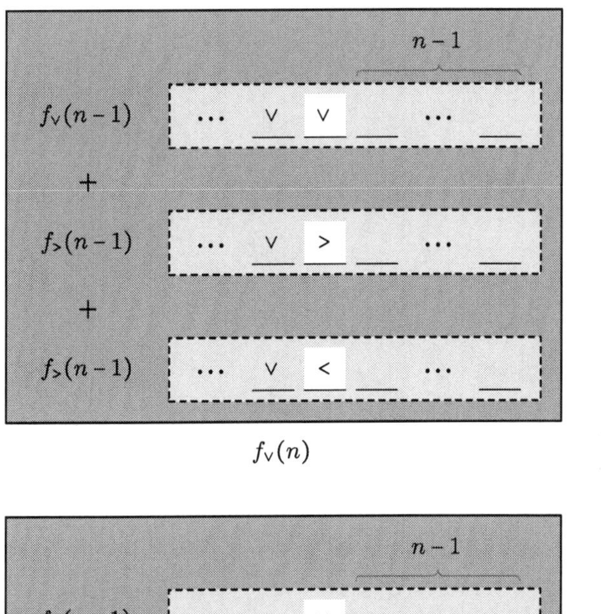

$f_\vee(n)$

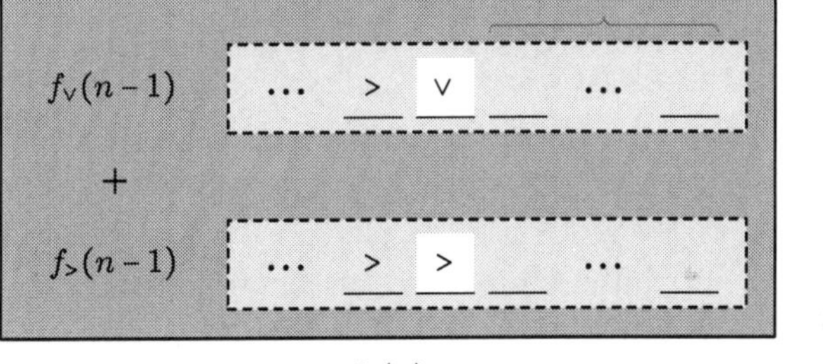

$f_>(n)$

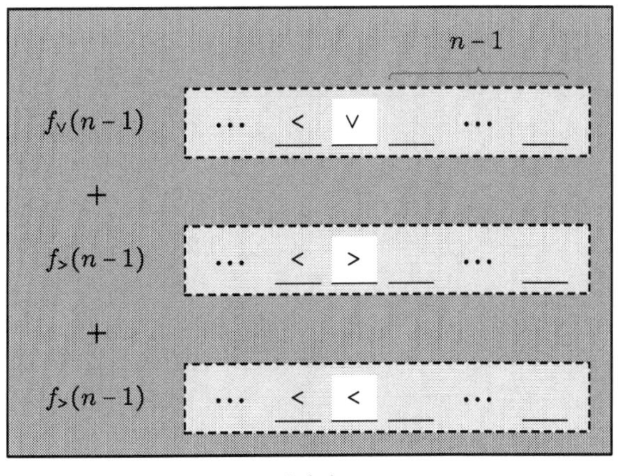

$f_<(n)$

Figure 9.9 Decompositions of the three problems for the mutually recursive solution of the water treatment plants puzzle.

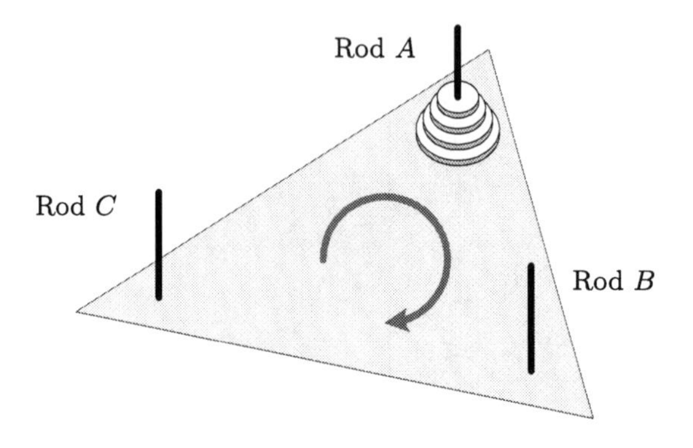

Figure 9.10 The cyclic towers of Hanoi puzzle.

Lastly, the solution $f(n)$ to the original problem can be expressed in different ways:

$$f(n) = f_{\vee}(n-1) + f_{>}(n-1) = f_{>}(n).$$

The first identity signifies that the first character of the string can either be 'v' or '>', in which case we would have to add the number of valid strings of length $n-1$ that start with those characters. The second identity follows from the recursive rule associated with $f_{>}(n)$. Finally, it is possible to show that $f_{\vee}(n) = f_{<}(n) = F_{2n+1}$, and $f_{>}(n) = F_{2n}$, where F is the Fibonacci function (see Exercise 9.5).

9.5 CYCLIC TOWERS OF HANOI

The following problem is known as the "cyclic" towers of Hanoi, which is a variant of the classical towers of Hanoi puzzle (see Section 7.2). Although it introduces a constraint that increases its complexity, the way to think about the problem and its solution (i.e., moving individual disks or stacks of several disks) is analogous. Thus, it constitutes an excellent companion to the original problem that could reinforce the reader's understanding of its solution. Moreover, it also supports the importance of recursion for problem solving in general, since (to the best of the author's knowledge) the only known iterative algorithm that solves the problem is a direct translation of the recursive solution by simulating the program stack (see Section 10.3).

The rules of the cyclic towers of Hanoi are identical to those of the original problem, but individual disks are constrained to move in a cyclical pattern. For example, if A, B, and C denote the three rods, then a

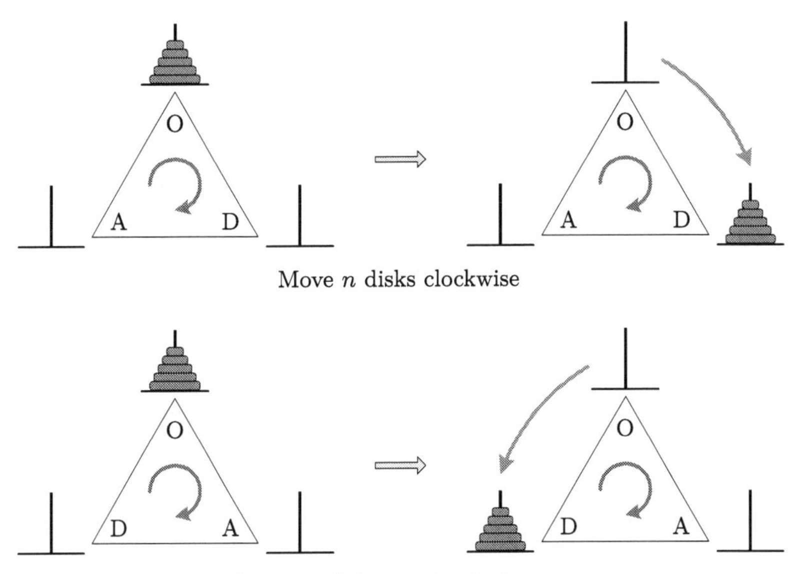

Figure 9.11 Illustration of two different problems comprised in the cyclic towers of Hanoi puzzle.

disk is only allowed to move from A to B, from B to C, or from C to A. The problem can be visualized by arranging the rods in a triangular configuration, and allowing moves only in a certain circular direction. In particular, we will assume that the disks must move in the clockwise direction, as shown in Figure 9.10.

The problem consists of describing how to move an entire stack of n disks from an origin rod, say A, to another destination rod, by using the remaining auxiliary rod. Notice that the previous statement does not mention whether the destination rod will be B or C. In fact, the key insight for solving the problem consists of realizing that although individual disks must move in a particular circular direction (we have assumed that it is clockwise), we can build different procedures in order to move an entire stack of n disks either clockwise or counterclockwise. In other words, the cyclic towers of Hanoi puzzle comprises two different problems: moving n disks clockwise, or counterclockwise, but where each single disk must move in only one circular direction. Both problems are illustrated in Figure 9.11, where O, D, and A denote the origin, destination, and auxiliary rods, respectively. Observe that the destination and auxiliary rods change their location depending on the specific problem.

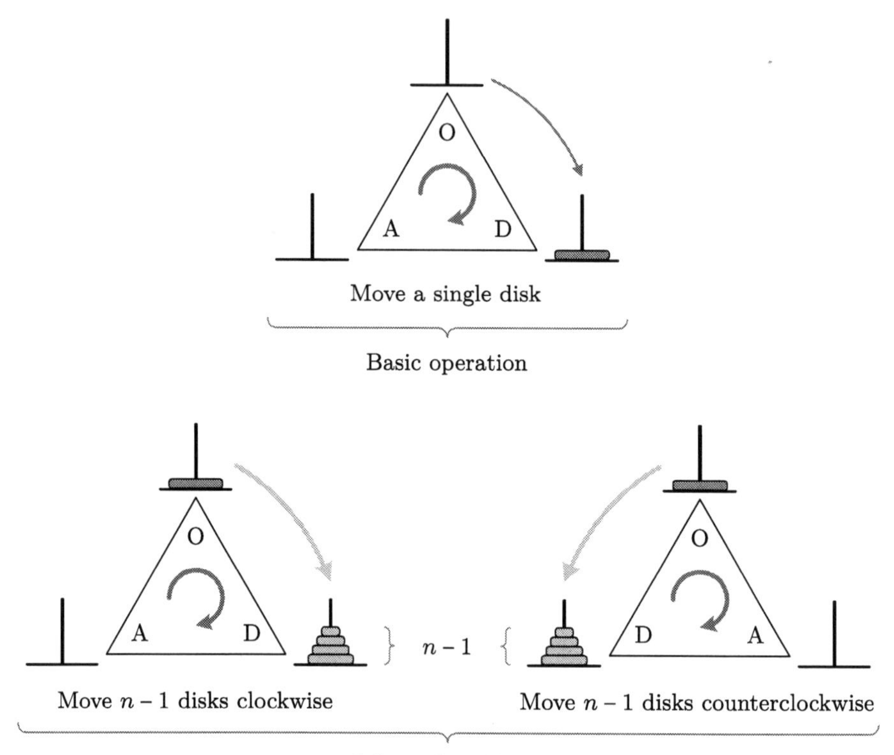

Figure 9.12 Three operations used to implement the recursive methods for solving the cyclic towers of Hanoi puzzle.

The size of both problems is naturally n. For the base case one option is to consider that it occurs when $n = 1$. This trivial solution involves a single move if the goal is to move the stack of one disk clockwise, or two moves if the disk must be moved counterclockwise. Another possibility is to avoid any action if $n = 0$, similarly to the function `towers_of_Hanoi_alt` in Listing 7.3.

For the recursive cases we can decompose the problems by reducing their size by a unit. This implies that there are three operations that we can use to build the recursive methods (see Figure 9.12). Firstly, we can move an individual disk clockwise, according to the rules of the problem. In addition, we can assume by induction that we can move stacks of $n-1$ disks either clockwise or counterclockwise.

Figure 9.13 illustrates the steps involved in solving the problem of moving a stack of n disks clockwise, from an origin rod to a destination

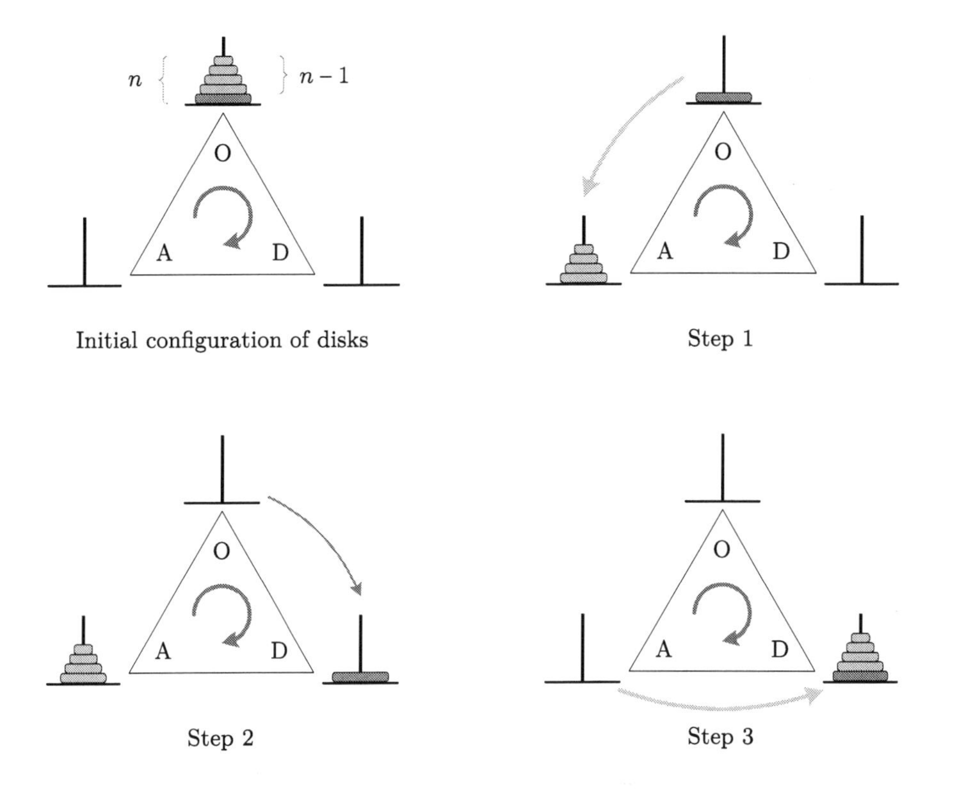

Figure 9.13 Operations for moving n disks clockwise.

rod. Firstly, we must move $n-1$ disks to the auxiliary rod, since otherwise it would be impossible to move the largest disk. The operation that accomplishes this task consists of moving the stack of $n-1$ disks counterclockwise, from the origin to the auxiliary rod, and with the help of the destination rod (step 1). The next step (2) involves the basic operation that inserts the largest disk into the destination rod. Finally, the $n-1$ disks on the auxiliary rod need to be moved into the destination rod. Again, this is carried out by moving them counterclockwise (step 3).

The procedure `clockwise` in Listing 9.6 implements the method. The way to reason about the problem is very similar to how we solved the original towers of Hanoi puzzle. Thus, the algorithm is almost identical to the code in Listing 7.3, with the exception that the recursive calls involve a second method. Observe that when programming the complete algorithm we assume that `counterclockwise` works correctly, even if we have not written a line of its code.

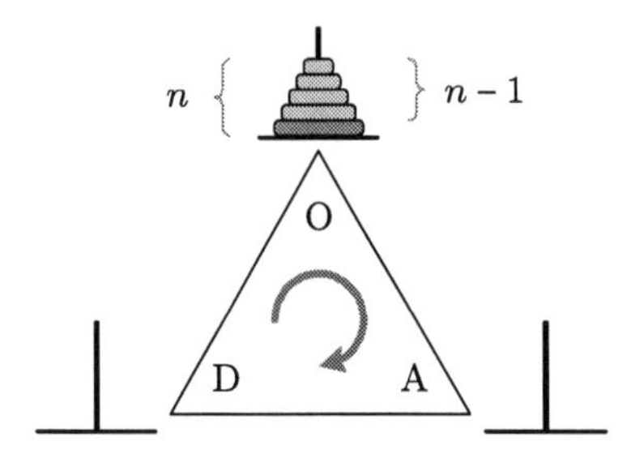

Initial configuration of disks

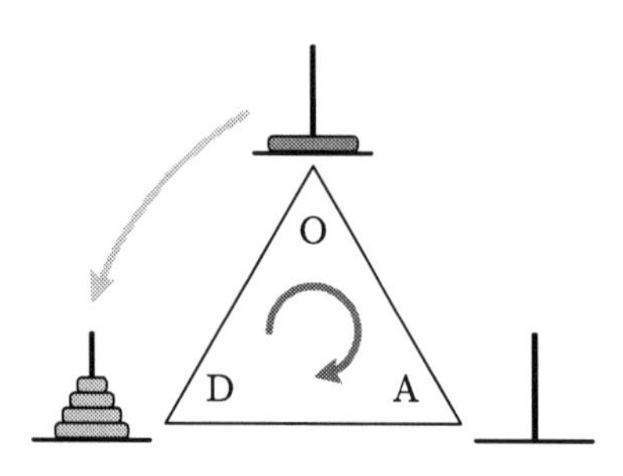

Step 1

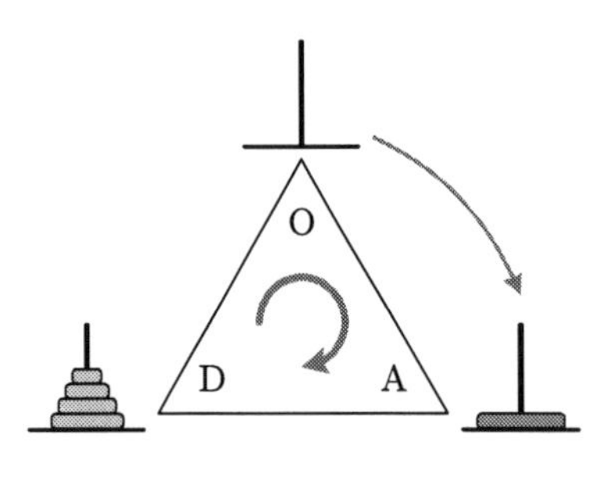

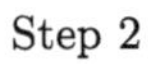

Step 2

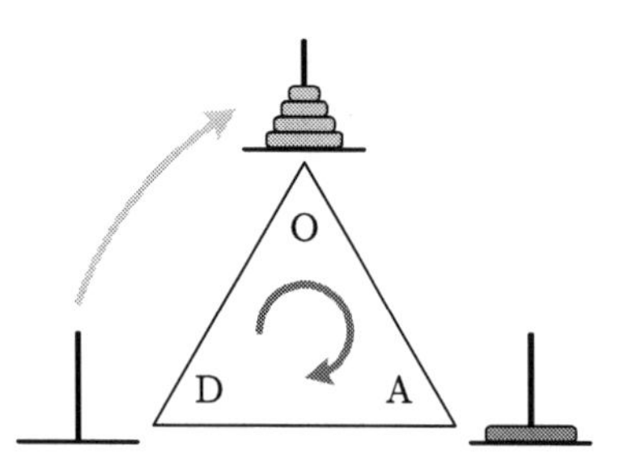

Step 3

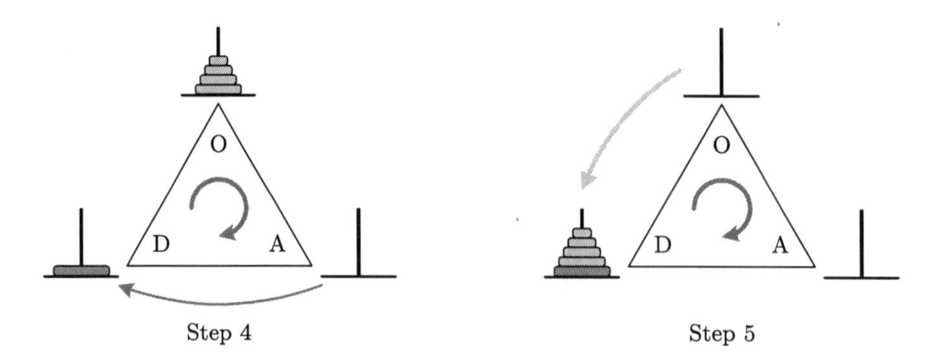

Step 4

Step 5

Figure 9.14 Operations for moving n disks counterclockwise.

Listing 9.6 Mutually recursive procedures for the cyclic towers of Hanoi puzzle.

```
def clockwise(n, o, d, a):
    if n > 0:
        counterclockwise(n - 1, o, a, d)
        print('Move disk', n, 'from rod', o, 'to rod', d)
        counterclockwise(n - 1, a, d, o)


def counterclockwise(n, o, d, a):
    if n > 0:
        counterclockwise(n - 1, o, d, a)
        print('Move disk', n, 'from rod', o, 'to rod', a)
        clockwise(n - 1, d, o, a)
        print('Move disk', n, 'from rod', a, 'to rod', d)
        counterclockwise(n - 1, o, d, a)
```

Lastly, Figure 9.14 shows the operations that solve the problem of moving a stack of n disks counterclockwise, and which are coded in procedure `counterclockwise` of Listing 9.6. This case requires more work since we need to be able to move the largest disk twice clockwise in order to simulate a single counterclockwise move. The first step consists of moving $n-1$ counterclockwise with a recursive call. This allows us to perform a basic move that inserts the largest disk into the auxiliary rod (step 2). In the third step the stack of $n-1$ disks needs to be moved clockwise back to the origin rod (by calling the previously implemented `clockwise` method) in order to allow the second basic move in the fourth step. Finally, having placed the largest disk in the destination rod, the last operation consists of moving the stack of $n-1$ disks counterclockwise into the destination rod.

9.6 GRAMMARS AND RECURSIVE DESCENT PARSERS

In computer science a **formal grammar** is a set of **production rules** that specify the syntax of a **formal language**. In other words, it describes how to combine and build sequences of elements of the formal language's **alphabet**, which is a set of **tokens** that can be words, symbols, numbers, punctuation marks, etc. In this section we will see a more complicated example involving these concepts that is usually covered in advanced courses related to compilers and language processors. In particular, we will implement a calculator that receives a string representing a mathematical expression, and outputs its value. The calculator will be able to add, subtract, multiply, and divide numbers. For simplicity these will correspond to integer inputs (although the result can be a rational number). In addition, it will be able to process parentheses and unary minus signs that precede left parentheses.

The construction of the calculator requires two steps. The first one breaks down an initial string **s**, which represents a mathematical expression, into a sequence of tokens. In the second step this list of tokens is analyzed according to a formal grammar by a **recursive descent parser** in order to evaluate the initial expression. In practice there are powerful tools and libraries to accomplish these tasks, but we will present low-level algorithms for solving them in order to provide more examples of mutual recursion. Naturally, a full coverage of the mentioned concepts and topic is well beyond the scope of this book.

9.6.1 Tokenization of the input string

The goal of this stage consists of splitting the input string that represents a mathematical expression into a list of consecutive tokens, which can be:

- Positive or negative integers: an optional '-' symbol followed by a sequence of digits.
- Operators: '+', '-', '*', and '/'.
- Parenthesis symbols: '(' and ')'.

For example, if the input string is "-(-6 / 3) - (-(4)) + (18-2)", this first algorithm should produce the following list:

['-', '(', '-6', '/', '3', ')', '-', '(', '-', '(', '4', ')', ')', '+', '(', '18', '-', '2', ')'].

Firstly, the algorithm ignores space and tab (whitespace) characters. Furthermore, observe that a minus sign can play different roles. Besides specifying a negative number, it can also be either a unary or binary operator. For instance, the algorithm interprets that -6 is a number, and therefore does not generate two separate tokens for the (unary) minus sign and the 6. The example also shows that unary minus signs can appear at the beginning of the input list, or in between two left parenthesis symbols. The mutually recursive functions that we will see shortly arise due to the necessity to handle unary minus signs in the

Listing 9.7 Mutually recursive functions for tokenizing a mathematical expression.

```
def tokenize_unary(s):
    if not s or s == ' ' or s == '\t':
        return []
    elif len(s) == 1:
        return [s]
    elif s[0] == ' ' or s[0] == '\t':
        return tokenize_unary(s[1:])
    elif s[0].isdigit() or s[0] == '-':
        if s[0].isdigit() and (s[1] == ' ' or s[1] == '\t'):
            return [s[0]] + tokenize(s[2:])
        else:
            t = tokenize(s[1:])
            if t == []:
                return [s[0]]
            else:
                if is_number(t[0]):
                    t[0] = s[0] + t[0]
                    return t
                else:
                    return [s[0]] + t
    else:
        if s[0] == '(':
            t = tokenize_unary(s[1:])
        else:
            t = tokenize(s[1:])
        return [s[0]] + t


def tokenize(s):
    if not s or s == ' ' or s == '\t':
        return []
    elif len(s) == 1:
        return [s]
    elif s[0] == ' ' or s[0] == '\t':
        return tokenize(s[1:])
    elif s[0].isdigit():
        if s[1] == ' ' or s[1] == '\t':
            return [s[0]] + tokenize(s[2:])
        else:
            t = tokenize(s[1:])
            if t == []:
                return [s[0]]
            else:
                if is_number(t[0]):
                    t[0] = s[0] + t[0]
                    return t
                else:
                    return [s[0]] + t
    else:
        if s[0] == '(':
            t = tokenize_unary(s[1:])
        else:
            t = tokenize(s[1:])
        return [s[0]] + t
```

Listing 9.8 Function that checks whether a string represents a number.

```
1 def is_number(s):
2     try:
3         float(s)
4         return True
5     except ValueError:
6         return False
```

mathematical expressions. Lastly, observe that a token can be a number composed of several digits (and a possible minus sign).

In order to tackle this problem we will use two similar functions, included in Listing 9.7, which receive an input string `s` of length n, and return the corresponding list of tokens. The difference between them is that one (denoted as `tokenize_unary`) contemplates receiving a unary minus sign as its first non-whitespace character, while the other one (called as `tokenize`) assumes that such a minus sign will not appear. Both methods also check whether the content of a string corresponds to a number through the function `is_number`, described in Listing 9.8.

The size of the problems is the length of the input string (n). There are several base cases. If the string is empty, or if it contains a single whitespace character, the methods must return an empty list. In addition, if the string is a single character (we know it is not a whitespace character since this has already been checked in the previous condition) then it should return a list containing that character. Both of these conditions guarantee that the input string will contain at least two characters in the recursive cases. Therefore, we will be able to access the second element of the string safely, avoiding a "string index out of range" error.

The recursive cases for this problem are more complex than in previous examples, since there are several situations that need to be taken into account. Firstly, if the first element of the input string is a whitespace character it has to be ignored. Thus, both methods can simply return the result of invoking themselves with the rest of the input string ($\mathbf{s}_{1..n-1}$).

9.6.1.1 Function tokenize()

In order to analyze the remaining recursive cases we will focus on `tokenize` first, which is slightly simpler. The next recursive case occurs if the first character of the input string (s_0) is a digit (line 36), which naturally belongs to some number. The algorithm must then perform different actions depending on the following character (s_1) or token associated with the remaining string ($\mathbf{s}_{1..n-1}$).

Firstly, if s_1 is a whitespace character then s_0 will constitute a token. Therefore, the method can return a list containing s_0, concatenated with the list of tokens related to $\mathbf{s}_{2..n-1}$. For example, this can be seen through the following diagram:

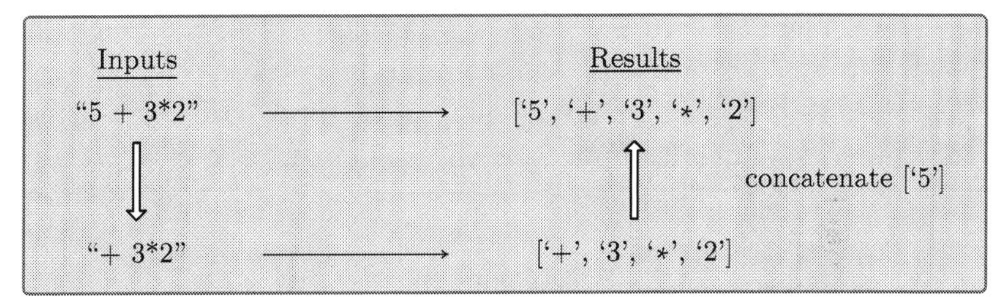

Otherwise, the method computes the tokens associated with $\mathbf{s}_{1..n-1}$ through a recursive call, and stores them in the list $\mathbf{t}$ (line 40). If this list is empty, then the method can simply return the token corresponding to s_0 (line 42). Instead, the following action will depend on whether the first token of $\mathbf{s}_{1..n-1}$ is a number or not. Consider the following diagram:

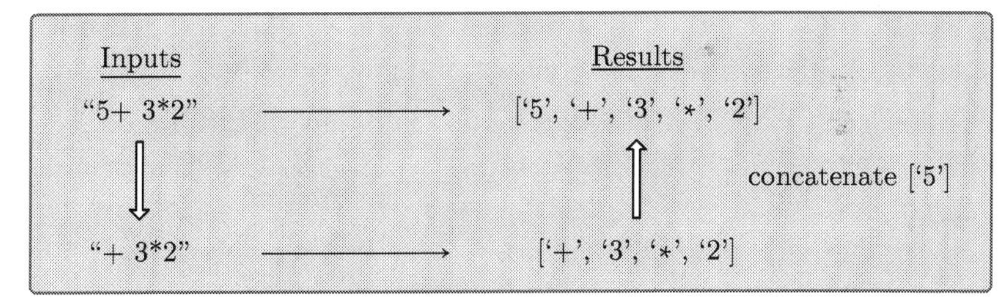

Since the first token of $\mathbf{t}$ ('+') is not a number, the algorithm must concatenate the list containing s_0 with $\mathbf{t}$ (line 48).

Alternatively, if the first token of $\mathbf{t}$ is a number the method will have to append the character s_0 to this first token (lines 45 and 46). The following diagram illustrates the idea:

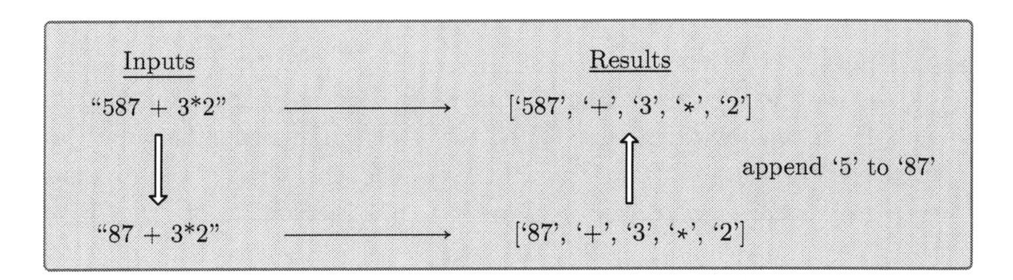

Observe that this is the way that the algorithm builds numerical tokens, one digit at a time.

Lastly, there are two more recursive cases when the first element of the string is not a digit. Since the function has already checked that it is not a whitespace character (in line 34), the character must be an operator or a parenthesis, and will constitute a token. Thus, the method will concatenate such token to the result of solving the problem for $\mathbf{s}_{1..n-1}$. However, if the character s_0 is a left parenthesis, then the first token of the substring could be a unary minus sign. In that case the method has to call function `tokenize_unary` (line 51). In any other situation the unary minus sign cannot appear as the first token, and the method can simply invoke itself (line 53).

9.6.1.2 Function *tokenize_unary()*

Function `tokenize_unary` is very similar to `tokenize`. There are two main differences. On the one hand, the method calls itself in line 7 if it encounters a whitespace character at the first position of the string. Naturally, the algorithm could still find a unary minus sign as the first token of the string.

On the other hand, if the first character is a minus sign, then it must be a unary operator. If it precedes a number, then the minus sign will form part of the number, similarly as if it were a digit. Consider the following diagram:

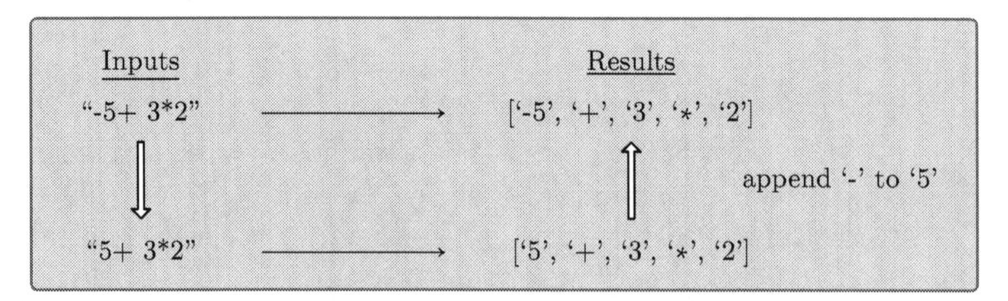

Note that the '-' is appended to the '5′ in order to generate the token '-5'. Therefore, line 8 also checks if s_0 is a minus sign, and lines 12–20 are identical to 40–48.

Lastly, line 9 is also different from line 37 due to the following subtlety. If there is a whitespace character after a digit, then this digit will constitute a token (lines 9 and 10). However, there may be whitespace characters after a unary minus sign, but this does not imply that it will be a token. For example, given the string "- 6", the algorithm must only return the token '-6' (instead of two tokens for each of the non-whitespace characters). Thus, the first condition in line 9 makes sure that the character is indeed a digit, before proceeding to line 10 where a token is generated. Finally, when processing a string like "- 6", the recursive call in line 12 returns the token '6', ignoring the whitespace. Therefore, the code continues and appends the minus sign to the 6 correctly, as if it were a digit, in line 17.

9.6.2 Recursive descent parser

Having generated a list of tokens, we are ready to process it in order to evaluate the initial mathematical expression. Naturally, these tokens cannot appear in any random order (e.g., the list of tokens cannot have two consecutive operators or numbers, unbalanced parentheses, etc.). They must conform to the production rules of a formal grammar, which describe how to combine the tokens in a valid way. Roughly speaking, the production rules are used to group or break up tokens hierarchically into fragments that belong to certain categories. For our calculator these classes will be "expressions," "terms," "factors," and a few others that we will describe shortly. To understand this idea, consider the following mathematical expression:

$$-(2+3) \;+\; 2*8/4 \;-\; 3*(4+2)/6 \;+\; 7 \;-\; (-(2+8)*3).$$

Figure 9.15 shows that within it there are five terms separated by plus or minus signs that are binary operators. These terms can be further decomposed into "multiplicative" factors, separated by '$*$' or '/'. The example only shows how the term $3*(4+2)/6$ is broken up into three factors. It is important to note that the plus in the middle factor does not separate terms in the initial expression, since it appears inside parentheses. In addition, the first minus sign of the expression is a unary operator and naturally does not separate terms. A unary minus sign can also appear in between two left parentheses, as illustrated in the last term.

The figure also shows that a multiplicative factor can be an expression enclosed in parentheses (i.e., a "parenthesized expression"). This implies that the definitions of an expression, a term, and a factor are mutually recursive. Notice that an expression is a sequence of additive terms, a term is a series of multiplicative factors, and a factor can be either a number, or it can contain yet another expression enclosed in parentheses. Naturally, an expression, term, or factor can be a single number.

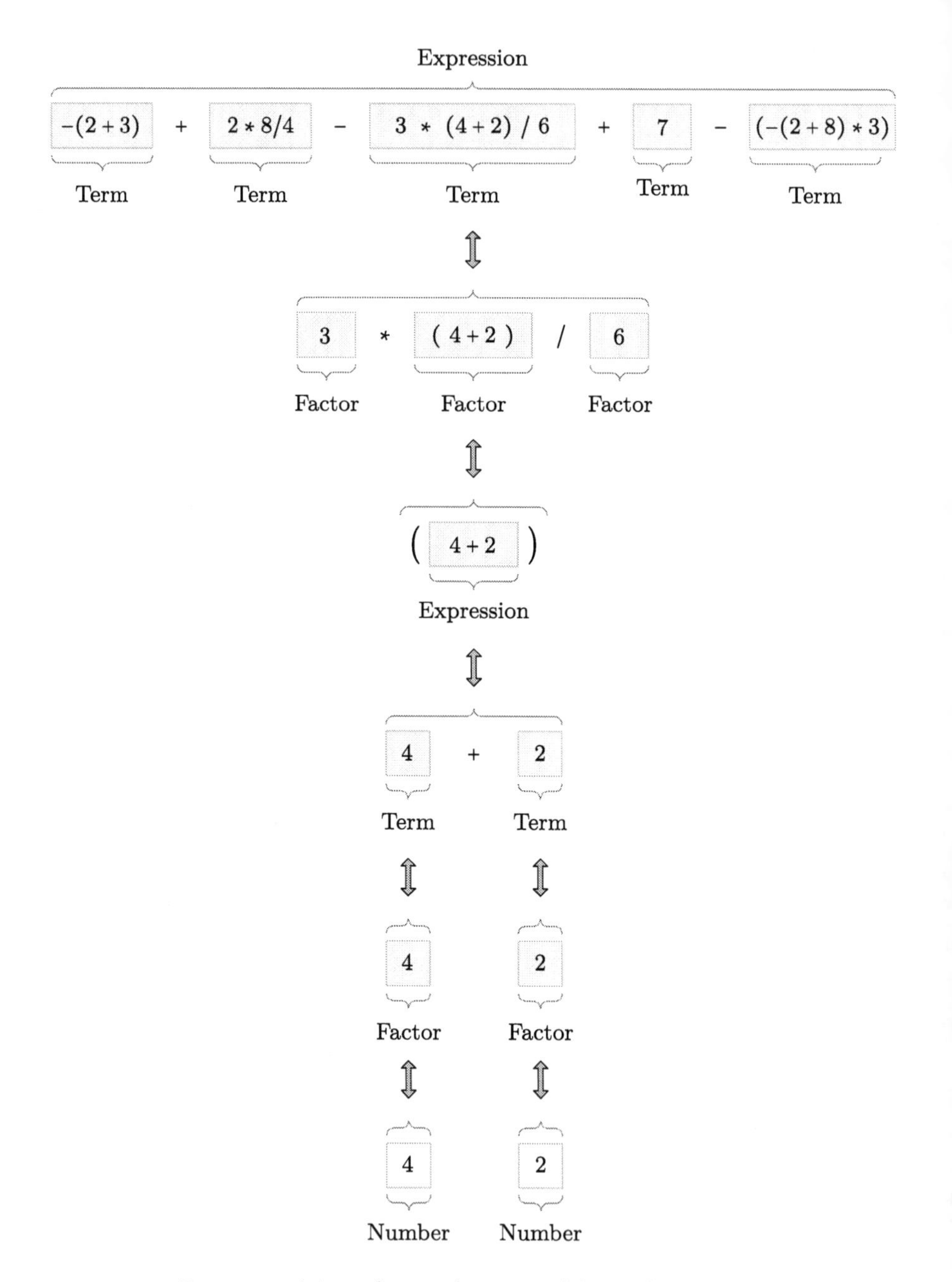

Figure 9.15 Decomposition of a mathematical formula into categories such as expressions, terms, factors, or numbers.

Consider defining an expression recursively. Its "size" could be the number of terms within it. The base case would occur when it consists of a single term. However, if there is more than one term then we can reduce the size of the recursive entity by a unit, and define it as a smaller expression plus or minus a term. These recursive definitions are precisely the production rules of a formal grammar. If the symbol E represents an expression, and T a term, the production rules for an expression can be described by using the following notation:

$$E \quad \longrightarrow \quad T \;\mid\; E+T \;\mid\; E-T, \tag{9.9}$$

where the plus and minus signs are tokens (binary operators) from the input list, and the vertical bars (|) can be read as "or," and are used for notational convenience, in order to define several production rules in a single line. In other words, (9.9) defines the three production rules: $E \rightarrow T$, $E \rightarrow E+T$, and $E \rightarrow E-T$.

Nevertheless, the definition of an expression in (9.9) is incomplete given the calculator that we wish to build. In particular, there is another type of base case related to a term that begins with a unary minus sign, continues with a parenthesized expression, and could be followed by several multiplicative factors. Let I denote an "inverted" term (because of the sign change associated with the unary operator) of these multiplicative factors, where the first one is a parenthesized expression. The production rules for E are:

$$E \quad \longrightarrow \quad T \;\mid\; E+T \;\mid\; E-T \;\mid\; -I. \tag{9.10}$$

Similarly, we can build recursive (or non recursive) definitions for other elements in the language. Let F denote a multiplicative factor, P a parenthesized expression, and $Number$ an integer token. The complete description of the grammar that we will use is:

$$\begin{array}{lcl} E & \longrightarrow & T \;\mid\; E+T \;\mid\; E-T \;\mid\; -I \\ T & \longrightarrow & F \;\mid\; T*F \;\mid\; T\,/\,F \\ I & \longrightarrow & P \;\mid\; P*T \;\mid\; P\,/\,T \\ F & \longrightarrow & P \;\mid\; Number \\ P & \longrightarrow & (E), \end{array}$$

where '+', '-', '*', '/', '(', and ')' are possible input tokens. Other formal grammars are also possible, but developing them is far beyond the scope of this book.

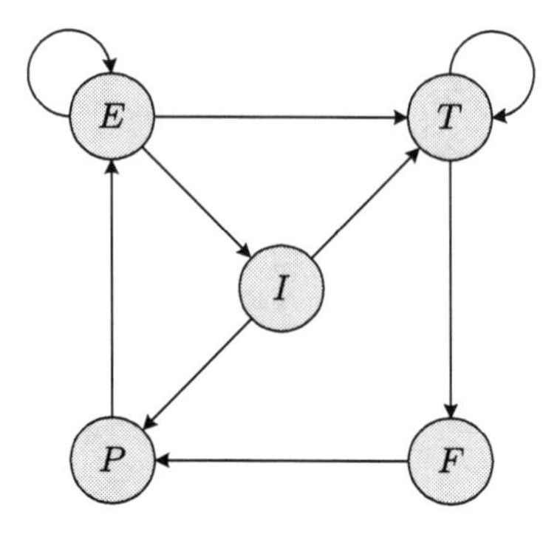

Figure 9.16 Method calls of the mutually recursive functions that implement the recursive descent parser associated with the calculator program.

A **recursive descent parser** is a program that analyzes a list of tokens by implementing the production rules of a grammar as methods, which are typically mutually recursive. Specifically, these systems define one method per symbol that appears on the left-hand side of the production rules, and which assumes that the input list of tokens is of the type specified by the symbol. For our calculator we will build a recursive descent parser based on five mutually recursive functions, one for each of the symbols E, T, I, F, and P (Figure 9.16 shows how the methods invoke each other). For example, the function related to E will process expressions, assuming that the input list of tokens is indeed an expression. The following subsections will show possible implementations of the functions.

9.6.2.1 Function `expression()`

Listing 9.9 includes a function that handles expressions (E). Firstly, all of the functions of the recursive descent parser begin by checking whether the list of tokens is empty, raising a syntax error exception if it is. Otherwise, the method must decide which production rule to apply, assuming that the input list of tokens represents an expression. The options are illustrated in Figure 9.17. On the one hand, the expression could be a single regular or inverted term, as shown in (a) and (b), respectively. On the other hand, if it contains more than one additive term the expression will be decomposed into a smaller one, plus or minus a term, as shown in (c).

Listing 9.9 Function that parses a mathematical expression of additive terms.

```
1  def expression(tokens):
2      if tokens == []:
3          raise SyntaxError('Syntax error')
4      else:
5          par_counter = 0
6          i = len(tokens) - 1
7          while i > 0 and ((tokens[i] != '+' and tokens[i] != '-')
8                                 or par_counter > 0):
9              if tokens[i] == ')':
10                 par_counter = par_counter + 1
11             if tokens[i] == '(':
12                 par_counter = par_counter - 1
13 
14             i = i - 1
15 
16         if i == 0:
17             if tokens[0] == '-':
18                 return - inverted_term(tokens[1:])
19             else:
20                 return term(tokens)
21         else:
22             e = expression(tokens[0:i])
23             t = term(tokens[i + 1:])
24             if tokens[i] == '+':
25                 return e + t
26             else:
27                 return e - t
```

In order to recognize these cases the function uses a while loop (lines 7–14) to search for the first plus or minus binary operator that appears outside of any parenthesis, by starting at the end of the list (i.e., it will search from right to left). The method keeps track of the number of left and right parenthesis symbols through the accumulator variable `par_counter`. After exiting the loop, the variable `i` (i in Figure 9.17(c)) will contain the index of the found '+' or '-' token, or 0 if it has not detected any, implying that there is only one term in the expression. In that case there are two possibilities. If the first token of the input list is a '-' symbol, then it must be an inverted term (I). The function can then return the result of evaluating the method `inverted_term(tokens[1:])` with the input list of tokens, but discarding the first '-' token (line 18). Otherwise it will be a regular term (T), and the function can invoke

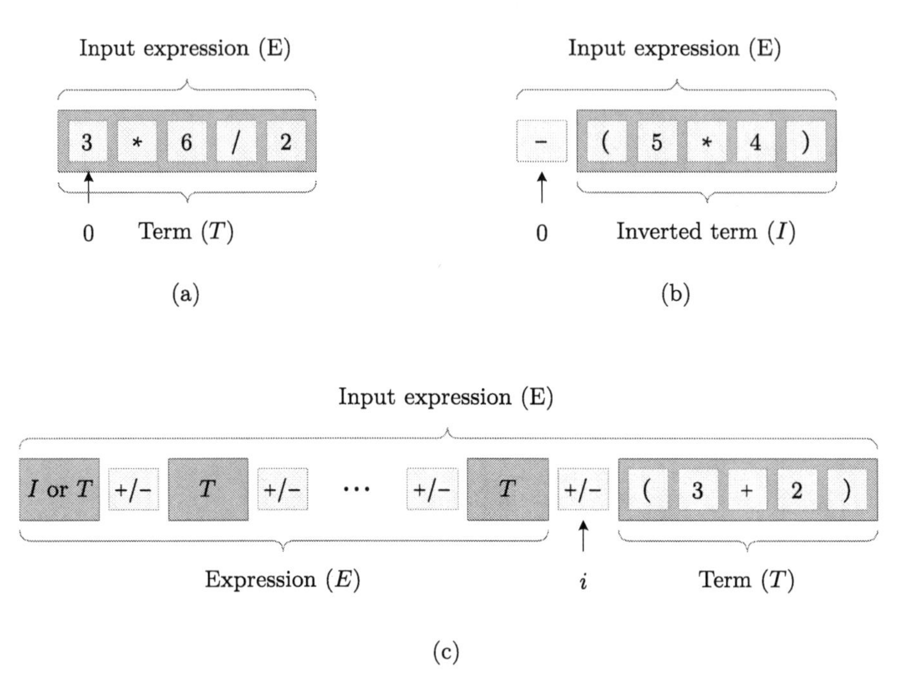

Figure 9.17 Possible decompositions of an expression.

`term(tokens)`, in line 20. Alternatively, if `i` is greater than 0, then `token[i]` will be an additive binary operator ('+' or '-'). In that case the function evaluates the entire expression that precedes the symbol, storing the result in `e`, and the term to the right of the symbol, saving the result in `t`. The method will then return the appropriate additive result, depending on whether it should add or subtract `e` and `t`, which are obviously computed through mutually recursive function calls.

9.6.2.2 Function *term()*

Listing 9.10 codes the method that parses a term of multiplicative factors. It is very similar to the code in Listing 9.9, where the multiplicative symbols, '*' and '/', play the roles of the '+' and '-' symbols. The while loop identifies the index of a multiplicative symbol that separates factors, and which is not inside parentheses. If `i`= 0 the term does not contain a separating symbol. Thus, it must be a single factor, and the function can call the `factor` method with the entire input list (line 17). Otherwise, the algorithm will multiply or divide the term of multiplicative

Listing 9.10 Function that parses a term of multiplicative factors.

```
def term(tokens):
    if tokens == []:
        raise SyntaxError('Syntax error')
    else:
        par_counter = 0
        i = len(tokens) - 1
        while i > 0 and ((tokens[i] != '*' and tokens[i] != '/')
                                  or par_counter > 0):
            if tokens[i] == ')':
                par_counter = par_counter + 1
            if tokens[i] == '(':
                par_counter = par_counter - 1

            i = i - 1

        if i == 0:
            return factor(tokens)
        else:
            t = term(tokens[0:i])
            f = factor(tokens[i + 1:])
            if tokens[i] == '*':
                return t * f
            else:
                return t / f
```

factors (`t`) that precede the separating symbol (located at index `i`), by the single factor to its right (`f`).

9.6.2.3 Function `inverted_term()`

Listing 9.11 shows the code that analyzes inverted terms. Since they must start with a left parenthesis, the function can also raise a syntax error exception if the first token is not '('. Otherwise, the function uses a loop to figure out the index (`i`) of the operator that goes immediately after the right parenthesis token that is balanced with the first left one. If `i` is equal to the length of the input list then the entire term is an expression enclosed between parentheses. Thus, the function can return the result of calling method `parenthesized_expression` with the entire input list (line 17). Otherwise, the function evaluates the expression enclosed within the first parentheses (line 19), and the term of multiplicative factors that appear after index `i` (line 20). The function

Listing 9.11 Function that parses a term of multiplicative factors, where the first one is a parenthesized expression.

```
def inverted_term(tokens):
    if tokens == [] or tokens[0] != '(':
        raise SyntaxError('Syntax error')
    else:
        par_counter = 1
        n = len(tokens)
        i = 1
        while i < n and par_counter > 0:
            if tokens[i] == '(':
                par_counter = par_counter + 1
            if tokens[i] == ')':
                par_counter = par_counter - 1

            i = i + 1

        if i == n:
            return parenthesized_expression(tokens)
        else:
            p = parenthesized_expression(tokens[0:i])
            t = term(tokens[i + 1:])
            if tokens[i] == '*':
                return p * t
            else:
                return p / t
```

finally multiplies or divides the previous values depending on the token at position `i`.

9.6.2.4 Function `factor()`

Listing 9.12 contains the code that parses a factor. If the input list contains a single token it must be numerical, and the function can return the corresponding value (line 6). If the token is not numerical then there is a syntax error in the initial expression. Finally, if the length of the input list is greater than one, the factor must be a parenthesized expression, and the method can call the corresponding function with the entire input list (line 10).

Listing 9.12 Function that parses a mathematical factor.

```
1 def factor(tokens):
2     if tokens == []:
3         raise SyntaxError('Syntax error')
4     elif len(tokens) == 1:
5         if is_number(tokens[0]):
6             return float(tokens[0])
7         else:
8             raise SyntaxError('Syntax error')
9     else:
10        return parenthesized_expression(tokens)
```

Listing 9.13 Function that parses a parenthesized expression.

```
1 def parenthesized_expression(tokens):
2     if tokens == [] or tokens[0] != '(' or tokens[-1] != ')':
3         raise SyntaxError('Syntax error')
4     else:
5         return expression(tokens[1:-1])
```

9.6.2.5 Function *parenthesized_expression()*

Listing 9.13 parses an expression enclosed in parentheses. Besides checking that the input list is not empty, the method also makes sure that the first and last tokens are left and right parenthesis symbols, respectively. If the method does not raise a syntax error exception it can proceed by evaluating the expression contained inside the parentheses (line 5), and returning the result. This method therefore calls the `expression` function, completing the cycle that makes all of the functions mutually recursive.

Finally, the calculator can be executed with the code in Listing 9.14.

Listing 9.14 Basic code for executing the calculator program.

```
1 s = input('> ')
2 print(expression(tokenize_unary(s)))
```

9.7 EXERCISES

Exercise 9.1 — Mary and John are going to play a game involving n pebbles placed on a board. Each player can remove one, three, or four pebbles at a time. The one who takes the last pebbles wins. Both players will win if there are one, three, or four pebbles left and it is their turn to play. If there are two pebbles left then they will lose since they can only pick up one pebble. When there are more than four pebbles Mary has decided that she will take four pebbles whenever she can, while John's strategy consists of removing only one pebble. Implement a pair of mutually recursive functions that simulate the game, and return a 1 in case John wins, and a 0 if Mary wins. Which player has adopted a better strategy in general, if $1 \le n \le 100$? Consider that any player can start the game.

Exercise 9.2 — Consider the functions $A(n)$ and $B(n)$ defined in (3.38) and (3.39). Show that $A(n) + B(n)$ corresponds to the n-th Fibonacci number, without solving the recurrences. In other words, do not use (3.40) or (3.41).

Exercise 9.3 — Determine the asymptotic order of growth of the running times of $A(n)$ and $B(n)$ defined in (3.38) and (3.39). In other words, obtain (3.40) and (3.41).

Exercise 9.4 — Show that $f(n)$, as described in (9.6), is equal to $F(2n)$, where F is the Fibonacci function. Prove the identity by: (a) mathematical induction, and (b) solving the recurrence relation in (9.6), obtaining a formula similar to that in (3.37).

Exercise 9.5 — Express the functions in (9.7) and (9.8) in terms of themselves (without mutual recursion).

Exercise 9.6 — Let $A(n)$ and $B(n)$ denote the number of individual disk operations carried out by the procedures `clockwise` and `counterclockwise`, respectively, in Listing 9.6. Provide recursive formulas for $A(n)$ and $B(n)$, and express each one in terms of itself, avoiding mutual recursion. Finally, solve the recurrences and indicate their order of growth.

Exercise 9.7 — Recall the "sideways" towers of Hanoi problem introduced in Exercise 7.4. Design mutually recursive procedures in order to move a stack of n disks from:

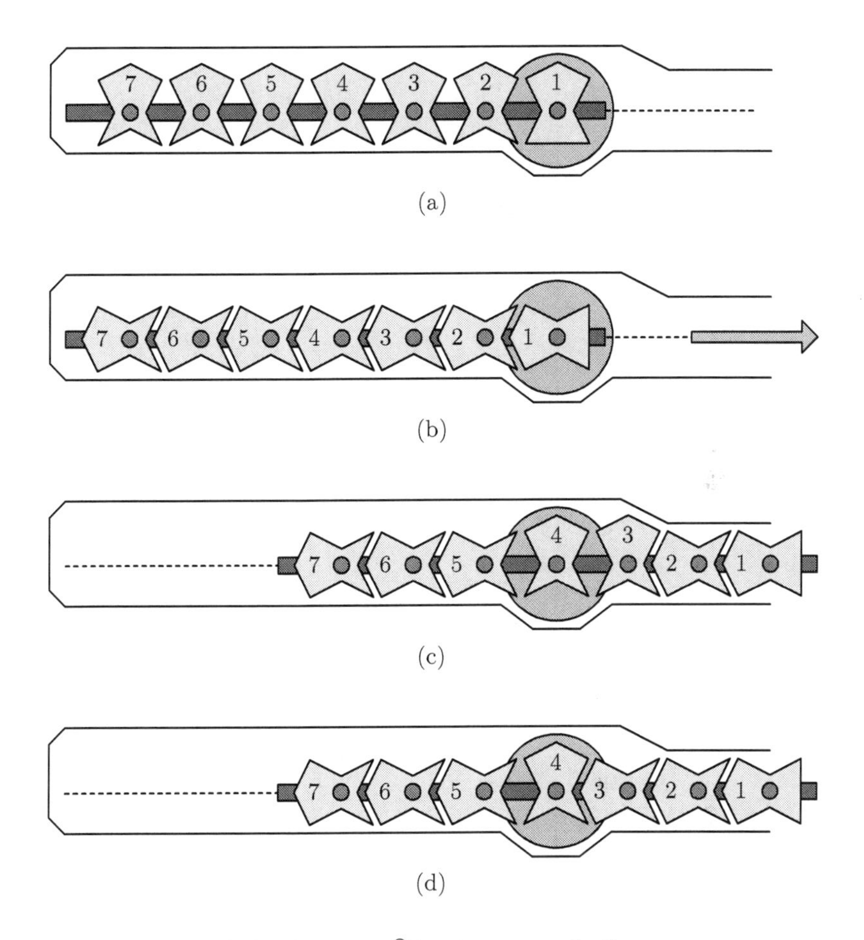

Figure 9.18 Rules of the Spin-out® brainteaser challenge.

a) the left or right rod to the middle rod,

b) the middle rod to the left or right rod.

Describe functions that determine the number of individual disk moves that the procedures carry out when moving a stack of n disks.

Exercise 9.8 — The following problem is known as the Spin-out® brainteaser challenge. We will use the illustrations in Figure 9.18 to explain the problem. Initially, there are n pieces (seven in the example) that appear contiguously along a rod, and inside a container that has a narrow opening on its right side. The pieces can appear in two orientations: either vertically, pointing upwards, as in (a); or horizontally, pointing towards the left, as in (b). The goal of the problem is to rotate the pieces from the initial "locked" configuration in (a), to the "unlocked" pattern in (b), where it would be possible to extract the rod and its pieces. Observe that a piece in a vertical orientation would hit the walls of the container, preventing the extraction of the rod. Only one of the pieces can be rotated at a time, from a vertical to a horizontal position, or vice versa. Furthermore, they can only be turned in the "rotation spot," which is the shaded circle in the illustrations. In (c), the fourth piece can be rotated in order to place it horizontally. In addition, observe that the walls of the container impede moving the rod any further to the right, since the piece to the right of the rotation circle is in a vertical position. Furthermore, while we can always turn the first piece, we can only rotate another if the one to its right is in a vertical position, as in (c). Thus, in (d), we cannot turn the fourth piece.

Implement two mutually recursive procedures in order to indicate the sequence of pieces that must be turned in order to "unlock" the rod, arriving at the configuration in (b), when starting from the layout in (a), for n pieces. Hint: use one procedure that "unlocks" the first n pieces of a rod, and another that "locks" them, where the goal would be the opposite: to reach (a) when starting from (b). Finally, specify a function that determines the number of turns needed to unlock a rod of n pieces.

CHAPTER 10

Program Execution

You don't really understand something if you only understand it one way.

— Marvin Minsky

PREVIOUS chapters have shown how to think and design programs recursively, by analyzing a wide variety of computational problems and the main types of recursion. The algorithms followed the declarative programming paradigm, which focuses on *what* programs compute from a high-level point of view. Instead, this chapter examines recursive programs from a low-level perspective, illustrating *how* they work, by showing the order in which instructions are carried out (i.e., the control flow).

The chapter describes several representations of the thought process regarding how recursive programs work, some of which are known as "mental models" of recursion. It introduces recursion trees, which are popular graphical representations of the sequence of calls carried out by recursive programs. Among their benefits, they are useful for debugging, help evaluate functions, allow us to visualize the computational (time and space) complexity of a recursive program, and can characterize the types of recursion covered so far. In addition, the chapter covers their relationship with the program stack, which is a data structure that programs use to implement general subroutine (i.e., method) calls.

The chapter ends with a brief introduction to memoization, which is a strategy for reducing the time complexity of recursive methods by storing results of potentially expensive calls. Memoization is closely related to dynamic programming, which is an important algorithm design technique.

Listing 10.1 Methods for computing the cosine of the angle between two vectors.

```
import math


def dot_product(u, v):
    s = 0
    for i in range(0, len(u)):
        s = s + u[i] * v[i]
    return s


def norm_Euclidean(v):
    return math.sqrt(dot_product(v, v))


def cosine(u, v):
    return dot_product(u, v) / norm_Euclidean(u) /
        norm_Euclidean(v)


# Example
a = [2, 1, 1]
b = [1, 0, 1]
print(cosine(a, b))
```

10.1 CONTROL FLOW BETWEEN SUBROUTINES

Regardless of whether the methods executed by a program are recursive or not, it is important to understand the order in which they are called. Consider the code in Listing 10.1, which computes the cosine of the angle between two general n-dimensional vectors, according to the definitions in (3.14), (3.15), and (3.16). Figure 10.1 shows the associated sequence of subroutine calls (darker arrows) and returns (lighter arrows) when executing the code. The boxes on the top-right corners of the functions indicate their concrete arguments, which end up producing the value $\sqrt{3}/2$. The numbers inside the circles indicate the order in which the operations are executed. The obvious, but key, observation is that a subroutine cannot terminate until all of the methods it calls have been completely processed.

The procedure is similar when running recursive code. Figure 10.2 shows the sequence of recursive calls of the function

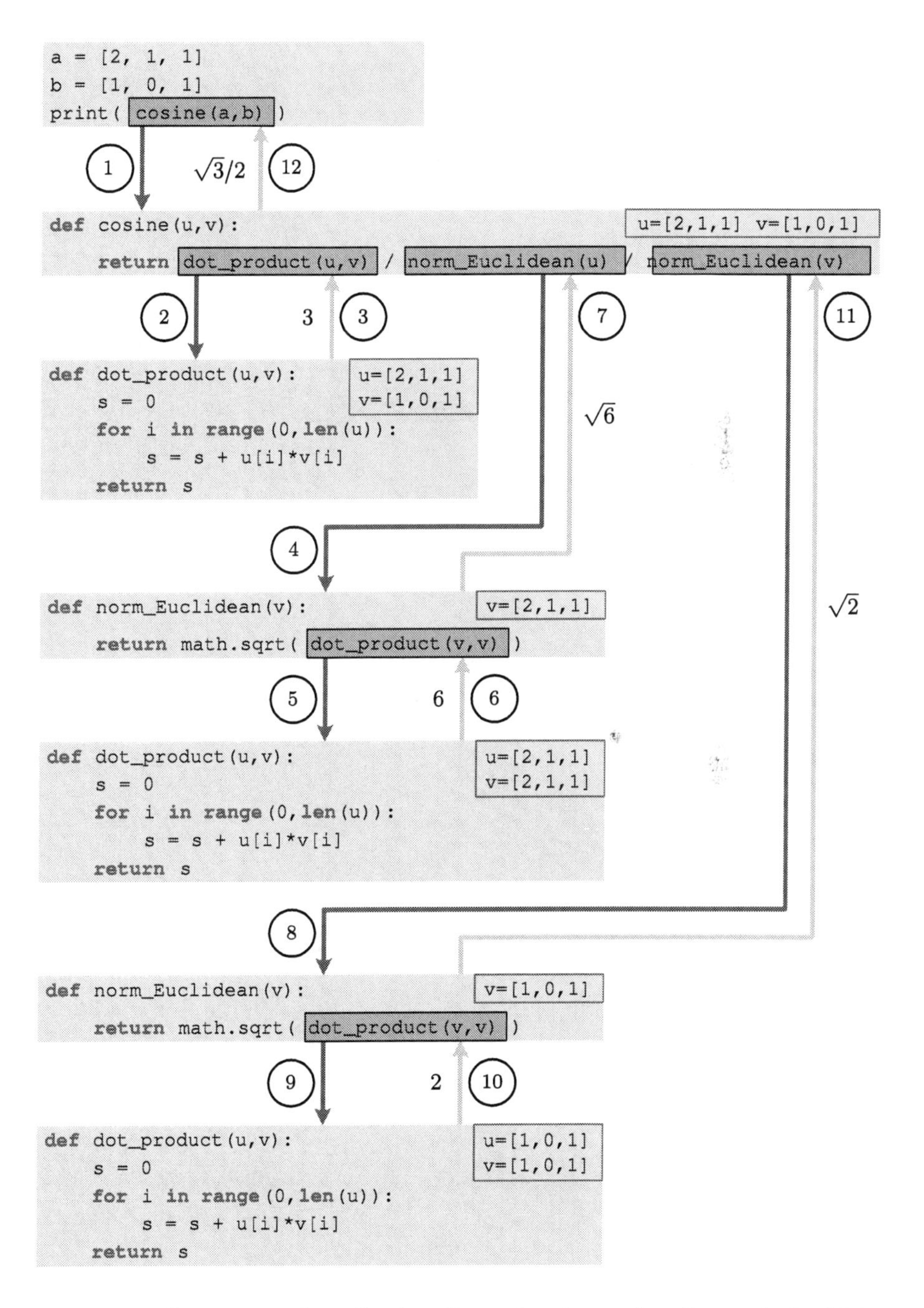

Figure 10.1 Sequence of method calls and returns for the code in Listing 10.1.

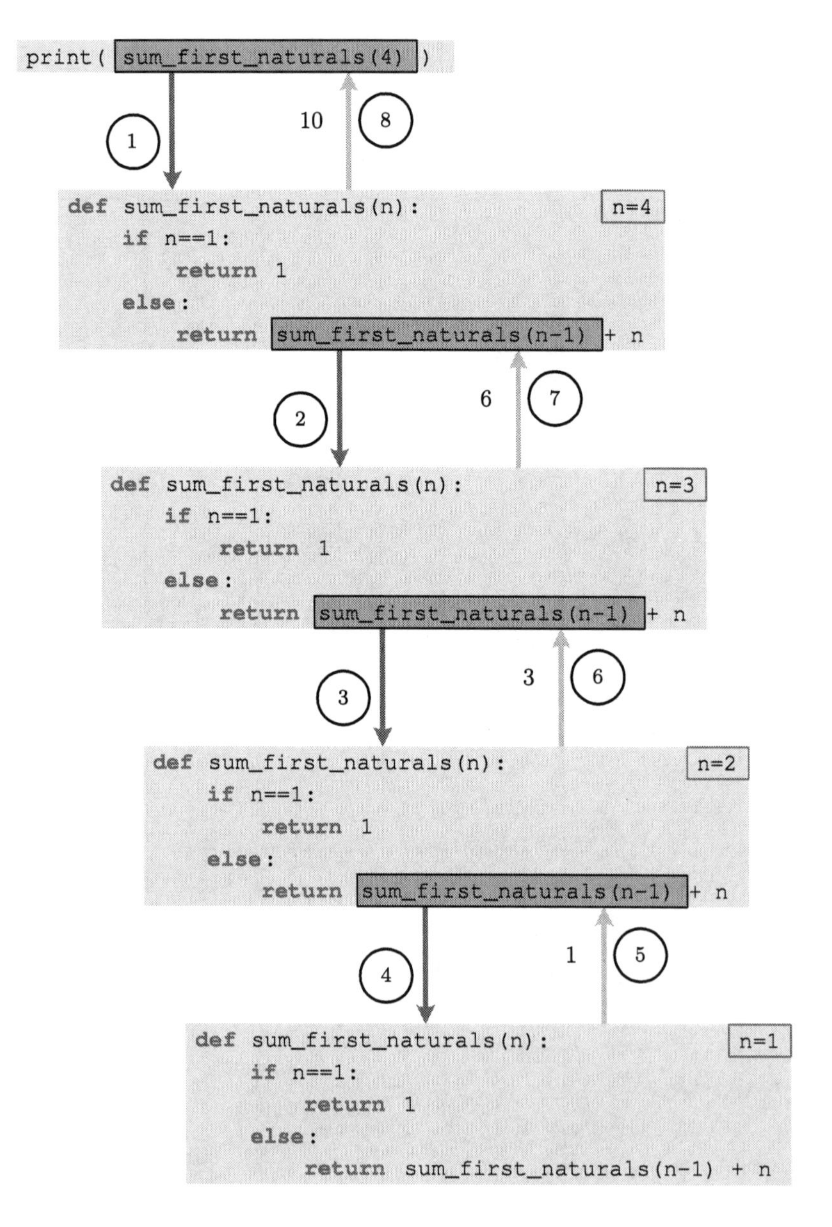

Figure 10.2 Sequence of calls and returns for `sum_first_naturals(4)`.

Listing 10.2 Similar recursive methods. What do they do?

```
def mystery_method_1(s):
    if s:
        print(s[0], end='')
        mystery_method_1(s[1:])


def mystery_method_2(s):
    if s:
        mystery_method_2(s[1:])
        print(s[0], end='')


s = 'Word'
mystery_method_1(s)
print()
mystery_method_2(s)
```

sum_first_naturals in Listing 1.1, for an initial input $n = 4$. When reaching a recursive case, the method calls itself (decreasing the input argument a unit), and jumps to the first instruction of the function. It is crucial to understand that the process does not terminate when reaching a base case ($n = 1$). The method naturally calls itself several times until reaching it, but the program must continue until every call has returned. Failing to acknowledge this last sequence of operations (also known as the "passive flow") is one of the major misconceptions regarding recursion. Lastly, the entire sequence of steps taken by the algorithm can be examined by running the code line by line, and stepping into the methods with a debugger. These tools are available in most integrated development environments.

The code in Listing 10.2 includes two procedures whose input parameter consists of a string, and simply print characters on the output console (they do not return values). They are very similar since the only difference between them is the order of the two instructions in the recursive case, which is executed if the input string is not empty. The first one prints a character and then makes a recursive call, while the second one invokes itself before printing a character. The enigmatic names of the methods have been purposely chosen in order to avoid revealing what they perform. The reader is encouraged to figure this out on his or her own before continuing to read.

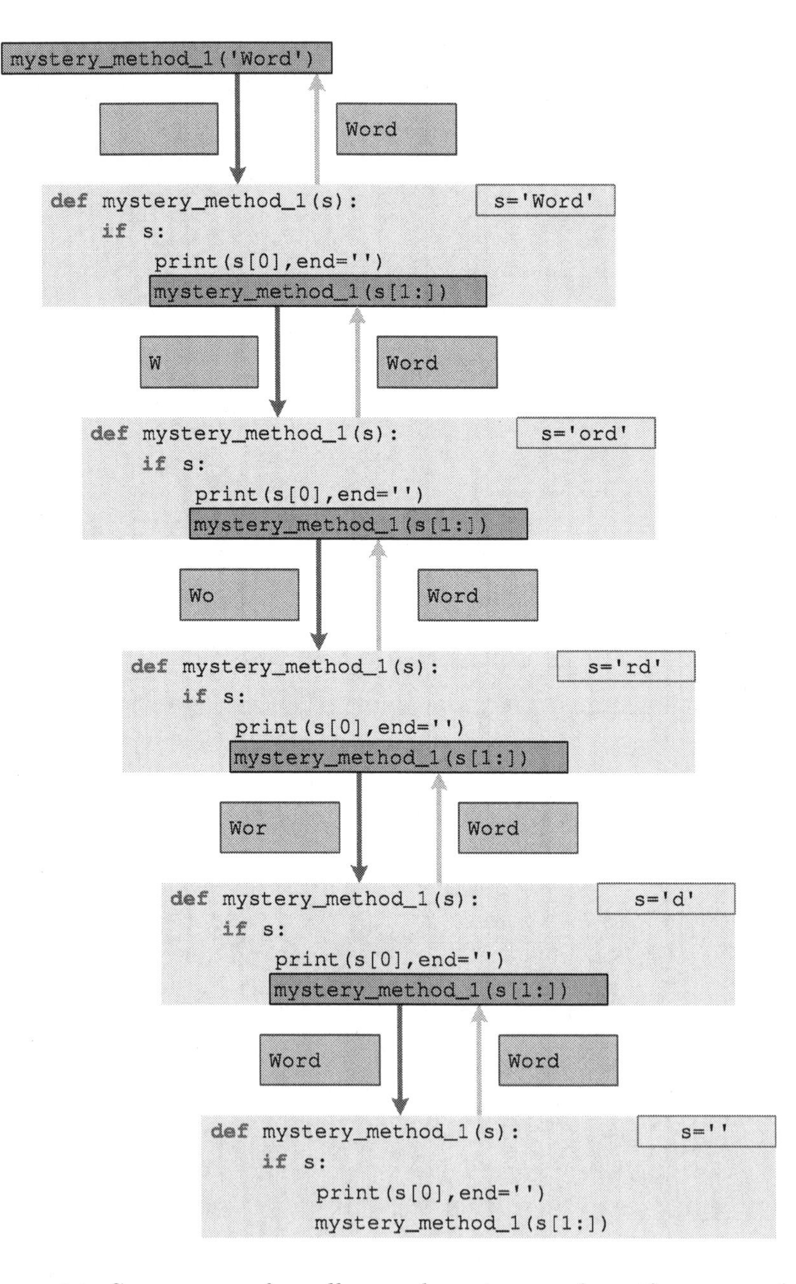

Figure 10.3 Sequence of calls and returns for the procedure `mystery_method_1('Word')`.

We will analyze the procedures by examining the sequence of operations they perform, similarly to the examples in Figures 10.1 and 10.2. Figure 10.3 illustrates the process for `mystery_method_1`, for the input string `s = 'Word'`, where the boxes next to the arrows represent the state of the output console when the recursive procedure is called, and after it terminates. The first time the method is called the console is empty. Subsequently, since `s` is not an empty string it prints its first character (`'W'`), and invokes the method with the string formed by the remaining characters (`'ord'`). These two steps are repeated until the argument to the method is an empty string. When reaching this base case the program will have printed, character by character, the initial input string `'Word'` on the console. Afterwards, the method terminates and returns the control to the previous call, which also terminates. This is repeated until every call has finished. The procedure therefore simply prints a string on the console. Lastly, observe that the method is tail-recursive, since the recursive call is the last operation in the recursive case. Thus, the algorithm will have performed its task when completing the base case. Note that the console is not modified as the recursive calls terminate.

Alternatively, method `mystery_method_2` prints the input string on the console, but in reversed order. Figure 10.4 shows the sequence of calls for the method. In this case, the method is linear-recursive, which implies that it performs an action (printing the first character of the input string) after completing a full recursive call. When the algorithm reaches the base case it will not have printed anything, since none of the `print` commands will have been executed. The base case does not perform any action either. However, as the methods terminate, the control jumps to the first instruction after the recursive call, which is the `print` statement. This ultimately prints the characters of the initial input string, character by character, in reverse order.

10.2 RECURSION TREES

A simpler way to visualize the sequence of recursive calls consists of illustrating a **recursion tree**. Its nodes correspond to method calls with specific arguments. If a particular method invokes other subroutines, its associated node will appear as the parent of the nodes related to the subroutine calls. Thus, the leaves are associated with base cases, while the internal nodes are related to recursive cases. In addition, the children nodes appear from left to right according to the order in which they are

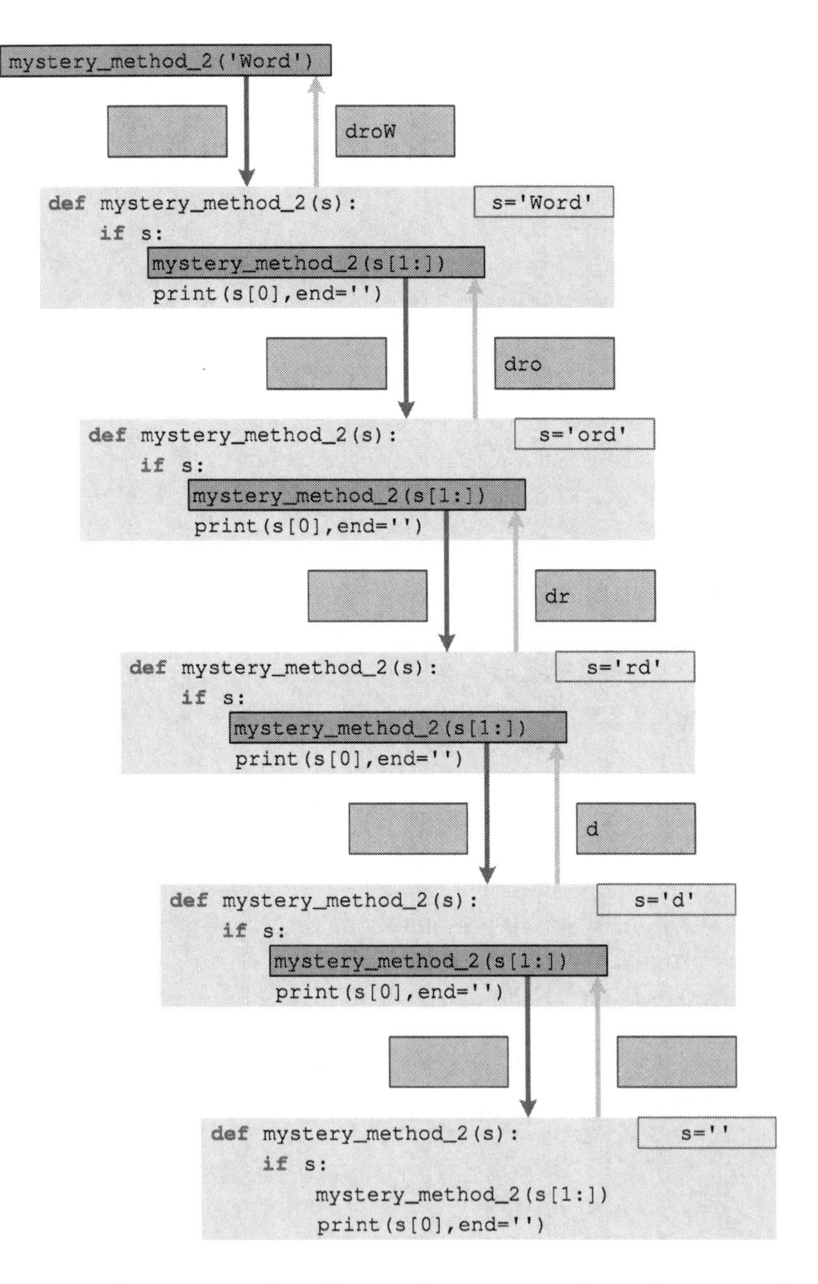

Figure 10.4 Sequence of calls and returns for the procedure `mystery_method_2('Word')`.

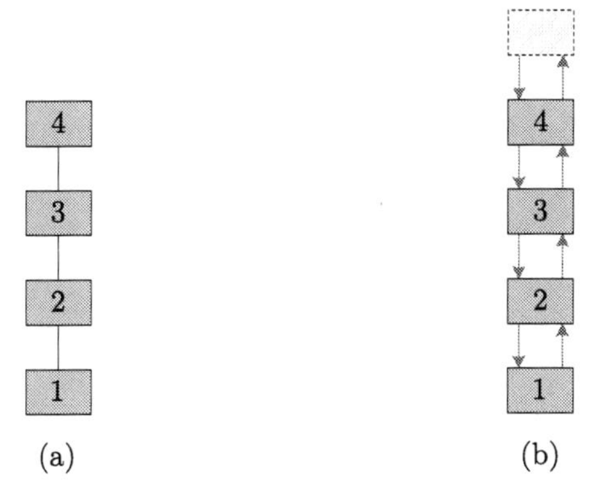

Figure 10.5 Recursion trees for `sum_first_naturals(4)`.

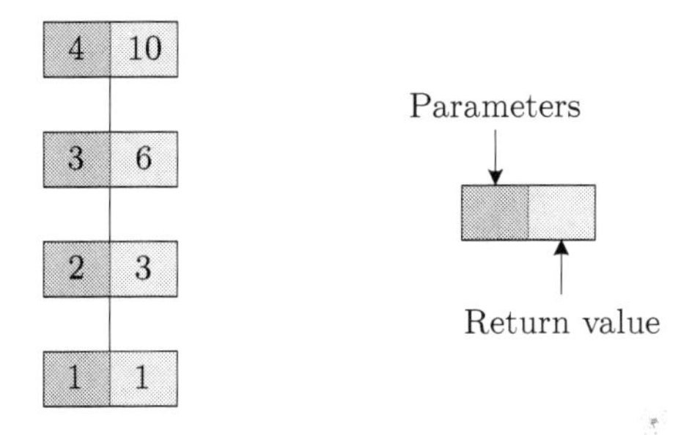

Figure 10.6 Activation tree for `sum_first_naturals(4)`.

called. Technically, a preorder traversal of a recursion tree indicates the order in which the subroutines are called, while a postorder traversal specifies the order in which the program finishes processing a complete recursive call (see Section 7.3).

Typically, the parameters that characterize a call are displayed next to the corresponding node. Figure 10.5(a) shows the recursion tree related to a call to `sum_first_naturals(4)`, where the number inside each node indicates the value of n associated with a call. Although the tree is undirected, it is important to remember that computing `sum_first_naturals(4)` implies both descending through the tree (recursive calls) towards the base case, and then ascending towards the root (method returns). Thus, an alternative representation of a recursion tree can use arrows to represent calls and returns, as shown in (b). Lastly, the graphic also shows that `sum_first_naturals(4)` returns its value to the method that called it (or to an interpreter's console).

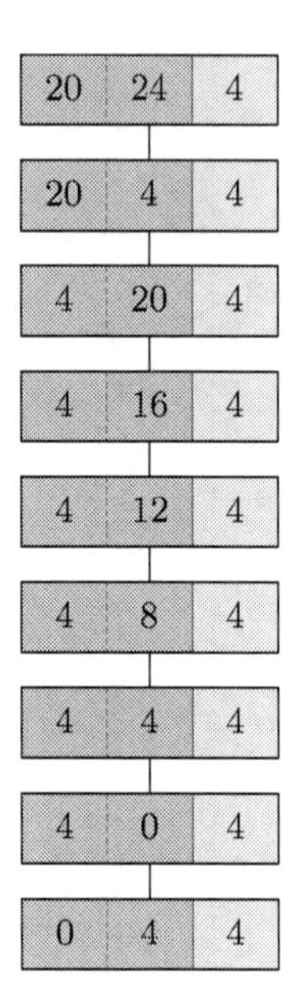

Figure 10.7 Activation tree for `gcd1(20,24)`.

Recursion trees can be extended in order to display the value of a function call, once it has been completed. In particular, an **activation tree** shows these values inside the nodes, next to the parameters of the method, as shown in Figure 10.6 for a call to `sum_first_naturals(4)`. Activation trees can be useful for evaluating and debugging the methods, since they help to keep track of the returned values.

Both linear-recursive and tail-recursive methods generate recursion trees that exhibit a linear structure, since they only invoke themselves once in the recursive case. However, while a linear-recursive function returns different values at each call, tail-recursive function calls always return the same value. Consider the function `gcd1` in Listing 5.18 that describes Euclid's algorithm. The activation tree for `gcd1(20,24)` is shown in Figure 10.7 (see (5.6)), where it is apparent that the method returns the value obtained in the base case. However, the procedure does not finish at the base case. Instead, every call must terminate returning the same value (4). Similarly, tail-recursive procedures complete their task when reaching the base case, as illustrated in the example involving method `mystery_method_1` (see Figure 10.3).

In addition, consider the `fibonacci` function in Listing 1.3. Its recursion tree, for computing the 6-th Fibonacci number, is shown in Fig-

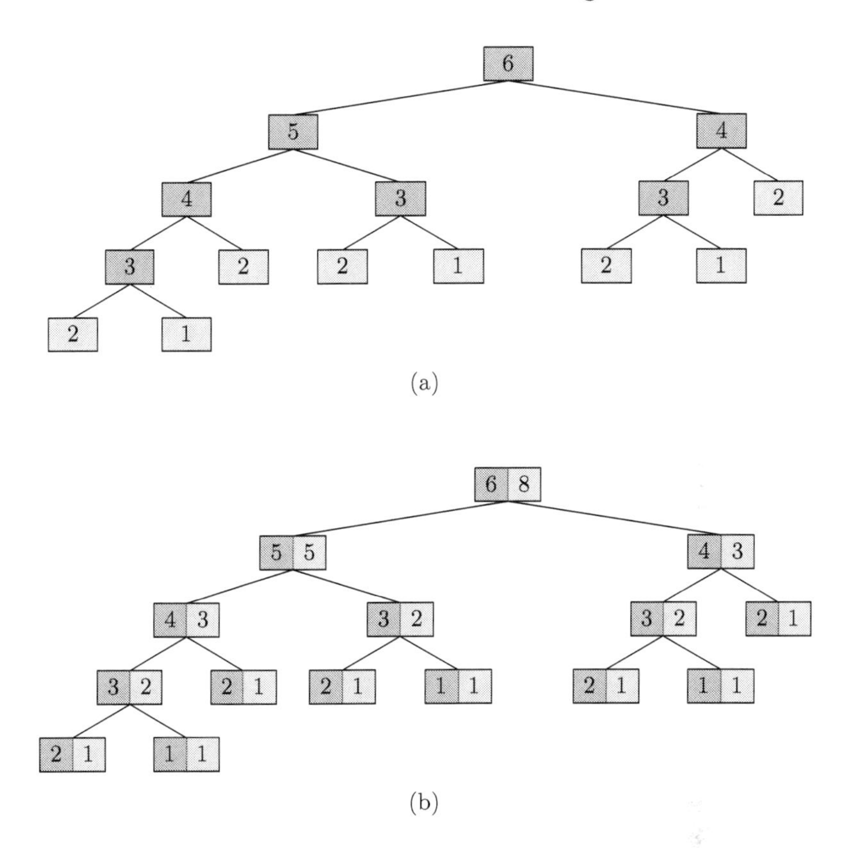

Figure 10.8 Recursion (a) and activation (b) trees for `fibonacci(6)`.

ure 10.8(a). It is more complex since the function is based on multiple recursion. In particular, it is a binary tree, since the recursive case involves two recursive calls. Lastly, Figure 10.8(b) illustrates the corresponding activation tree. The result of the method is the number of leaves in these trees, since the definition states that the method returns 1 at a base case, and the sum of the two calls in the recursive case. Thus, the algorithm simply counts the number of times it reaches a base case.

Finally, if the method is a procedure that does not return a value, the recursion tree can be extended by depicting additional information regarding its action next to arrows that indicate the sequence of calls. For example, Figures 10.3 and 10.4 are analogous to recursion trees (where each node also includes the source code), but also print the state of the console before and after the calls (see Exercise 10.2).

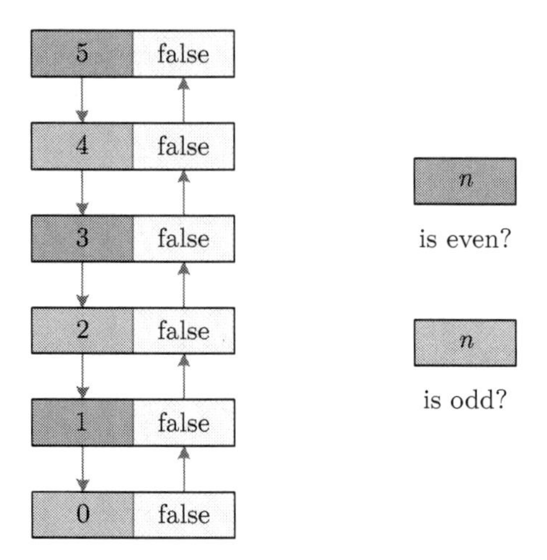

Figure 10.9 Activation tree for the mutually recursive functions in Listing 9.1.

It is possible to generate recursion trees when working with other types of recursion. Firstly, mutually recursive methods can also produce recursion trees with a linear structure. The methods in Listing 9.1, which determine the parity of a nonnegative integer, provide a simple example. Figure 10.9 shows an activation tree for `is_even(5)`, whose result is false. Notice that the methods are tail-recursive. Thus, the Boolean value found at the base case is simply propagated upwards as the functions return.

Naturally, the recursion trees can be more complex if the algorithm also incorporates multiple recursion. Figure 10.10 shows an example where the recursion tree is associated with the mutually recursive functions in (1.17) and (1.18), and where the root node corresponds to $A(6)$.

Building a recursion tree for nested recursive functions is considerably more complex, since it is necessary to evaluate all of the recursive calls in order to specify the arguments of the function. Thus, for nested-recursive functions, activation trees are more appropriate than the simpler recursion trees. Figure 10.11(a) illustrates the activation tree for the function in (1.19), associated with a call to $f(6,0)$. The particular rules related to the nested-recursive function are shown in (b). The left subtree corresponds to a call to $f(4,0)$, which returns some value x. The right subtree, associated with a call to $f(5,0+x)$, cannot be expanded until we obtain a concrete value of x, which requires evaluating $f(4,0)$

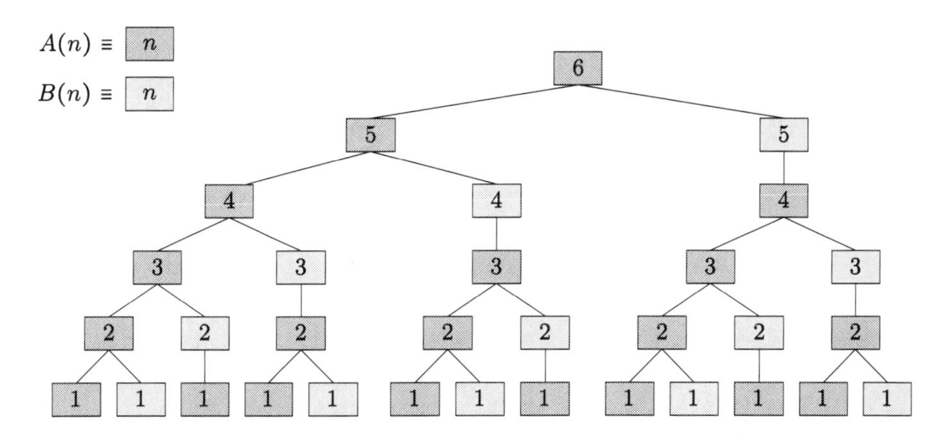

Figure 10.10 Recursion tree for the mutually recursive functions in (1.17) and (1.18).

(generating the entire left subtree). In this case, $f(4,0) = x = 3$, and therefore the root node of the right subtree is related to a call to $f(5,3)$.

10.2.1 Runtime analysis

The number of nodes (i.e., of recursive calls) of a recursion tree also provides insight regarding the computational time complexity of a recursive algorithm. If the tree is linear, then clearly the height of the tree contributes a linear factor to the runtime of the method (the cost of a node must be multiplied by the tree's height). It is also possible to determine the number nodes when the tree is not linear. For example, the following function defines the number of recursive calls performed by the `fibonacci` method:

$$Q(n) = \begin{cases} 1 & \text{if } n = 1, \\ 1 & \text{if } n = 2, \\ 1 + Q(n-1) + Q(n-2) & \text{if } n \geq 3. \end{cases}$$

It is similar to the Fibonacci function. However, besides counting the number of leaves of the recursion tree, it adds a unit in its recursive case in order to also count the number of internal nodes associated with the recursive case. In this case, it can be shown that $Q(n) = 2F(n) - 1$, where $F(n)$ is the n-th Fibonacci number. Due to (3.37), the number of calls is (approximately) on the order of 1.618^n. In this regard, observe that the number of nodes of a full binary tree of height n is on the order

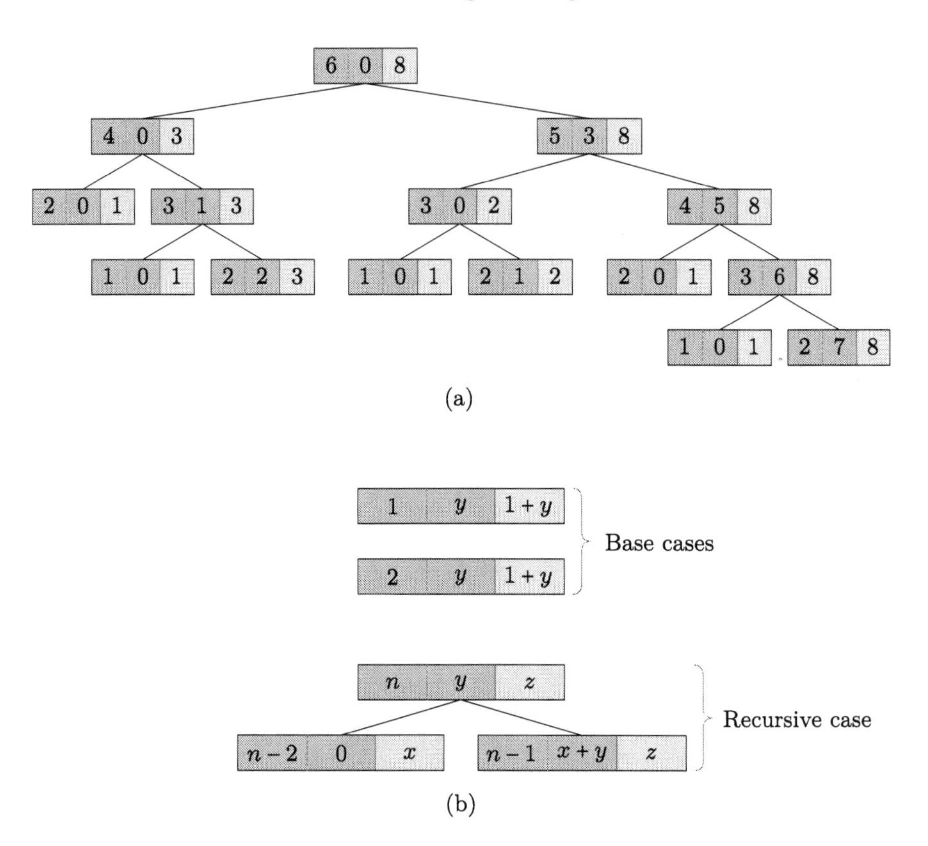

Figure 10.11 Activation tree for the nested recursive function in (1.19).

of 2^n. The Fibonacci recursion tree contains a lower number of nodes since it is a pruned binary tree. Thus, it makes sense that the base of the related exponential $(1.618\ldots)$ is smaller than 2.

In addition, recursion trees can be used to analyze the runtime of divide and conquer algorithms, through an approach denoted as the "tree method," which is related to the expansion method and the master theorem. The idea consists of forming an alternative recursion tree whose nodes contain the cost of the operations carried out within a call to a recursive method, but ignoring the ones associated with future calls. Figure 10.12 shows this tree for the recurrence $T(n) = 2T(n/2) + n^2$. A first call to a related method requires n^2 operations, plus the cost $2T(n/2)$ associated with future calls. Thus, the first node simply contains the term n^2. The nodes in the next level of the tree contain the number of operations related to $T(n/2)$, which is $(n/2)^2 = n^2/4$. Since there are

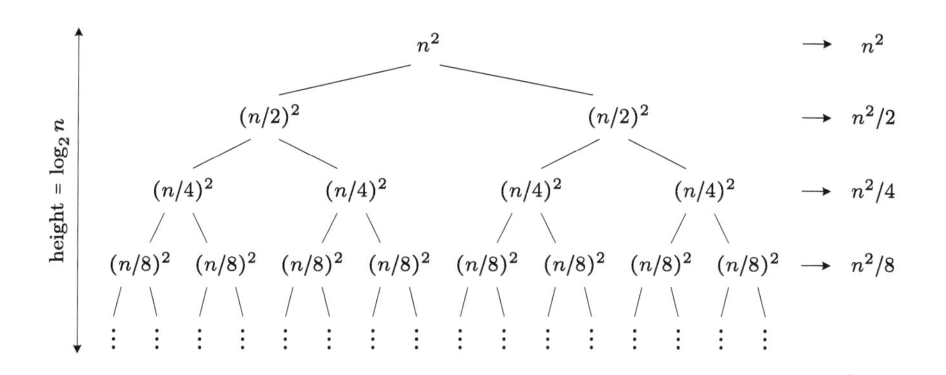

Figure 10.12 Recursion tree associated with the recurrence $T(n) = 2T(n/2) + n^2$.

two nodes, the total cost for that level is $n^2/2$. The same reasoning is applied in the following levels, where the sum of the terms on a level leads to the expressions on the right of the figure. Their sum will end up describing the cost of the algorithm. If we assume that n is a power of two and that the base case is reached when $n = 1$, then the height of the tree will be $\log_2 n$. In this example, assuming $T(1) = 1$ (the last level would contain n terms equal to 1), the final nonrecursive expression of the recurrence would be:

$$T(n) = n^2 \sum_{i=0}^{\log_2 n} \frac{1}{2^i} = n^2 \left(2 - \frac{1}{n}\right) = 2n^2 - n.$$

10.3 THE PROGRAM STACK

Recursion is especially beneficial when an algorithm requires managing a **stack**, which is a linear data structure that simulates a pile of objects, where elements can be added or deleted from the stack, but only in a certain order. In particular, when an element is added it is placed on the top of the stack, while the only element that can be removed is the one on the top of the stack, as illustrated in Figure 10.13(a). Thus, the data structure is also denoted as LIFO (last in, first out). The operations that introduce and extract an element from the stack, which in this case grows downwards, are called **push** and **pop**, respectively. Alternatively, a **queue** is a similar data structure that uses a FIFO (first in, first out) policy, simulating a waiting line, as shown in Figure 10.13(b). When an element is added to the queue it is placed at the last position in the

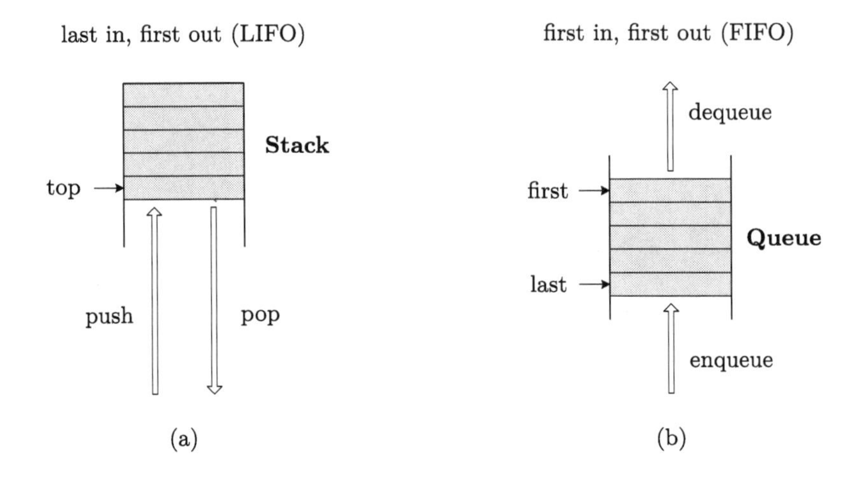

Figure 10.13 The stack and queue data structures.

queue, while only the first element can be extracted from the queue. These operations are denoted as **enqueue** and **dequeue**, respectively.

From a low-level perspective, subroutine calls and returns are implemented by using a special stack in memory, which is referred to as the **program stack**, call stack, control stack, or simply as "the stack." Although it is managed automatically when working with high-level programming languages, understanding how it is used is important for comprehending how recursion works.

10.3.1 Stack frames

Firstly, every subroutine call creates a **stack frame**, which is a block of data that is placed on top of the program stack. In particular, it contains the method's parameters and local variables (or pointer references to them), together with the return address, and other low-level information. The return address indicates the instruction to which a method should return the control when it finishes executing. Lastly, when the method terminates, its stack frame is removed from the program stack, and the freed memory can be used for other method calls. Thus, at any moment, the program stack only contains information relative to the "active" subroutines, which are those that have been called but have not yet returned (stack frames are also called active frames, or activation records).

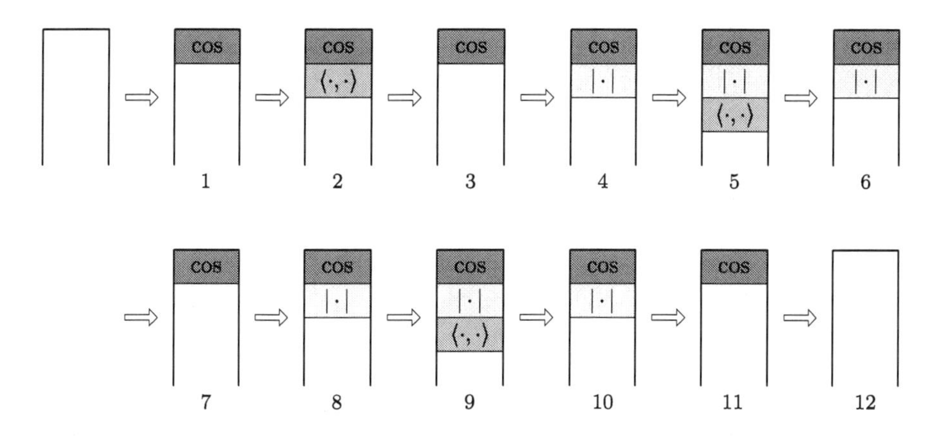

Figure 10.14 Evolution of stack frames when running the code in Listing 10.1, where cos, $|\cdot|$, and $\langle\cdot,\cdot\rangle$ represent the methods `cosine`, `norm_Euclidean`, and `dot_product`, respectively.

Figure 10.14 shows the states of the program stack when running the code in Listing 10.1, and essentially represents the sequence of method calls and returns in Figure 10.1. After initializing the lists `a` and `b` in lines 16 and 17, the call to `cosine` generates a first stack frame (represented by a shaded rectangle) that is pushed onto the program stack, which stores references to the two input lists (step 1). The control then jumps to the called method, where the first operation consists of computing the dot product of its input arguments (`dot_product(u,v)`) in line 13. Thus, the program calls this new method, which pushes another stack frame onto the stack (step 2), allocating memory for storing two references to the input lists, and the local variables `s` and `i`. When the method returns, the entire stack frame is deleted from the stack (step 3). The next step consists of evaluating `norm_Euclidean(u)`, also in line 13. The call to this function pushes another stack frame onto the program stack (step 4). The method computes the Euclidean norm of its input list (which in this case is `a`) by calling `dot_product(v,v)`, which generates a new stack frame (step 5). Figure 10.15 shows the parameters and local variables stored in the program stack when computing this dot product, together with the program data associated with the two input lists in the example code. The next two stages delete stack frames since the program returns from functions `dot_product` and `norm_Euclidean`. Afterwards, the algorithm computes the Euclidean norm of the second input list (`b`).

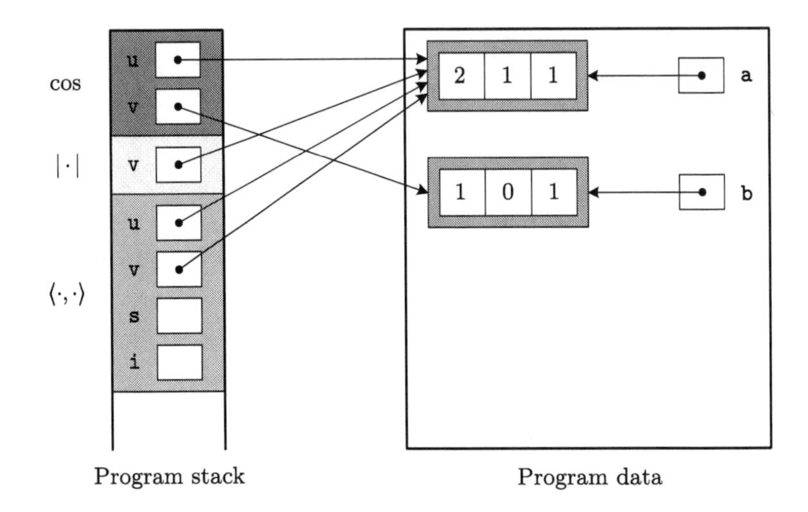

Figure 10.15 Program stack and data at step 5 in Figure 10.14.

Thus, steps 8–11, are analogous to stages 4–7. Finally, function `cosine` finishes by returning its result, which empties the program stack.

The program stack is an appropriate data structure for implementing sequences of nested subroutine calls, and recursion in particular. Consider the recursive calls to the function `sum_first_naturals` in Figure 10.2. Although the graphic may seem to give the impression that there exist multiple copies of the code, there is only one in memory. The mechanism for differentiating each call, and keeping track of the sequence of calls, is precisely the program stack. In particular, each new stack frame, associated with an additional call, holds the value of the input parameter n for the specific call. Thus, when running the code, the concrete value of n will be the one stored on the frame located at the top of the program stack. Figure 10.16 shows the state of the program stack when running `sum_first_naturals(4)`, where each of the stack frames is obviously associated with the method. The evolution of the stack frames provides the sequence of calls and returns, similarly to a recursion tree (see Figure 10.5). Finally, the darker rectangle represents a set of stack frames related to methods that may have been called prior to the call to `sum_first_naturals(4)`.

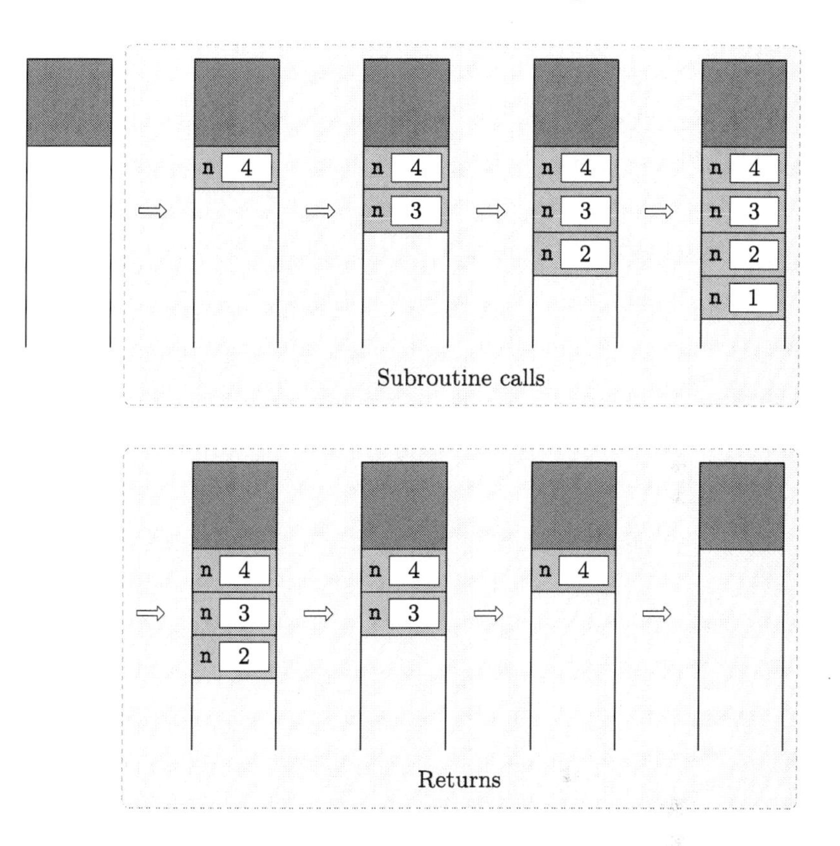

Figure 10.16 Evolution of stack frames for `sum_first_naturals(4)`.

10.3.2 Stack traces

Another important concept useful for debugging is a **stack trace**, which is a description of the program stack at a particular moment during the execution of a program. Most programming languages provide functionality for displaying a stack trace. For instance, stack tracing can be carried out in Python through the traceback module. In addition, stack traces are often printed when a program generates a runtime error. Consider replacing the addition operator in line 6 of Listing 10.1 by a subtraction: `s = s - u[i]*v[i]`, which is obviously incorrect (the other functions would also be erroneous since they rely on `dot_product`). For instance, trying to compute the Euclidean norm of a (nonzero) vector leads to a runtime error, since the program would try to compute the square root of a negative number. When running the code the standard Python interpreter would print a message similar to:

```
Traceback (most recent call last):
  File "traceback_example.py", line 18, in <module>
    print(cosine(a,b))
  File "traceback_example.py", line 13, in cosine
    return dot_product(u,v)/norm_Euclidean(u)/norm_Euclidean(v)
  File "traceback_example.py", line 10, in norm_Euclidean
    return math.sqrt(dot_product(v,v))
ValueError: math domain error
```

It indicates that there has been a math domain (runtime) error in line 10 within the `norm_Euclidean` method, which was called in line 13 within the `cosine` function. The message also indicates the line (18) where the method `cosine` is called (in some programming languages the program stack can hold a stack frame associated with the "main" method). Lastly, note that the structure of the message is in accordance with the active frames in step 5 of Figure 10.14.

10.3.3 Computational space complexity

The memory storage requirements of a recursive algorithm depend on the largest number, say h, of stack frames placed on the program stack at any given moment when running it. This quantity can also be understood as the height of the recursion tree. Naturally, for linear or tail-recursive methods h is also the number of nodes of the recursion tree, since its structure is linear. Assuming that each call to the recursive method requires the same amount M of memory (this may not always be the case), the total storage space needed by the algorithm would be Mh, since at some point there would be h stack frames related to the method on the program stack. For example, each stack frame related to the function `sum_first_naturals` requires a constant amount of memory c (i.e., $\Theta(1)$ storage space), since it only needs to store the input argument and a constant amount of low-level information. In addition, since the height of the recursion tree is n, the total amount of memory needed is cn, and we can conclude that the algorithm requires $\Theta(n)$ storage space.

For other algorithms whose recursion tree does not have a linear structure it is important to understand that the memory required by a recursive algorithm is the height of the recursion tree, and not the number of nodes that it contains. Consider the recursion tree for the `fibonacci` function in Figure 10.8(a). It contains an exponential number of nodes with respect to the input parameter n, but its height is simply $n-1$, and it is this quantity that determines the method's storage needs. This

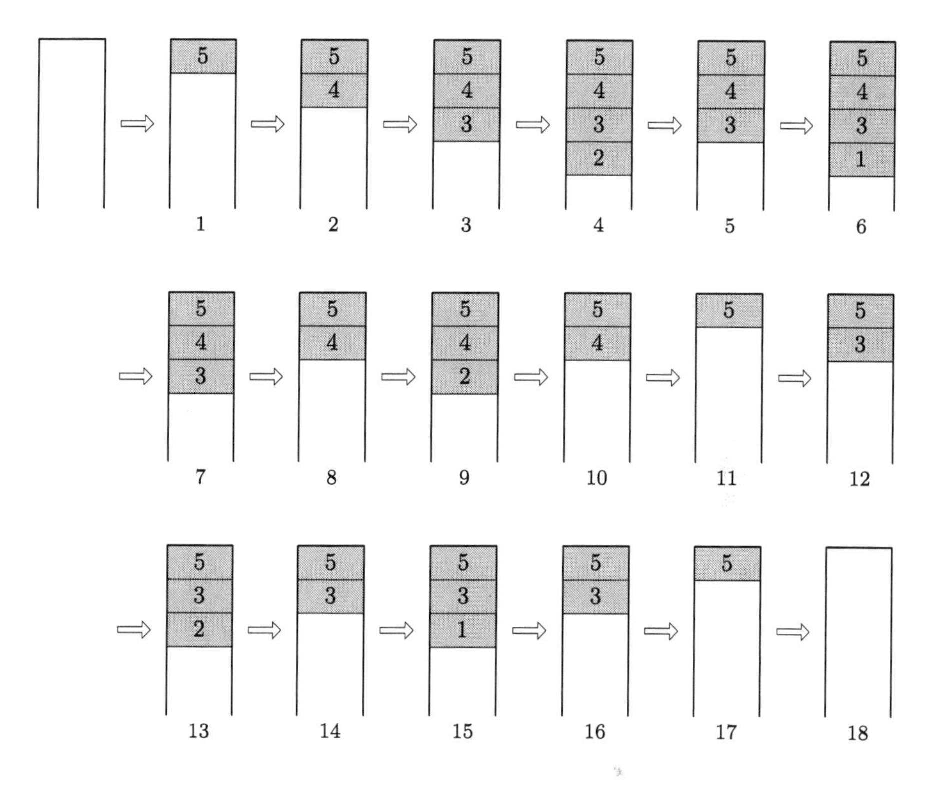

Figure 10.17 Evolution of stack frames for `fibonacci(5)`.

can be seen by analyzing the different states of the program stack when running `fibonacci(5)`, as shown in Figure 10.17. Since stack frames are deleted from the stack when the method returns, the maximum number that it contains at any moment is only 4 (in the fourth and sixth steps). Lastly, since each call only needs to store a constant amount of data, the computational space complexity for the algorithm is linear with respect to n.

Since recursion is implemented through the program stack, it always requires at least h units of memory (i.e., $\Omega(h)$). In this regard, recursive algorithms generally use more memory than iterative versions. For example, basic iterative algorithms that compute the sum of the first positive integers, or Fibonacci numbers, only use a constant amount of memory, while the recursive versions necessarily require on the order of n units of storage space. In addition, the push and pop operations on the program stack lead to an extra computational overhead. Thus, considering algorithms that share the same computational time complexity, iterative ones will generally be (slightly) more efficient than recursive ones (see Section 11.1).

10.3.4 Maximum recursion depth and stack overflow errors

When executing programs there is a limit on the amount of memory they can use, which depends on a number of factors including the programming language, machine architecture, or multi-threading. In particular, there is a limit on the amount of data or number of stack frames that we can store in the program stack. For instance, some Python implementations allow 1000 frames (i.e., nested recursive calls) to be stored in the stack by default. The concrete amount, say M, can be obtained by invoking `sys.getrecursionlimit()`. Thus, the maximum height of a recursion tree is M. A program that attempts to perform more recursive calls would crash, generating a message similar to "RecursionError: maximum recursion depth exceeded."

This type of runtime error is also called a "stack overflow." In other languages, instead of fixing the maximum number of recursions, the total memory available for the program stack and the "heap" (a zone of memory that stores dynamically allocated data) is limited. As the heap and program stack grow in size, the program can run out of available memory to use, leading to a runtime error. Thus, if the heap is large, a stack overflow could occur even with a small number of recursive calls. In this regard, programmers should try to limit the size of each stack frame, avoiding parameters passed by value or local variables that store large amounts of data (for instance, programmers should consider using dynamic allocated memory that allows them to work with pointer references, as in Figure 10.15).

The most common cause of a stack overflow error is an infinite recursion, originated by recursive code that does not terminate properly at a base case, such as method `factorial_no_base_case` in Listing 2.1. However, even for properly coded algorithms, a limit on the number of stack frames has important implications regarding the applicability of recursion, especially when the recursion tree has a linear structure. For example, most linear-recursive methods that use lists will not work if their size is greater than the maximum recursion depth (M). Observe that when a decomposition reduces the size of a problem n by a unit, it will usually generate n recursive calls until reaching the base case (assuming it occurs when $n = 1$ or $n = 0$). In these cases, if $n > M$, the program will crash due to a stack overflow runtime error. This is one

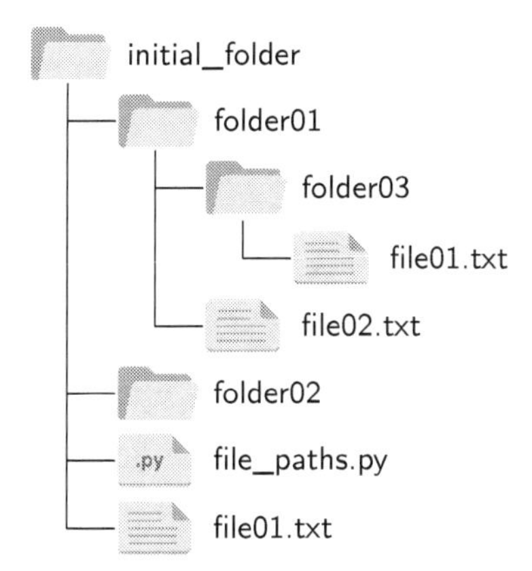

Figure 10.18 File system tree example.

of the main issues of using linear or tail-recursive algorithms. In these cases, if the size of the problem is large, iterative methods should be considered instead.

10.3.5 Recursion as an alternative to a stack data structure

Recursion is a clear alternative to iteration when it is the programmer's responsibility to explicitly manage a stack (or similar) data structure. Consider the problem of finding a file in a file system. In particular, given the name of a file f, and an initial folder F, the problem consists of printing the path of every file contained in F, or any of its nested folders, whose name is f. For example, consider the files and folders in Figure 10.18. Assuming file file_paths.py contains the method that solves the problem, when calling it with `'file01.txt'`, and the initial folder `'.'`(which represents the directory where file_paths.py lies), it should produce an output similar to:

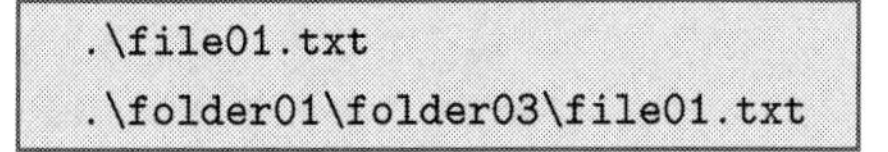

```
.\file01.txt
.\folder01\folder03\file01.txt
```

Since the file system has a tree structure, an iterative algorithm would need to use a stack or a queue in order to perform a depth or breadth-first search for the file f. Listing 10.3 shows a possible implementation that uses a stack, leading to a depth-first search. Firstly, it imports the module `os` in order to use the method `os.scandir`, which allows us to obtain a description of the contents of a particular folder. Specifically, it returns an iterator of `DirEntry` objects (which are yielded

Listing 10.3 Iterative algorithm for finding a file in a file system.

```
import os


def print_path_file_iterative(file_name, folder):
    stack = []
    for entry in os.scandir(folder):
        if entry.is_file() and file_name == entry.name:
            print(entry.path)
        elif entry.is_dir():
            stack.append(entry)

    while len(stack) > 0:
        entry = stack.pop()
        for entry in os.scandir(entry.path):
            if entry.is_file() and file_name == entry.name:
                print(entry.path)
            elif entry.is_dir():
                stack.append(entry)


print_path_file_iterative('file01.txt', '.')
```

in arbitrary order) in a given folder. We can use it within a loop in order to recover the names of the files and folders in the indicated path (through the attribute `name`), together with other information, such as whether an entry is a file, by using the `is_file` method. Finally, we will use the attribute `path` to print the location of the searched file in the file system.

The defined method first declares an empty list that will represent the stack data structure, and which will contain folder entries. Subsequently, it uses a loop in order to analyze all of the files and folders in the directory indicated by the `folder` parameter. For each entry provided by `os.scandir`, the algorithm checks if it corresponds to a file, and prints its path only if its name is the same as the argument `file_name`. Otherwise, the procedure determines whether the entry is a folder by calling the `is_dir` method. If the result is true, it pushes the entry onto the stack (through the `append` method). Afterwards, the algorithm performs the same operations on the folder entries that are contained in the stack. This process is repeated while the stack is not empty.

Figure 10.19 shows several states of the stack when running the code in Listing 10.3, for the files and folders in Figure 10.18. Initially the stack

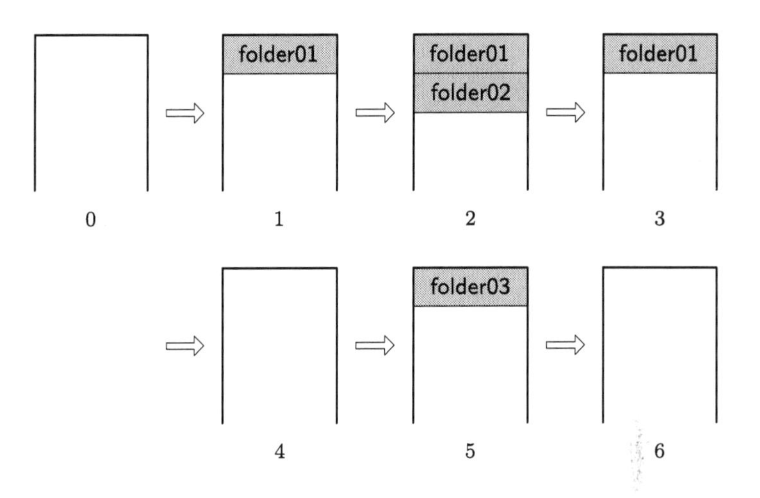

Figure 10.19 State of a stack data structure when running the iterative code in Listing 10.3, for the files and folders in Figure 10.18.

is empty (step 0). Afterwards, the first loop of the algorithm pushes the two directories contained immediately under the initial folder (steps 1 and 2), and prints the path `.\file01.txt`. Subsequently, the method executes the second loop (lines 11–17). Firstly, it pops the directory folder02 (step 3), and analyzes it. Since it is empty the method does not print new paths or insert new subfolders in the stack. In the second iteration of the while loop the procedure pops folder01 (step 4). In this case the folder contains a file, but it is not file01.txt. In addition, it encounters one subfolder, which it pushes onto the top of the stack (step 5). In the next iteration of the loop the method pops folder03 (step 6), finds a file named file01.txt, and prints its complete path. In addition, it does not introduce any more entries in the stack, since folder03 does not contain subdirectories. Thus, the stack ends up being empty, and the method terminates. Lastly, the reader can test the algorithm on larger file systems in order to verify that indeed the method performs a depth-first search. In this brief example, the algorithm begins by searching for the file in folder02. If it were to have any subfolders the method would have continued analyzing them before searching for the file in folder01.

Alternatively, Listing 10.4 shows a simpler recursive solution that captures a natural way to reason about the solution to the problem. In particular, the idea consists of determining whether the file is located in a given folder (without considering nested subfolders), and then checking if it appears in its subfolders (considering nested subdirectories). Observe

Listing 10.4 Recursive algorithm for finding a file in a file system.

```
import os


def print_path_file_recursive(file_name, folder):
    for entry in os.scandir(folder):
        if entry.is_file() and file_name == entry.name:
            print(entry.path)
        elif entry.is_dir():
            print_path_file_recursive(file_name, entry.path)


print_path_file_recursive('file01.txt', '.')
```

that a search in these subdirectories is a smaller instance of the original problem. The method also uses a loop to analyze every entry in a given folder (therefore, it uses multiple recursion). However, when it encounters a folder, it generates a recursive call in order to search for the file in that folder, or any of its nested subfolders.

Figure 10.20 shows the state of the program stack when executing Listing 10.4, for the files and folders in Figure 10.18. The label within a stack frame corresponds to the folder argument (the name of the file is the same for every method call), where the first call to the method uses the initial directory. As soon as the algorithm encounters a folder F it generates a new method call, which will print every path associated with the supplied `file_name` argument in F and all of its nested subfolders. Thus, the algorithm implements a depth-first search for the file in the file system, but does not use and manage a stack explicitly. Instead, all of the operations on the program stack are transparent to the programmer.

The example shows that for certain problems a recursive solution can be more concise, simpler (in the sense that it does not use a stack), and easier to understand, since it may reflect a natural way of thinking about the solution to a problem. Moreover, in this example the iterative algorithm is not more efficient than the recursive one. Readers are encouraged to evaluate the running times of both approaches on their own computers. Lastly, recursive algorithms can be easier to maintain and debug, due to their superior readability and simplicity. For certain problems, these aspects can outweigh time and space efficiency issues.

Finally, any recursive program can be converted into an iterative one (and vice versa). A general idea consists of using a stack data structure

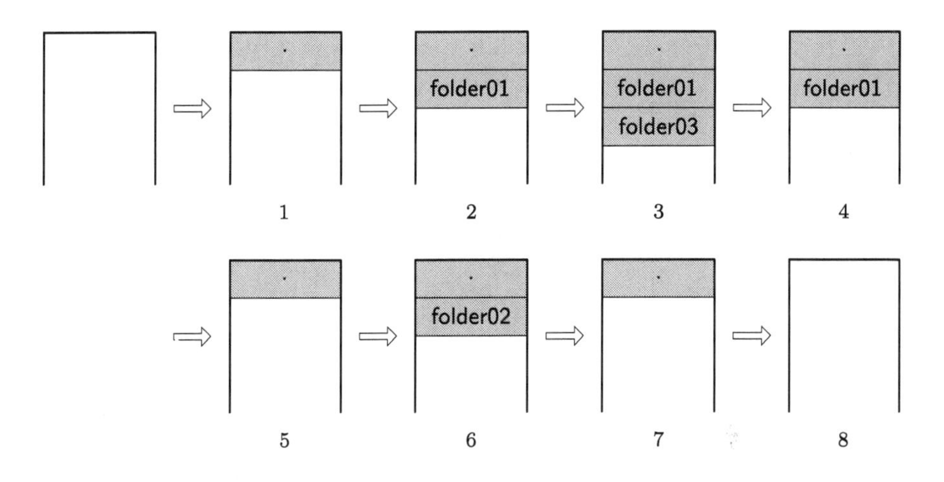

Figure 10.20 Evolution of stack frames when running the code in Listing 10.4, for the files and folders in Figure 10.18.

in order to simulate the operations carried out in the program stack. For example, it would contain the parameters passed to the recursive functions. There are standard methods for converting recursive code into iterative versions. However, this topic is beyond the scope of the book.

10.4 MEMOIZATION AND DYNAMIC PROGRAMMING

This section presents a technique known as **memoization** that is used for increasing the speed of some recursive algorithms substantially. The approach is related to **dynamic programming**, which is an important and advanced algorithm design technique.

10.4.1 Memoization

When using recursion, some algorithms perform the same recursive calls several times (see Section 7.4). In other words, they have to solve identical **overlapping subproblems** more than once. Memoization is an approach for avoiding recalculating the solutions to these subproblems, which allows us to speed up the algorithms considerably. The idea consists of storing the results of recursive calls in lookup tables. In particular, before performing a recursive call the method checks whether it has already calculated, and stored, the result. If it has, then the algorithm can simply return the stored result, discarding the associated recursive call. Instead, if the result has not been computed, then the algorithm proceeds as usual by invoking the recursive method.

One of the simplest examples that illustrates the approach involves the algorithm related to the `fibonacci` function where $F(n) = F(n-1)+F(n-2)$. Consider the associated recursion tree in Figure 10.21, for $F(6)$. In (a) there are two identical subtrees related to `fibonacci(4)` (i.e., the root node corresponds to $n = 4$), which correspond to two overlapping subproblems. Naturally, they perform the same calculations and return the same output. Therefore, an algorithm could store the result in a lookup table the first time `fibonacci(4)` terminates, and use it afterwards in order to avoid calling the method again, and repeating the same redundant computations. This strategy allows us to omit all of the calls related to the recursion tree's right subtree, as shown in (b). Moreover, we can continue to prune the recursion tree by noticing more overlapping subproblems. In the remaining tree there are again two identical subtrees related to $n = 3$ (observe that there are three of these subtrees in the entire recursion tree). Thus, we can store the result of `fibonacci(3)` the first time it is calculated, and use the value when the algorithm needs it in the future. Applying this idea in general leads to an algorithm whose recursion tree is shown in (d), which only contains n nodes. Thus, the final algorithm will run in linear time, at the expense of using more memory in order to store the n intermediate results associated with $F(n)$, for $n = 1,\dots,n$. The increase in speed is therefore remarkable, since the original function runs in exponential time.

Listing 10.5 shows one way to implement the approach for the Fibonacci function. Besides the parameter n, the method receives an initial list **a**, initialized with zeros. Its length is $n+1$ since indices begin at 0 in Python. The method begins by checking the base case. Since it only requires returning the constant 1, it is more efficient to simply return it than to look up its value in a table. Thus, this version of the algorithm does not store the values of the base cases in the lookup list. For the recursive case, if $a_n > 0$ then the value of $F(n)$ has been previously stored in a_n (i.e., $a_n = F(n)$). If the result is true the method can simply return a_n, without the need to perform recursive calls. Instead, if the result is false the algorithm proceeds as in function `fibonacci`, but stores the result in a_n before returning it. Lastly, the method is able to compute Fibonacci numbers in linear time, but needs an additional list of size n. However, this does not affect the computational space complexity, since

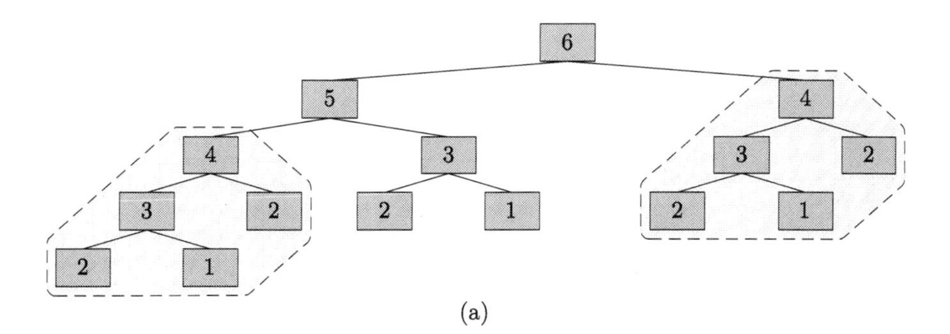

(a)

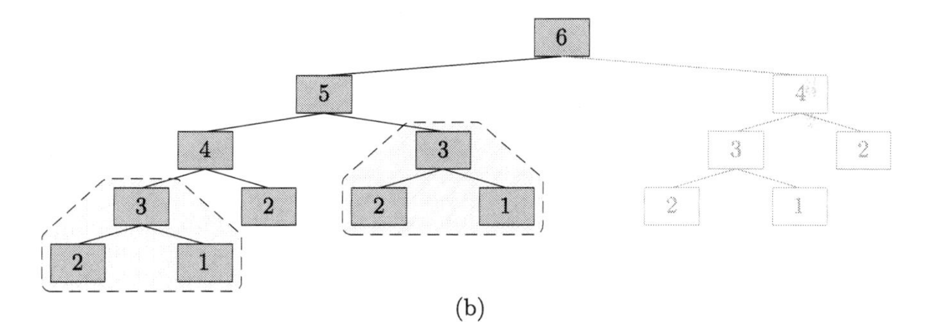

(b)

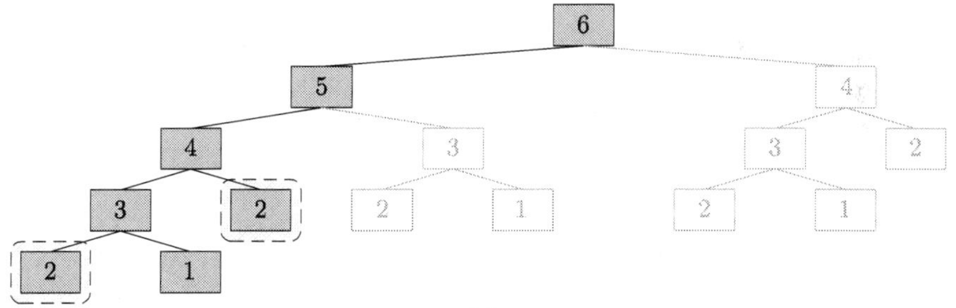

(c)

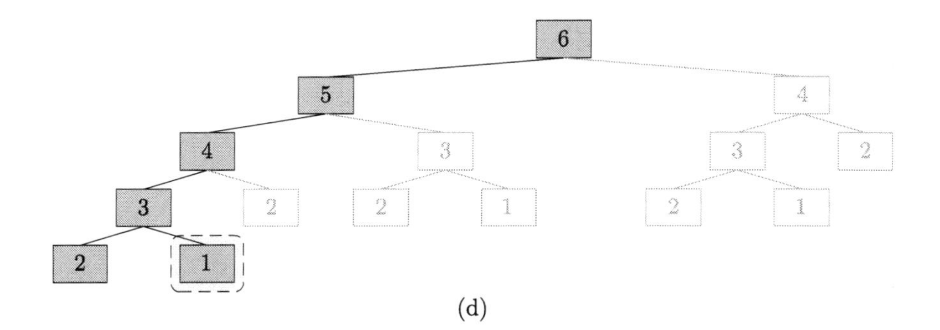

(d)

Figure 10.21 Overlapping subproblems when computing Fibonacci numbers through $F(n) = F(n-1) + F(n-2)$.

Listing 10.5 Recursive algorithm for computing Fibonacci numbers in linear time, by using memoization.

```
def fib_memoization(n, a):
    if n == 1 or n == 2:
        return 1
    elif a[n] > 0:
        return a[n]
    else:
        a[n] = fib_memoization(
            n - 1, a) + fib_memoization(n - 2, a)
        return a[n]


n = 100
print(fib_memoization(n, [0] * (n + 1)))
```

function `fibonacci` also requires on the order of n units of storage space on the program stack, as described in Section 10.3.3.

Section 7.4 described two recursive algorithms for solving the longest palindrome substring problem, whose runtime is exponential with respect to the length of the input string. We will now see a solution based on memoization that runs in quadratic time. In particular, Listing 10.6 contains a memoized version of Listing 7.7. Besides the initial string $\mathbf{s}$ of length n, the method incorporates three additional parameters. Firstly, it uses lower (i) and upper (j) indices in order to specify the substring where the function will search for a longest palindrome substring. In addition, it will use an $n \times n$-dimensional matrix $\mathbf{L}$ of strings that will contain the longest palindrome substrings found by the method (naturally, it is possible to implement a more efficient version that, instead of using that entire matrix, only stores the longest palindrome substring found by the method at any given time). In particular, the element $l_{i,j}$ will be the longest palindrome substring found within the substring $\mathbf{s}_{i..j}$, where $i \leq j$. Initially, the method is called with the string $\mathbf{s}$, matrix $\mathbf{L}$ where every entry is initialized to the empty string, and 0 and $n-1$ as the lower and upper indices, respectively.

The method starts by checking whether the problem has already been solved. In that case the element $l_{i,j}$ will contain a nonempty string, which the method can directly return (lines 2 and 3). The next condition checks $i > j$, returning an empty string if the condition is `True`. In lines 6–8 the algorithm checks if the substring determined by the indices

Listing 10.6 Memoized version of Listing 7.7.

```
def lps_memoization(s, L, i, j):
    if len(L[i][j]) > 0:
        return L[i][j]
    elif i > j:
        return ''
    elif i == j:
        L[i][j] = s[i]
        return s[i]
    else:
        if len(L[i + 1][j - 1]) > 0:
            s_aux_1 = L[i + 1][j - 1]
        else:
            s_aux_1 = lps_memoization(s, L, i + 1, j - 1)
            L[i + 1][j - 1] = s_aux_1

        if len(s_aux_1) == j - i - 1 and s[i] == s[j]:
            L[i][j] = s[i:j + 1]
            return s[i:j + 1]
        else:
            if len(L[i + 1][j]) > 0:
                s_aux_2 = L[i + 1][j]
            else:
                s_aux_2 = lps_memoization(s, L, i + 1, j)
                L[i + 1][j] = s_aux_2

            if len(L[i][j - 1]) > 0:
                s_aux_3 = L[i][j - 1]
            else:
                s_aux_3 = lps_memoization(s, L, i, j - 1)
                L[i][j - 1] = s_aux_3

            if len(s_aux_2) > len(s_aux_3):
                return s_aux_2
            else:
                return s_aux_3


s = 'bcaac'
L = [['' for i in range(len(s))] for j in range(len(s))]
print(lps_memoization(s, L, 0, len(s) - 1))
```

consists of a single character. In that case it simply stores and returns this character. The rest of the code processes the recursive cases. In lines 10–14 the method searches for the longest palindrome substring in $\mathbf{s}_{i+1..j-1}$,

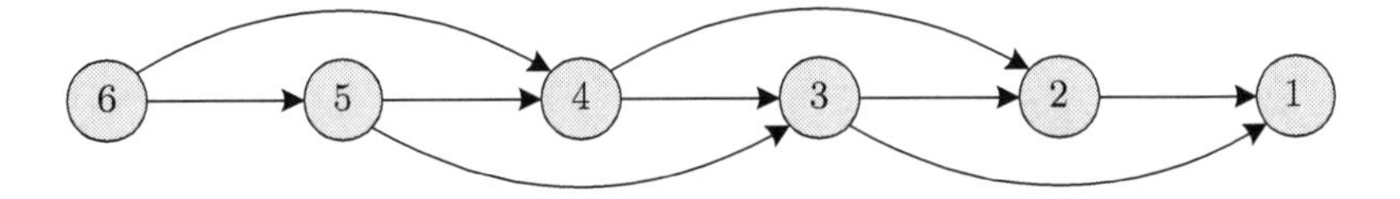

Figure 10.22 Dependency graph for $F(n) = F(n-1) + F(n-2)$.

but only invokes the recursive method if the solution has not yet been calculated and stored in **L**. Subsequently, if the substring $\mathbf{s}_{i+1..j-1}$ is a palindrome (`len(s_aux_1)==j-i-1`), then the method checks whether $s_i = s_j$ (line 16). If the result is `True`, then $\mathbf{s}_{i..j}$ will be the longest palindrome, and the method proceeds by storing and returning it (lines 17 and 18). If $\mathbf{s}_{i..j}$ is not a palindrome the method carries out the two recursive calls analogous to the ones in lines 10 and 11 of Listing 7.7, and stores their results, but only if the respective subproblems have not been solved previously (lines 20–30). Finally, the method returns the longest of the output strings associated with the two subproblems.

The algorithm has quadratic computational complexity since it essentially constructs the 2-dimensional matrix **L**, and does not solve a subproblem more than once. It is therefore much more efficient than the method in Listing 7.7, which runs in exponential time.

10.4.2 Dependency graph and dynamic programming

Dynamic programming is an algorithm design paradigm that is used for solving optimization problems that can be broken up recursively, and where the decomposition leads to overlapping subproblems. The technique can be applied in a top-down manner by using recursion together with memoization. However, a more common approach consists of using a bottom-up strategy. The idea is to implement an iterative algorithm that fills a lookup table, analogous to the one used in a memoized recursive algorithm, in a bottom-up manner, starting with the simplest instances, and progressively storing the values that will allow us to solve the original problem.

Even though the bottom-up approach is iterative, the recursive solution provides insight regarding how the lookup table should be filled. In particular, analysts can construct a **dependency graph** (also called a "subproblem graph") that indicates the dependencies between the different subproblems (i.e., method calls) needed to be solved when applying

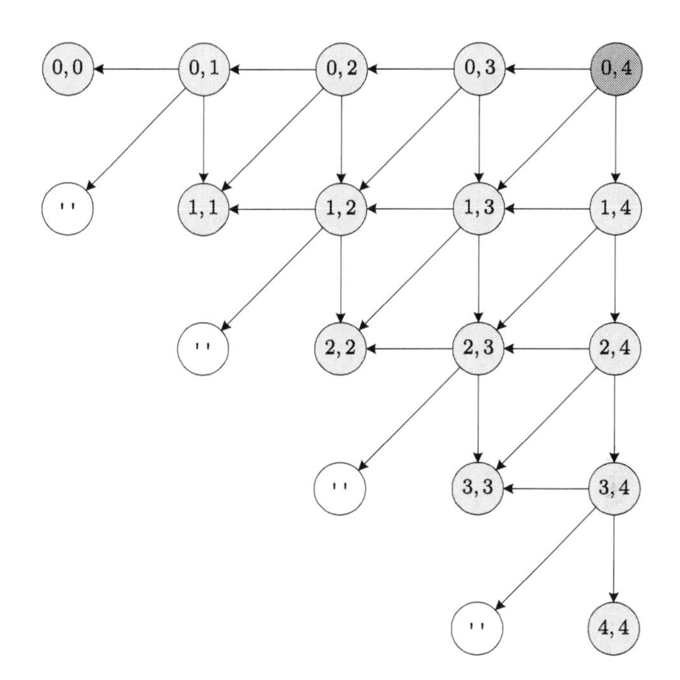

Figure 10.23 Dependency graph for Listing 10.6, which solves the longest palindrome substring problem.

the recursive algorithm. Specifically, it is a directed graph where each node represents a particular subproblem, and whose edges indicate dependencies. We will consider that if subproblem A needs the solution to subproblem B, then there will be an edge from A to B. Figure 10.22 shows the dependency graph for a recursive algorithm based on the recurrence $F(n) = F(n-1) + F(n-2)$, for $F(6)$, and base cases at $F(2)$ and $F(1)$. Note that there is an edge from node n to nodes $n-1$ and $n-2$. The dependency graph can then be analyzed in order to implement an iterative algorithm that solves the problem. For Fibonacci numbers it is clear that the algorithm can compute a result in n steps by solving and storing $F(1)$, $F(2)$, and so on, until reaching $F(n)$. Moreover, the dependency graph illustrates that it is not necessary to use a list of length n to store intermediate results, since at each step i the only information needed is $F(i-1)$ and $F(i-2)$, which can be stored in two variables.

Regarding the longest palindrome substring problem, Figure 10.23 shows the dependency graph related to Listing 10.6, for an initial input string containing five characters. The nodes only indicate the lower and

Listing 10.7 Code based on dynamic programming that computes the longest palindrome substring within a string **s**.

```
def lps_dynamic_programming(s):
    n = len(s)
    L = [['' for i in range(n)] for j in range(n)]

    for i in range(n):
        L[i][i] = s[i]

    k = 1
    while k < n:
        i = 0
        j = k
        while j < n:
            if (len(L[i + 1][j - 1]) == j - i - 1
                    and s[i] == s[j]):
                L[i][j] = s[i:j + 1]
            elif len(L[i][j - 1]) >= len(L[i + 1][j]):
                L[i][j] = L[i][j - 1]
            else:
                L[i][j] = L[i + 1][j]

            i = i + 1
            j = j + 1

        k = k + 1

    return L[0][n - 1]
```

upper indices i and j (if $i > j$ the result is an empty string, and the node is labeled as `''`). Observe that a node (i, j) only depends on the solutions associated with the pairs $(i+1, j)$, $(i, j-1)$, and $(i+1, j-1)$, which follows from the recursive decomposition of the problem. Each node (i, j) is related to a different subproblem, and the solution to the initial problem is obtained when the call related to the (darker) node $(0, n-1)$ terminates.

Considering that the results are stored in some matrix **L**, the example illustrates that $l_{i,j}$, associated with node (i,j), can be obtained after computing and storing $l_{i+1,j}$, $l_{i,j-1}$, and $l_{i+1,j-1}$. Thus, the dependency graph indicates an order or direction (towards the top-right corner) in which we can solve the subproblems and store their solutions. In particular, it is possible to develop an algorithm that begins by solving the subproblems associated with the nodes (i, i), for $i = 0, \ldots, n-1$, on the

i	0	1	2	3	4
0	'b'	'b'	'b'	'aa'	'caac'
1	''	'c'	'c'	'aa'	'caac'
2	-	''	'a'	'aa'	'aa'
3	-	-	''	'a'	'a'
4	-	-	-	''	'c'

j

Figure 10.24 Matrix $\mathbf{L}$ after running Listing 10.7 with `s = 'bcaac'`.

large diagonal. Then, it would continue with the next diagonal (nodes $(i, i+1)$, for $i = 0, \ldots, n-2$), and so on, until reaching node $(0, n-1)$. Listing 10.7 contains a possible iterative implementation. Note that it calculates the solutions to the subproblems in a bottom-up manner, starting with the smaller instances, until it is capable of returning the solution to the full initial problem. In particular, it first initializes matrix $\mathbf{L}$ with empty strings (line 3), and also stores s_i in $l_{i,i}$, for $i = 0, \ldots, n-1$ (lines 5 and 6). The two following loops progressively fill matrix $\mathbf{L}$, diagonal by diagonal, until they reach the top-right node in the dependency graph. Lines 13–15 check if the substring $\mathbf{s}_{i..j}$ is a palindrome, and store it in case that it is. Otherwise, the following four lines store in $l_{i,j}$ the longest substring that is either in $l_{i+1,j}$ or $l_{i,j-1}$. Figure 10.24 shows the state of matrix $\mathbf{L}$ when the method returns, with all of the solutions to the subproblems, for the initial input string `s = 'bcaac'`. The algorithm runs in $\Theta(n^2)$ time, since it also builds the matrix $\mathbf{L}$, and where each entry is obtained in $\Theta(1)$ time (it is possible to develop more efficient algorithms that do not use a matrix of strings).

10.5 EXERCISES

Exercise 10.1 — Guess what each of the following functions f compute by evaluating them with different (nonnegative integer) input arguments. In addition, verify that a guessed function, say g, is correct by designing a recursive algorithm for it, and by using the decomposition in the corresponding description of f. The recursive definition of g should therefore be identical to f.

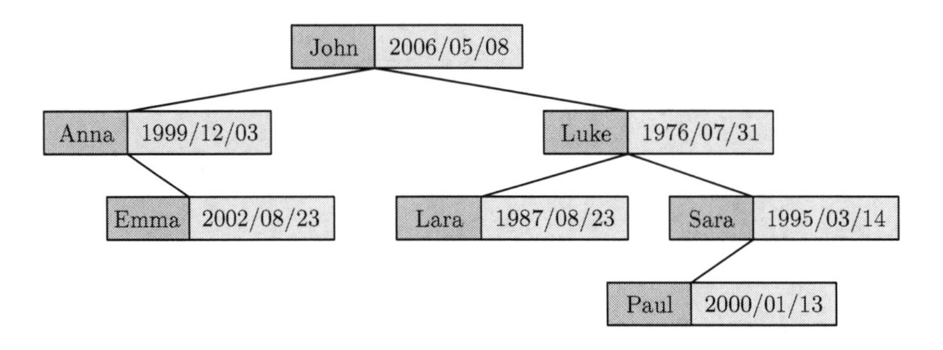

Figure 10.25 Alternative binary search tree that stores information about a birthday calendar.

a)

$$f(n) = \begin{cases} True & \text{if } n = 0, \\ \neg f(n-1) & \text{if } n > 0. \end{cases}$$

b)

$$f(n) = \begin{cases} 3 & \text{if } n = 0, \\ n \cdot f(n-1) & \text{if } n > 0. \end{cases}$$

c)

$$f(n) = \begin{cases} 0 & \text{if } n = 0, \\ f(n-1) + 2n - 1 & \text{if } n > 0. \end{cases}$$

d)

$$f(m,n) = \begin{cases} 0 & \text{if } n = 0 \text{ or } m = 0, \\ f(m-1, n-1) + m + n - 1 & \text{otherwise.} \end{cases}$$

Exercise 10.2 — Draw recursion trees as in Figure 10.5(b), for the mystery methods in Listing 10.2, when called with the string `'Word'`. In particular, indicate the state of the output console next to each of the drawn arrows.

Exercise 10.3 — Consider a list **a** of tuples of the form (key, item). Design methods that use the procedure `insert_binary_tree` in Listing 5.10 in order to store the tuples of **a** in a binary search tree, which

Listing 10.8 Methods that, supposedly, add and count the digits of a nonnegative integer. Are they correct?

```
1  def add_digits(n):
2      if n == 0:
3          return 0
4      else:
5          return add_digits(n // 10) + (n % 10)
6
7
8  def count_digits(n):
9      if n == 0:
10         return 0
11     else:
12         return count_digits(n // 10) + 1
```

will be a list of four components, as described in Section 5.3. Implement different methods that produce the binary search tree in Figure 5.2, and the one in Figure 10.25, for the input list **a** = [('John', '2006/05/08'), ('Luke', '1976/07/31'), ('Lara', '1987/08/23'), ('Sara', '1995/03/14'), ('Paul', '2000/01/13'), ('Anna', '1999/12/03'), ('Emma', '2002/08/23')].

Exercise 10.4 — Consider the code in Listing 7.3. Draw the recursion tree for `towers_of_Hanoi(3,'O','D','A')`. Specify all of the parameters inside the nodes, and indicate the disk moves.

Exercise 10.5 — Draw activation trees for computing 2^7 described in: (a) Listing 4.1, (b) Listing 4.3, and (c) Listing 4.4.

Exercise 10.6 — Consider the functions in Section 9.6.2 that define a recursive descent parser for a calculator. Draw the activation tree when evaluating the expression $(5 - 3) * 2 + (-(-7))$. Indicate the name of the methods (e.g., E, T, I, F, and P), and express the list of tokens as mathematical expressions.

Exercise 10.7 — Listing 10.8 contains a function designed to add the digits of a nonnegative integer n, and another to count its number of digits. However, one of them is not correct. Explain the error and the wrong design decision that leads to it.

Listing 10.9 Erroneous code for computing the number of times that two adjacent elements in a list are identical.

```
1 def count_consecutive_pairs(a):
2     if len(a)==2:
3         return int(a[0]==a[1])
4     else:
5         return int(a[0]==a[1]) + count_consecutive_pairs(a[1:])
```

Listing 10.10 Code for computing the smallest prime factor of a number n, which is greater than or equal to m.

```
1 # Preconditions: n>=m, n>=2, m>=2
2 def smallest_prime_factor(n, m):
3     if n % m == 0:
4         return m
5     else:
6         return smallest_prime_factor(n, m + 1)
```

Exercise 10.8 — The function in Listing 10.9 counts the number of times that two adjacent elements are identical in a list of length $n \geq 0$. However, the method contains a bug. Find it and fix the function.

Exercise 10.9 — The code in Listing 10.10 computes the smallest prime factor of a number n, which is greater than or equal to m. The preconditions are: $n \geq m$, $n \geq 2$, and $m \geq 2$. If it returns n when $m = 2$ then n is a prime number. What problem will we encounter when trying to calculate the first 200 prime numbers?

Exercise 10.10 — The code in Listing 10.11 pretends to compute the floor of a logarithm: $\lfloor \log_b x \rfloor$, where $x \geq 1$ is a real number, and $b \geq 2$

Listing 10.11 Erroneous code for computing the floor of a logarithm.

```
1 def floor_log(x, b):
2     if x == 1:
3         return 0
4     else:
5         return 1 + floor_log(x / b, b)
```

Listing 10.12 Erroneous code for determining if a list contains an element that is larger than the sum of all of the rest.

```
1 def contains_greatest_element(a):
2     if a == []:
3         return False
4     else:
5         return (2 * a[0] > sum(a)
6                 or contains_greatest_element(a[1:]))
```

is an integer. The idea consists of counting the number of times it is possible to divide x by b. However, the code is incorrect. Find the bugs and modify the code in order to fix them.

Exercise 10.11 — Consider a list **a** of n numbers. We say that it contains a "greatest" element if one of the entries of the list is greater than the sum of all of the rest. In particular, a_i would be the greatest element if:

$$a_i > \sum_{\substack{j=1 \\ j \neq i}}^{n} a_j = \sum_{j=1}^{n} a_j - a_i$$

which is equivalent to:

$$2a_i > \sum_{j=1}^{n} a_j,$$

where the right-hand term is the sum of the elements of the list. Listing 10.12 shows a possible solution, but is incorrect. Find the bug and fix the function.

Exercise 10.12 — Consider a nonempty list **a** of n numbers that appear in increasing order until a certain index i $(0 \le i \le n-1)$, and then continue in decreasing order until the end of the list. The element a_i, denoted as the "peak element," and which we can consider to be unique, will therefore be the largest one on the list. The code in Listing 10.13 tries to find this index, but contains an error. Find the bug and fix the function.

Exercise 10.13 — Consider a variant of the Koch curve that replaces a line segment by five smaller segments of a third of the length of the original one, as shown in Figure 10.26. Listing 10.14 shows a function that

Listing 10.13 Erroneous code for finding the location of the “peak element.”

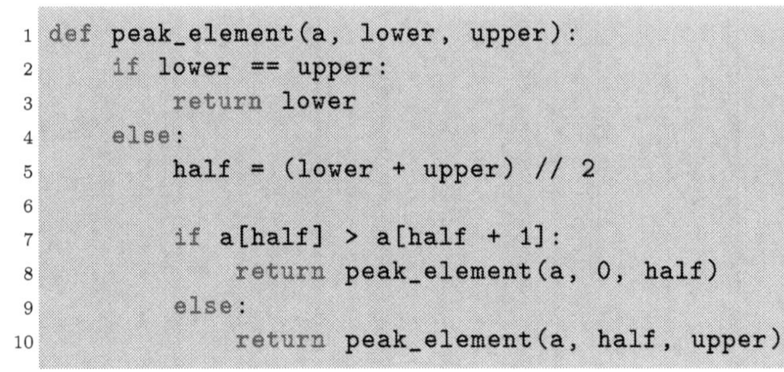

```
def peak_element(a, lower, upper):
    if lower == upper:
        return lower
    else:
        half = (lower + upper) // 2

        if a[half] > a[half + 1]:
            return peak_element(a, 0, half)
        else:
            return peak_element(a, half, upper)
```

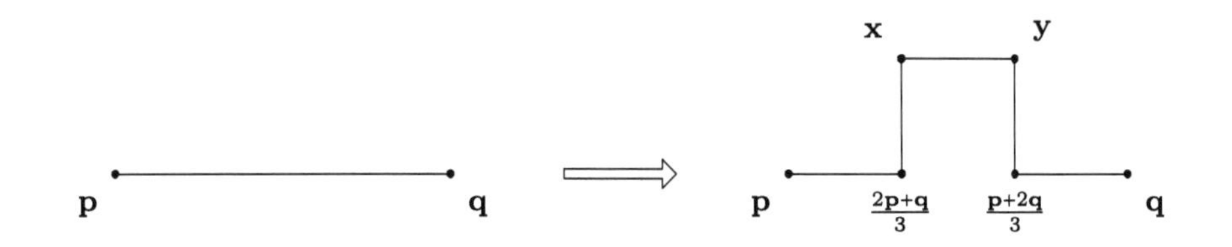

Figure 10.26 Transformation of a line segment into five smaller ones for a Koch curve variant.

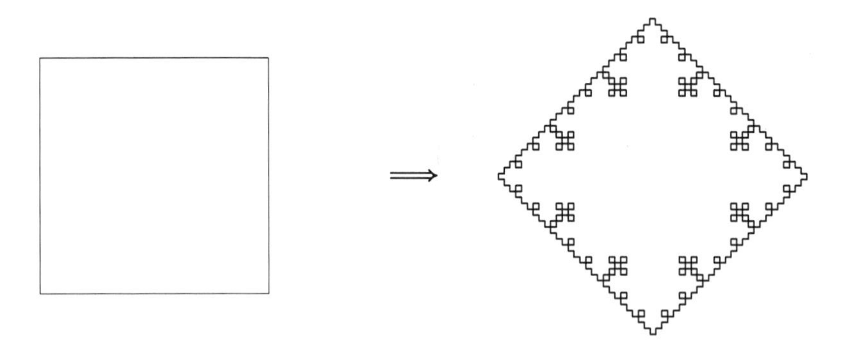

Figure 10.27 “Koch square” variant for $n = 4$.

produces a Koch fractal similar to the Koch snowflake (see Section 7.5.1), based on this transformation, and starting with four line segments forming a square. The method should produce the image in Figure 10.27, for $n = 4$, and for a square with vertices $(0,0)$, $(1,0)$, $(1,1)$, and $(0,1)$. However, it does not generate it. Find the bug and fix the code.

Listing 10.14 Code for generating a Koch fractal based on the transformation in Figure 10.26.

```
1  import numpy as np
2  import matplotlib.pyplot as plt
3
4
5  def koch_curve(p, q, n):
6      if n == 0:   # The base case is just a line segment
7          plt.plot([p[0, 0], q[0, 0]], [p[1, 0], q[1, 0]], 'k-')
8      else:
9          v = q - p
10         koch_curve(p, p + v / 3, n - 1)
11         R_90 = np.matrix([[0, -1], [1, 0]])
12         x = p + v / 3 + R_90 * v / 3
13         koch_curve(p + v / 3, x, n - 1)
14         y = x + v / 3
15         koch_curve(x, y, n - 1)
16         koch_curve(y, p + 2 * v / 3, n - 1)
17         koch_curve(p + 2 * v / 3, q, n - 1)
18
19
20 def koch_square(n):
21     p = np.array([[0], [0]])
22     q = np.array([[1], [0]])
23     r = np.array([[1], [1]])
24     s = np.array([[0], [1]])
25
26     koch_curve(p, q, n)
27     koch_curve(q, r, n)
28     koch_curve(r, s, n)
29     koch_curve(s, p, n)
30
31
32 fig = plt.figure()
33 fig.patch.set_facecolor('white')
34 koch_square(4)
35 plt.axis('equal')
36 plt.axis('off')
37 plt.show()
```

Exercise 10.14 — Implement an efficient version of the function in (3.2), which computes binomial coefficients, by using memoization. In addition, draw a dependency graph related to the function (for example, for nodes where $n \leq 4$).

Listing 10.15 Code for computing the length of the longest palindrome subsequence of a list.

```
def length_longest_palindrome_subseq(a):
    n = len(a)
    if n <= 1:
        return n
    else:
        if a[0] == a[n - 1]:
            return 2 + length_longest_palindrome_subseq(
                a[1:n - 1])
        else:
            l1 = length_longest_palindrome_subseq(a[1:n])
            l2 = length_longest_palindrome_subseq(a[0:n - 1])
            return max(l1, l2)
```

Exercise 10.15 — Consider the code in Listing 10.15, which computes the length of the longest palindrome subsequence within a list **a**. Recall that while a substring is a sequence of contiguous characters in an original string, a subsequence does not necessarily contain contiguous elements of the list. Implement an efficient version of the algorithm by using memoization, similarly to Listing 10.6.

CHAPTER 11

Tail Recursion Revisited and Nested Recursion

One should always look for a possible alternative, and provide against it. — Sherlock Holmes

— Arthur Conan Doyle

TAIL recursion is special due to its relationship with iteration. Firstly, it is fairly straightforward to transform tail-recursive algorithms into analogous iterative versions that are not only more efficient, but will not be liable to stack overflow errors. Thus, for some programmers, tail recursion is actually considered to be a bad practice. Furthermore, it is not uncommon to encounter tail-recursive algorithms that have been designed by using an imperative approach that is closer to iterative thinking than to problem decomposition and induction. In these cases iterative versions are clearly superior. This chapter will examine this connection between tail-recursion and iteration. In addition, it will provide a brief introduction to nested recursion (see (1.19)), and a strategy for deriving simple tail-recursive algorithms, often analogous to iterative methods, but by using a declarative approach.

11.1 TAIL RECURSION VS. ITERATION

Iterative programs can be converted into recursive ones, and vice versa. Moreover, the relationship between iteration and tail recursion is particularly apparent. This section analyzes this connection, and explains why iteration is usually preferred over tail recursion.

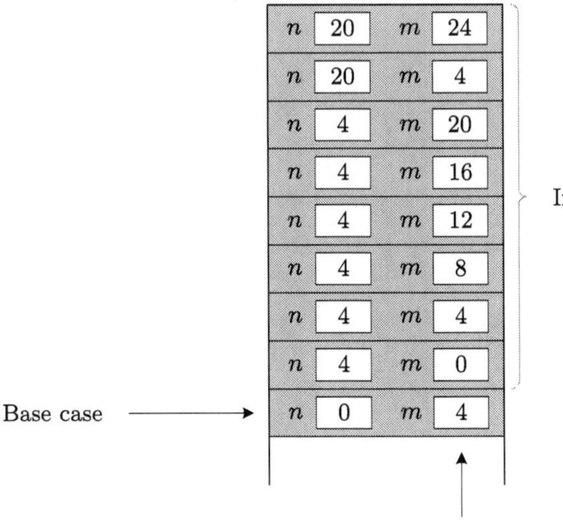

Figure 11.1 State of the program stack when running the base case relative to a call to `gcd1(20,24)`.

In general, recursive algorithms can be converted into equivalent iterative versions by using a stack data structure that plays the role of the program stack. However, for tail-recursive methods this conversion is considerably simpler since it does not require an explicit stack. The key observation is that tail-recursive algorithms do not really need to store information on the program stack, since they never use it when returning from method calls.

Again, consider the linear-recursive function `sum_first_naturals` in Listing 1.1. It requires saving the values of the parameters on the program stack (see Figure 10.16), which are later used and processed together with the results of the recursive calls in order to build the final output. For example, `sum_first_naturals(4)` recovers the argument 4 from the stack, and adds it to the result of `sum_first_naturals(3)`. In contrast, the function `gcd1` in Listing 5.18 obtains its final result after processing the base case, since it is tail-recursive. In particular, Figure 11.1 illustrates the program stack when executing the base case after a call to `gcd1(20,24)`. The graphic is analogous to the activation tree in Figure 10.7, or to (5.6), where the result of the method (4) is achieved at the base case, and directly returned by every recursive call. This implies that all of the stored data on the program stack is never

Listing 11.1 Iterative version of Euclid's method (gcd1).

```
1  def gcd1_iterative(m,n):
2      while m>0:
3          if m>n:
4              aux = n
5              n = m
6              m = aux
7          else:
8              n = n-m
9
10     return n
```

used, and is therefore unnecessary. Indeed, having obtained the result in the base case, the program basically propagates it towards the first recursive call, deleting each stack frame progressively as the recursive calls return.

This example illustrates that it is possible to design an equivalent iterative algorithm that computes the greatest common divisor, but without using a stack. In particular, it would require a single loop that would have to implement how the parameters n and m change in function gcd1 as it invokes itself. In other words, the loop would have to simulate how the parameters are updated from one stack frame to the next in Figure 11.1. Therefore, each iteration of the loop would essentially perform the same parameter updates carried out by the tail-recursive call (note that the recursive case of a tail-recursive method simply defines how arguments change from one call to the next). Lastly, the loop termination condition would be equivalent to the one used in the base case.

Listing 11.1 shows this alternative iterative variant. The base case in gcd1 is reached when $m = 0$. Thus, the recursive cases occur when $m > 0$. We can therefore use a loop that runs while $m > 0$, and which will simulate all of the recursive calls that gcd1 performs until it reaches the base case. There are two recursive cases. If $m > n$, the recursive call simply swaps the parameters. The iterative code accomplishes this task in lines 4–6 by using an auxiliary variable. If $m \leq n$ then the second recursive case replaces n by $n - m$, which gives rise to the trivial code in line 8. Finally, when $m = 0$ the method simply returns the value in n, in accordance with the base case of gcd1.

Analogous iterative versions of tail-recursive algorithms provide two main benefits. On the one hand, they are more efficient since they do not have to manage a (program) stack, and can return the result as

Listing 11.2 Iterative version of the bisection method.

```
1 def bisection_iterative(a, b, f, epsilon):
2     z = (a + b) / 2
3 
4     while f(z) != 0 and b - a > 2 * epsilon:
5         if (f(a) > 0 and f(z) < 0) or (f(a) < 0 and f(z) > 0):
6             b = z
7         else:
8             a = z
9 
10        z = (a + b) / 2
11
12    return z
```

soon as it is computed. Note that an iterative algorithm would only return its value once, unlike the tail-recursive function, which returns the same value multiple times. On the other hand, they do not cause stack overflow errors, even for large inputs. While the number of recursive calls is limited, the number of iterations of a loop (which essentially performs the same computations as the recursive calls) is not restricted. In many applications the loss in efficiency associated with tail-recursive methods is not relevant. However, the constraint on the number of recursive calls is an important factor regarding the scalability of recursive algorithms, representing a far more serious issue.

Nevertheless, when a recursive algorithm divides the size of a problem by two (or another constant), the limit on the number of recursion calls does not usually represent a drawback. For these algorithms the height of the recursion tree is logarithmic with respect to the size of the problem. Thus, it would be very unlikely to need a large number (e.g., 1000) of nested recursive calls in order to accomplish a task. If the efficiency of the algorithm is also not crucial, then the tail-recursive algorithm can be a perfectly valid option. For example, consider the tail-recursive function `bisection` in Listing 5.15, related to the bisection method. When applied to calculate a square root, the number of recursive calls that it performs is relatively small. For instance, a call to `bisection(0,4,f,10**(-10))` only requires 36 recursive calls in order to approximate $\sqrt{2}$ with an accuracy of up to nine decimal digits to the right of the decimal point. Moreover, due to the small number of recursive calls, the difference in running time with respect to the analogous iterative algorithm, shown in Listing 11.2, would be negligible. The method also uses a while loop in order to simulate the recursive calls, where its condition is the negated expression used to detect the base case in the recursive function. The `if` statement distinguishes the two scenarios related to the two recursive cases. In both situations only one parameter needs to be updated. Finally, the variable `z` is initialized and updated in every iteration, and stores the final result. Given the practical equivalence of both algorithms, the decision to use one or another can be based on which is more readable. In this example, both algorithms are easy to understand given the problem they are designed to solve. However, tail-recursive methods generally provide a greater emphasis on problem decomposition.

Finally, in order to improve efficiency, many modern compilers are able to detect tail-recursive methods and transform them into iterative versions automatically. Nevertheless, Python does not support this feature, which is often referred to as "tail recursion elimination." One of the arguments of Guido van Rossum, who created the Python language, is that it would make debugging harder. In addition, he recommends using iteration over tail recursion. The opinion of the author of the present book is that, if efficiency and the limitation on the size of the program stack are relevant issues, then clearly iterative methods should be used over tail-recursive ones. However, if those aspects are not problematic, then programmers should consider the implementation that better reflects the thinking process involved in solving the problem. In other words, the decision should depend on which version is more readable and maintainable.

11.2 TAIL RECURSION BY THINKING ITERATIVELY

Tail recursive algorithms can obviously be derived by thinking recursively, as illustrated in Chapter 5. However, it is possible to develop them by thinking iteratively. This section examines this idea through the factorial function and the base conversion problem.

11.2.1 Factorial

Consider the iterative version of the factorial function in Listing 11.3. The method computes the final value progressively by repeating the operations inside the while loop. In this algorithm, the initial parameter n controls the number of iterations of the loop, while p can be regarded as an accumulator variable that stores an intermediate product, and will

Listing 11.3 Iterative factorial function.

```
def factorial_iterative(n):
    p = 1
    while n > 0:
        p = p * n
        n = n - 1

    return p
```

Listing 11.4 Tail-recursive factorial function and wrapper method.

```
def factorial_tail(n, p):
    if n == 0:
        return p
    else:
        return factorial_tail(n - 1, p * n)


def factorial_tail_recursive_wrapper(n):
    return factorial_tail(n, 1)
```

end up holding the final factorial. Table 11.1 shows the values of the variables and parameters (i.e., the program state) of the method before every iteration of the loop when computing 4! (in other words, each entry in the table shows the values of n and p when the program evaluates the condition in line 3).

It is possible to build an equivalent tail-recursive factorial function that performs the same operations as the iterative algorithm. In particular, the method (say, f) would require both parameters n and p, and would need to perform function calls in which the arguments take

Table 11.1 Program state related to the iterative factorial function when computing 4!.

n	p
4	1
3	4
2	12
1	24
0	24

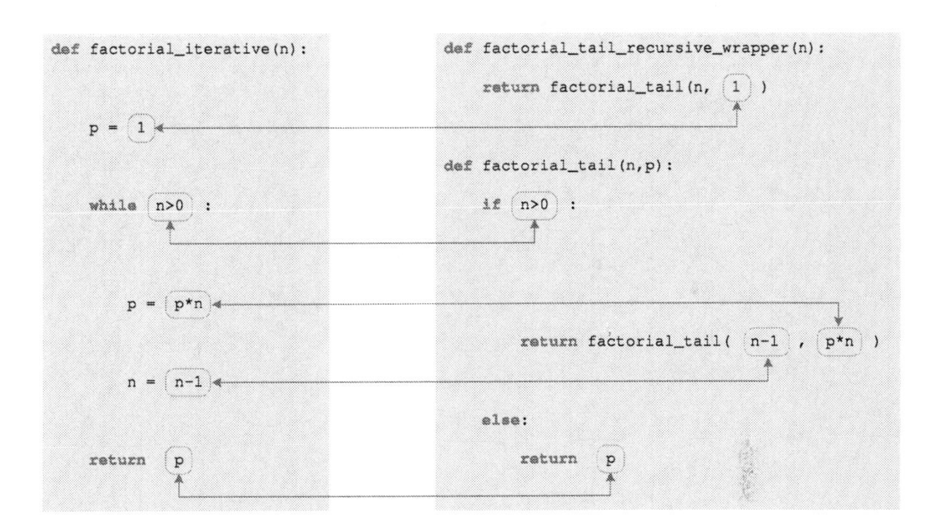

Figure 11.2 Similarities between the iterative and tail-recursive codes that compute the factorial function.

the values in the rows of Table 11.1. In other words, the tail-recursive method would need to generate the following recursive calls:

$$\begin{aligned} & f(4,1) \\ = \; & f(3,4) \\ = \; & f(2,12) \\ = \; & f(1,24) \\ = \; & f(0,24) \; = \; 24. \end{aligned}$$

This generates a correspondence between a tail-recursive call and an iteration of the iterative algorithm. Specifically, the recursive rule is clearly $f(n,p) = f(n-1, p \cdot n)$, which essentially carries out the (variable update) operations inside the loop of the iterative version. Observe that n controls the number of function calls, while p will contain the intermediate and final results. The last call corresponds to the base case (when $n = 0$), where the method must return its second argument. Finally, the factorial of n is achieved by calling $f(n,1)$, where the parameters n and p take the values of the corresponding variables in the iterative version before running the loop. Listing 11.4 shows a possible implementation, where a wrapper function is in charge of calling the recursive method with its second argument set to one. The iterative and tail-recursive functions are essentially equivalent. Figure 11.2 points out the similari-

Listing 11.5 Iterative conversion of a nonnegative integer n into its representation in base b.

```
def decimal_to_base_iterative(n, b):
    s = 0
    p = 1
    while n > 0:
        s = s + (n % b) * p
        p = p * 10
        n = n // b
    return s
```

Table 11.2 Program state related to the iterative base conversion function in Listing 11.5, when obtaining the base-5 representation of 142, which is 1032.

n	b	p	s
142	5	1	0
28	5	10	2
5	5	100	32
1	5	1000	32
0	5	10000	1032

ties between both methods (the recursive case appears before the base case in order to use the condition $n > 0$ in the recursive method).

11.2.2 Decimal to base b conversion

This iterative way of thinking can be applied in order to build tail-recursive algorithms for many problems. Consider the more complex base conversion problem addressed in Section 4.2.2. We could envision an iterative algorithm that takes the steps shown in Table 11.2 in order to compute the base-5 representation of 142 (which is 1032). Besides the parameters, the method requires a variable s that will hold intermediate results and will end up storing the final value. In addition, an extra variable p contains a power of 10 needed to update s. In each step n is modified by performing an integer division by the base ($n//b$), p is multiplied by 10, and s is an accumulator variable that is updated by adding $(n\%b) \cdot p$. Listing 11.5 implements this iterative algorithm, where initially $p = 1$ and $s = 0$. Finally, when n reaches 0 the result is stored in s and the method can return it.

Listing 11.6 Tail-recursive base conversion function and wrapper method.

```
1  def decimal_to_base_tail(n, b, p, s):
2      if n == 0:
3          return s
4      else:
5          return decimal_to_base_tail(n // b, b, 10 * p,
6                                      s + (n % b) * p)
7
8
9  def decimal_to_base_tail_wrapper(n, b):
10     return decimal_to_base_tail(n, b, 1, 0)
```

Listing 11.7 The Ackermann function implemented in Python.

```
1  def ack(m, n):
2      if m == 0:
3          return n + 1
4      elif n == 0:
5          return ack(m - 1, 1)
6      else:
7          return ack(m - 1, ack(m, n - 1))
```

An analogous tail-recursive function would have to simulate the iterations of the while loop by using function calls. The variables (and parameters) of the iterative version would be parameters in the recursive algorithm, and the recursive function call would need to capture how they are updated inside the iterative loop (lines 5–7). Finally, the base case is reached when $n = 0$, where the method can return s, since it will hold the final result. Listing 11.6 contains the tail-recursive function, together with a wrapper method in order to initialize p to one and s to zero. Note how the arguments in the recursive call reflect the variable updates of the iterative algorithm inside the loop.

Although the tail-recursive functions in Listings 11.4 and 11.6 are obviously recursive, this way of constructing them is in accordance with the imperative programming paradigm, which focuses on the variables and parameters that determine the state of the program. Observe that the approach does not rely on problem decomposition or induction in order to develop the methods. Instead, the recursive algorithms simply mimic the iterative versions. The resulting algorithms can therefore be confusing to other programmers who might try to understand the code from a problem decomposition perspective. In addition, due to the disadvantages mentioned in Section 11.1, this type of design strategy should clearly be avoided (stack overflow errors would be rare for Listings 11.4 and 11.6, since the input to a factorial is usually small, and the height of the recursion tree of the base conversion algorithm is logarithmic with respect to n). In summary, if we tackle a problem by thinking iteratively, then it is more appropriate to develop an iterative algorithm.

11.3 NESTED RECURSION

Nested recursion is a rare type of recursion that appears when the argument of a recursive call involves another recursive call. Similarly to tail-recursive procedures, in many nested-recursive functions (in particular, all those presented in this book) a recursive call is the last action carried out by the algorithm. However, nested-recursive methods must necessarily call themselves several times in some recursive case. Thus, they cannot be tail-recursive. This section describes two well-known methods based on nested recursion, and presents a problem that can be solved through it. Section 11.4 will provide additional examples.

11.3.1 The Ackermann function

One of the most popular examples of nested recursion is the Ackermann function, defined as follows:

$$A(m,n) = \begin{cases} n+1 & \text{if } m = 0, \\ A(m-1,1) & \text{if } m > 0 \text{ and } n = 0, \\ A(m-1,A(m,n-1)) & \text{if } m > 0 \text{ and } n > 0, \end{cases} \tag{11.1}$$

where the second argument in the last recursive case is a recursive call. The function grows extremely rapidly, even for small inputs, and is used in benchmarks related to the ability of compilers to optimize recursion. Lastly, Listing 11.7 shows the code associated with the function, which is straightforward to implement.

11.3.2 The McCarthy 91 function

Another well-known nested-recursive method is the "mysterious" McCarthy 91 function, defined as follows:

$$f(n) = \begin{cases} n-10 & \text{if } n > 100, \\ f(f(n+11)) & \text{if } n \le 100, \end{cases}$$

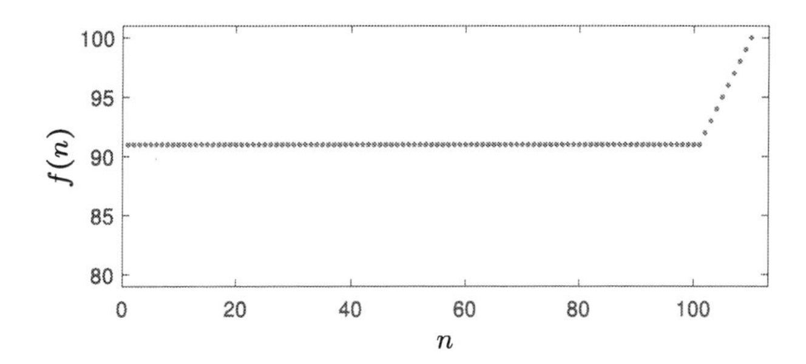

Figure 11.3 McCarthy 91 function.

where n is a positive integer. It is mysterious since it is not easy to picture its behavior when $n \leq 100$ intuitively. Figure 11.3 plots the function for the first 110 positive integers. In particular, the function can be rewritten as:

$$f(n) = \begin{cases} n - 10 & \text{if } n > 100, \\ 91 & \text{if } n \leq 100. \end{cases}$$

11.3.3 The digital root

The digital root, denoted here as $d(n)$, of a nonnegative integer n, is the result of repeatedly adding the digits of n until the remaining value is a single digit. For example, $d(79868) = d(7+9+8+6+8) = d(38) = d(3+8) = d(11) = d(1+1) = d(2) = 2$. The size of the problem is related to the number of digits of n, and the base case for the problem clearly occurs when n contains a single digit.

For the recursive case we will present two alternatives. Firstly, the statement of the problem implies that it can be simplified by replacing n with the sum of its digits. This leads to the following definition of the digital root:

$$d(n) = \begin{cases} n & \text{if } n < 10, \\ d(s(n)) & \text{if } n \geq 10, \end{cases}$$

where $s(n)$ denotes the sum of the digits of a nonnegative integer, whose code is described in Listing 2.2. Note that this function is tail-recursive.

Another idea consists of reducing the size of the problem by discarding one of the digits of n. Consider the following concrete recursive diagram:

Listing 11.8 Nested-recursive method for finding the digital root of a non-negative integer.

```
def digital_root(n):
    if n < 10:
        return n
    else:
        return digital_root(digital_root(n // 10) + n % 10)
```

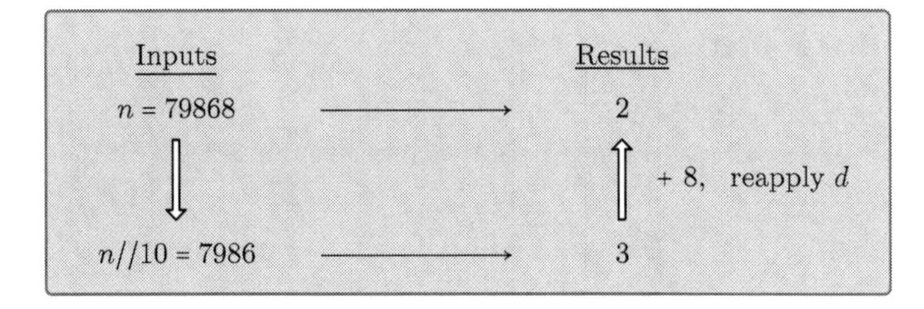

The recursive case requires adding the last digit of n to the result of the subproblem, and applying the entire method again. Thus, the function can be defined exclusively in terms of itself as:

$$d(n) = \begin{cases} n & \text{if } n < 10, \\ d(d(n//10) + n\%10) & \text{if } n \geq 10, \end{cases}$$

which represents an example of nested recursion. The function can be coded easily as in Listing 11.8. Lastly, d has a number of properties that involve nested recursion. For example, $d(n + m) = d(d(n) + d(m))$, or $d(n \cdot m) = d(d(n) \cdot d(m))$.

11.4 TAIL AND NESTED RECURSION THROUGH FUNCTION GENERALIZATION

We have seen that tail-recursive functions can be derived by thinking iteratively, or by relying on implemented iterative versions. In this section we will see that, for some simple computational problems that involve formulas, it is possible to design the same tail-recursive functions, but by using a purely declarative approach. The strategy is related to formal methods for developing software, which are beyond the scope of the book.

The idea stems from noticing that some tail-recursive functions are generalizations of the functions they are designed to solve. For example, the method `factorial_tail(n,p)` in Listing 11.4 is a generalization of the factorial function. Clearly, when $p = 1$ the method computes the factorial of n, but for other values of p it will calculate something else. In this case, it is not hard to see that the function computes $p \cdot n!$. The algorithm design strategy described in this section consists of developing these general tail-recursive functions by using recursive concepts and basic algebraic manipulation. Lastly, this methodology can also lead to inefficient, but correct, nested-recursive algorithms.

11.4.1 Factorial

The natural recursive definition of the factorial function, $n! = f(n) = n \cdot f(n-1)$, leads to a linear-recursive method. If we wish to build a tail-recursive function, then we must use additional parameters that will contain the information necessary to recover the final factorial in a base case. For example, we can propose building a new function $g(n,p)$ that incorporates the new parameter p. Function $g(n,p)$ will have its own expression that must allow us to compute the factorial of n, and which will also contain the parameter p. Some examples of generalizations of the factorial function are:

$$g(n,p) = p \cdot n!, \tag{11.2}$$

$$g(n,p) = p + n!, \tag{11.3}$$

$$g(n,p) = (n!)^p. \tag{11.4}$$

Of course, $f(n)$ can be recovered by setting $p = 1$ in (11.2) and (11.4), and $p = 0$ in (11.3).

11.4.1.1 Appropriate generalization

The first step consists of choosing a particular generalization to work with. Among the multiple possibilities for the factorial function, we will begin analyzing $g(n,p) = p \cdot n!$, since it leads to a simple tail-recursive algorithm. This choice turns out to be crucial, since other expressions usually yield inefficient and complex nested-recursive functions.

We can proceed as in previous chapters in order to build a recursive algorithm for $g(n,p)$. The size of the problem is n. Therefore, the base case is obtained when $n = 0$, where the function simply returns p (i.e., $g(0,p) = p$). For the recursive case we can try to decrease n by a unit, which would lead to the following recursive diagram:

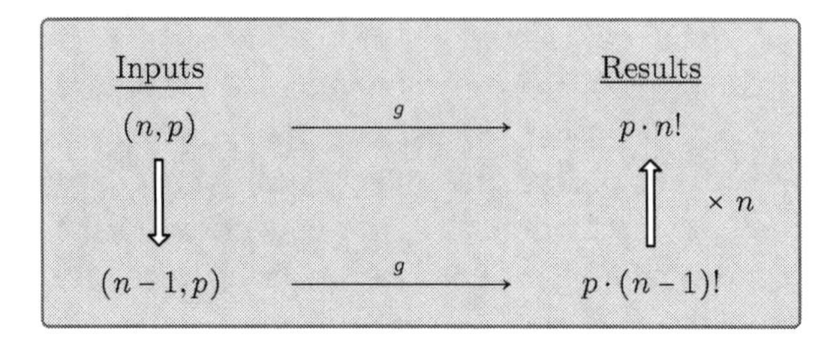

However, this decomposition leads to a linear-recursive algorithm instead of a tail-recursive one, since the result of the subproblem needs to be multiplied by n. The previous diagram is not useful in this case because it has not taken into account the fact that the goal is to develop a tail-recursive method, where the result of a subproblem needs to be identical to the output of the original problem. Firstly, the decomposition needs to be modified since $g(n,p) \neq g(n-1,p)$. If we wish to decrease the size of the problem by a unit, then we must use another expression for the second argument. For example, $g(n,p) = g(n-1,q)$, where q is just a dummy variable that represents an expression involving the input parameters n and p. The goal will be to determine q, which we can find by building an alternative, but similar, diagram:

Recursive rule		Expressions		
$f(n,p)$	$=$	$p \cdot n!$		
$\parallel$		$\parallel$	$\Rightarrow$	$q = p \cdot n$
$f(n-1,q)$	$=$	$q \cdot (n-1)!$		

The first thing to notice is that all of the terms in the diagram are equal: $g(n,p) = p \cdot n!$ is the definition of the general function, while $g(n,p) = g(n-1,q)$ would be the tail-recursive rule. Applying the definition to $g(n-1,q)$ naturally yields $g(n-1,q) = q \cdot (n-1)!$. Finally, the formulas on the right column must also be equal, and can be used to solve for q. In this case we have:

$$p \cdot n! = q \cdot (n-1)!,$$

which implies that $q = p \cdot n$. Finally, the function is defined as follows:

$$g(n,p) = \begin{cases} p & \text{if } n = 0, \\ g(n-1, p \cdot n) & \text{if } n > 0, \end{cases}$$

which corresponds to the method `factorial_tail(n,p)` in Listing 11.4 ($f(n) = g(n,1)$ would be the wrapper function). In this case, the parameter p arises from employing a more general function, and not by thinking of an accumulator variable as in the iterative algorithm.

11.4.1.2 *Inappropriate generalizations*

Generally, if the definition of a function involves a product, then the generalization should also include a product times the new parameter. Similarly, if the function involves a sum, then the generalization should add the extra parameter. This facilitates solving for the extra parameter (q, in the previous example) in the subproblem.

However, not all generalizations will lead to tail-recursive algorithms. For example, consider $g(n,p) = p + n!$, which uses an addition instead of a product. Firstly, the base case would be $g(0,p) = p + 0! = p + 1$. For the recursive case we can form the following diagram:

Recursive rule		Expressions		
$g(n,p)$	$=$	$p + n!$		
$\parallel$		$\parallel$	$\Rightarrow$	$q = ?$
$g(n-1,q)$	$=$	$q + (n-1)!$		

In this case solving for q is trickier. Starting with $q + (n-1)! = p + n!$ we have:

$$\begin{aligned} q &= p + n! - (n-1)! \\ &= p + (n-1) \cdot (n-1)! \\ &= p + (n-1) \cdot g(n-1,0), \end{aligned}$$

where the last equality follows from the definition of the generalization. The function can therefore be defined as:

$$g(n,p) = \begin{cases} p+1 & \text{if } n = 0, \\ g(n-1, p + (n-1) \cdot g(n-1,0)) & \text{if } n > 0, \end{cases} \tag{11.5}$$

which is an example of nested recursion. Although the algorithm is correct, it is not only more complex, but much more inefficient. Its recursion tree is a full binary tree, which implies that it runs in $\Theta(2^n)$ time.

Moreover, $g(n,p) = (n!)^p$ is even more problematic. Considering $((n-1)!)^q = (n!)^p$, it is possible to solve for q by first taking logarithms (the base is not relevant) on both sides. This yields $\log(((n-1)!)^q) =$

$\log((n!)^p)$, and consequently: $q\log((n-1)!) = p\log(n!)$. Thus, we have:

$$\begin{aligned} q &= p\left[\frac{\log(n!)}{\log((n-1)!)}\right] \\ &= p\left[\frac{\log(n)+\log((n-1)!)}{\log((n-1)!)}\right] \\ &= p\left[\frac{\log(n)}{\log((n-1)!)}+1\right] \\ &= p\left[\frac{\log(n)}{\log(g(n-1,1))}+1\right]. \end{aligned}$$

This expression is only valid for $n > 2$ (observe that it is not possible to obtain q for $(1!)^q = (2!)^p$). Thus, we need several base cases in order to generate a correct function. In particular, it can be defined as follows:

$$g(n,p) = \begin{cases} 1 & \text{if } n \le 1, \\ 2^p & \text{if } n = 2, \\ g\left(n-1, p\left[\frac{\log(n)}{\log(g(n-1,1))}+1\right]\right) & \text{if } n > 2, \end{cases}$$

which is also based on nested recursion. Similarly to (11.5), it runs in exponential time with respect to n. In addition, it operates with real numbers instead of integers due to the logarithmic computations. Thus, its output should be rounded to the nearest integer.

11.4.2 Decimal to base b conversion

We have already seen two ways of constructing algorithms for the base conversion problem. Listings 4.10 and 11.6 contain linear and tail-recursive solutions, respectively. We will now develop the same tail-recursive algorithm by using the generalization approach. The solution is more difficult since it will require two additional parameters. In addition, the expression for the base conversion function is more complex. For instance, it can be written as follows:

$$f(n,b) = \sum_{i=0}^{m-1} (d_i \cdot 10^i),$$

where n is a decimal number that we wish to represent in base b. Furthermore, m is the number of base-b digits in the new description of n, while d_i, for $i = 0,\ldots,m-1$, represent those digits. Fortunately, it is not necessary to provide a more concrete specification for these digits.

Since the formula contains a sum, we can begin by adding a new parameter s. This generates the following more general function:

$$g(n, b, s) = s + \sum_{i=0}^{m-1} (d_i \cdot 10^i).$$

The decomposition should consider performing the integer division of n by b, which decreases the size of the problem by a unit. This leads to the following diagram:

Recursive rule		Expressions		
$g(n,b,s)$	$=$	$s + \sum_{i=0}^{m-1} (d_i \cdot 10^i)$		
$\parallel$		$\parallel$	$\Rightarrow$	$q = ?$
$g(n//b,b,q)$	$=$	$q + \sum_{i=1}^{m-1} (d_i \cdot 10^{i-1})$		

We therefore have:

$$s + \sum_{i=0}^{m-1} (d_i \cdot 10^i) = q + \sum_{i=1}^{m-1} (d_i \cdot 10^{i-1}).$$

If we try to solve for q we will immediately see that the solution is complex. In particular:

$$q = s + d_0 + \sum_{i=1}^{m-1} d_i(10^i - 10^{i-1}). \tag{11.6}$$

However, we can notice that the formula would be considerably simpler if the sum related to the subproblem were multiplied by 10, since the sum in (11.6) would vanish. Therefore, we can include an additional parameter in order to create an even more general function, which will multiply the sum:

$$h(n, b, p, s) = s + p \cdot \sum_{i=0}^{m-1} (d_i \cdot 10^i). \tag{11.7}$$

With this new function we can form the following diagram:

Recursive rule		Expressions		
$h(n,b,p,s)$	$=$	$s + p \cdot \sum_{i=0}^{m-1} (d_i \cdot 10^i)$		
$\parallel$		$\parallel$	$\Rightarrow$	$q = ?,\quad r = ?$
$h(n//b,b,r,q)$	$=$	$q + r \cdot \sum_{i=1}^{m-1} (d_i \cdot 10^{i-1})$		

In this case we have:

$$s + p \cdot \sum_{i=0}^{m-1} (d_i \cdot 10^i) = q + r \cdot \sum_{i=1}^{m-1} (d_i \cdot 10^{i-1}),$$

which involves the dummy variables r and q. If we set $r = 10p$, then the expression reduces to:

$$s + p \cdot \sum_{i=0}^{m-1} (d_i \cdot 10^i) = q + p \cdot \sum_{i=1}^{m-1} (d_i \cdot 10^i),$$

where we can now solve for q easily. In particular, $q = s + p \cdot d_0 = s + p \cdot (n\%b)$.

Finally, assuming that the base case occurs when $n < b$, where $f(n,b) = n$, the base case for $h(n,b,p,s)$ should be $s + pn$ (this follows from (11.7), where $m = 1$ and $d_0 = n$). Thus the function can be defined as follows:

$$h(n,b,p,s) = \begin{cases} s + pn & \text{if } n < b, \\ h(n//b, b, 10p, s + p \cdot (n\%b)) & \text{if } n \geq b, \end{cases}$$

where $f(n,b) = h(n,b,1,0)$. Finally, another option is to consider that the base case is reached when $n = 0$. This would lead to the function implemented in Listing 11.6.

11.5 EXERCISES

Exercise 11.1 — Implement a tail-recursive algorithm that computes the sum of the first n positive integers by thinking iteratively, where $n \geq 0$. In particular, consider how an iterative algorithm would solve the problem by using an accumulator variable that stores intermediate sums (and the final result). Lastly, derive a tail-recursive algorithm by using the approach based on function generalization.

Exercise 11.2 — Implement a tail-recursive version of the function in Listing 5.16, and write an analogous iterative function.

Exercise 11.3 — Consider the problem of adding the elements of a list **a** of n numbers. Design a tail-recursive function by applying the generalization approach. Finally, transform it into an iterative algorithm.

Exercise 11.4 — Design a tail-recursive function that computes b^n, where b is a real number and n is an nonnegative integer, by applying the generalization approach. Finally, transform it into an iterative algorithm.

Exercise 11.5 — Design a tail-recursive function that computes the n-th Fibonacci number (for $n > 0$) by applying the generalization approach. In particular, use a general function $f(n,a,b) = G(n)$ that represents a sequence that follows the recursive rule $G(n) = G(n-1) + G(n-2)$, where $G(1) = a$, and $G(2) = b$. Finally, transform the code to an iterative version.

Exercise 11.6 — Design a function that computes the n-th Fibonacci number $F(n)$ (for $n > 0$) by applying the generalization approach with the following general function: $f(n,s) = s + F(n)$. In addition, use a decomposition that reduces the size of the problem by a unit.

Exercise 11.7 — Consider the Ackermann function in (11.1). Draw the activation tree for a call to $A(2,1)$.

CHAPTER 12

Multiple Recursion III: Backtracking

He who would search for pearls must dive below.

— John Dryden

BACKTRACKING is one of the most important algorithm design paradigms. It can be regarded as an "intelligent" brute-force strategy that performs exhaustive searches for solutions to constraint satisfaction and discrete optimization problems. The approach can be used to solve numerous puzzles and problems, including the eight-queens problem, finding paths through mazes, the sudoku puzzle, the 0-1 knapsack optimization problem, and many more.

Backtracking methods generally combine recursion and iteration, contain several parameters, and are not usually designed by strictly thinking about problem decomposition and induction. Thus, they appear to be complex to students who study the material for the first time. Fortunately, backtracking algorithms often share a similar structure, which can be exploited in order to ease their design. This structure is captured in what are know as "backtracking templates," which vary depending on coding styles and programming languages. In this book the methods will share a particular structure where they receive a relatively large number of parameters, and barely rely on additional methods. However, the reader should be aware that there are other possibilities. In any case, students can master backtracking with relative ease by studying examples, and by applying very similar algorithms to different problems.

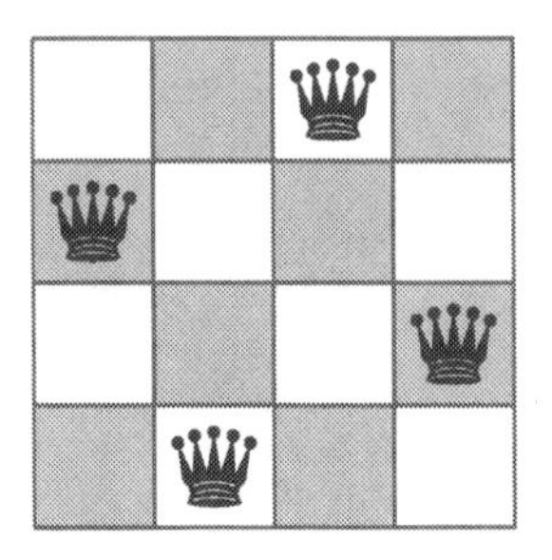

Figure 12.1 One solution to the four-queens puzzle.

12.1 INTRODUCTION

This section introduces fundamental concepts related to backtracking, and provides an overview of how it works by examining the simple four-queens puzzle. Its goal consists of placing four chess queens on a 4×4 chessboard so that they do not threaten each other. Since queens can move horizontally, vertically, and diagonally, two (or more) queens cannot appear in the same row, column, or diagonal on the board. Figure 12.1 illustrates one of the two possible solutions to the puzzle. Naturally, the problem can be generalized to placing n queens on an $n \times n$ chessboard (see Section 12.3).

12.1.1 Partial and complete solutions

Backtracking is a general strategy for finding solutions to computational problems among a finite set of possibilities. These solutions are collections of discrete items. Thus, technically we say that backtracking is a strategy that searches for solutions in a "discrete state space." It is also a brute-force method in the sense that the search is exhaustive. In other words, if a solution exists, a backtracking algorithm is guaranteed to find it.

The methods search for solutions by constructing and updating a **partial solution**, which could eventually become a valid **complete solution** to the problem. Partial solutions are built by incorporating **candidate** elements incrementally, step by step, in successive recursive calls. Thus, there is an implicit order in which the candidates appear in the solutions. In this regard, partial solutions can be understood as "prefixes" of complete solutions. When a solution is a list, a partial solution is simply a sublist that contains the first elements, as illustrated in Figure 12.2(a) through the lighter-shaded region. Instead, if it is a two-

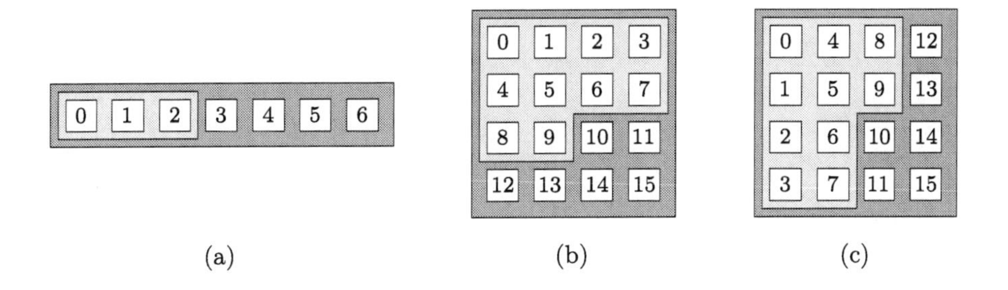

Figure 12.2 Partial solutions within complete solutions that are coded as lists or matrices.

dimensional matrix, we need to consider some linear arrangement of its elements. The two most natural choices consist of ordering the entries of a matrix in row or column-major order, as shown in (b) and (c), respectively. In those cases a partial solution is the collection of the first elements of the matrix according to one of these linear arrangements.

The key to using backtracking effectively is the possibility of efficiently analyzing whether a partial solution satisfies the constraints of the problem. If it does, then it could potentially be expanded into a complete valid solution. However, if it is not possible to create a valid partial solution by including a new item, the algorithm "backtracks" to another previous valid partial solution, and continues exploring other possibilities not contemplated yet. Backtracking algorithms therefore perform exhaustive searches for solutions, but can solve some problems efficiently by avoiding incrementing partial solutions that do not lead to valid complete solutions.

The first task when tackling a problem through backtracking consists of selecting a particular form or data structure for the solutions. Typically, they are implemented as lists or matrices. For the n-queens puzzle a first idea that could come to mind is to use a Boolean $n \times n$ matrix to encode the solutions, where (n) `True` values mark the squares where the queens lie on the board. A partial solution would have the same structure, but it would contain less than n queens (as soon as you place the n-th queen the solution is either not valid, or it is a complete solution to the problem). However, although it is possible to use a matrix, this choice of data structure does not lead to an efficient algorithm. Since backtracking methods perform exhaustive searches, an algorithm would traverse the entire matrix in row or column-major order, deciding whether to place a queen in each of the chessboard's squares. This would

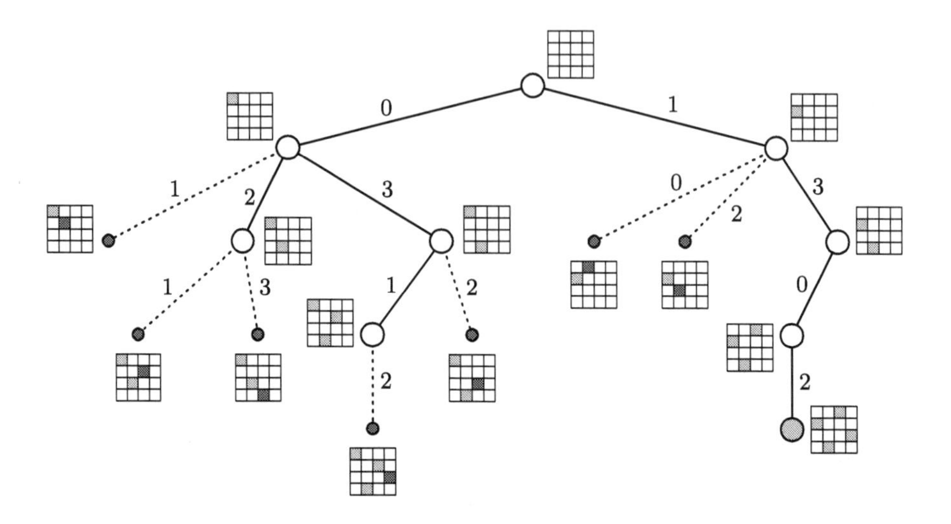

Figure 12.3 Recursion tree of a backtracking algorithm that finds one solution to the four-queens puzzle.

be time-consuming, since, for example, if a queen is placed in a particular column, then the algorithm does not need to try to place other queens on that same column. Instead, the method could simply move on to the next column.

A more appropriate choice consists of using a simple list of length n. In that case, the indices of the list could represent the columns of the chessboard, and the particular value could denote the row where a queen is placed. In other words, if $\mathbf{x}$ is a list that represents a solution, and there is a queen in column i and row j, then we have $x_i = j$. For example, if columns are numbered from left to right, and rows from top to bottom, both starting at 0, the solution in Figure 12.1 corresponds to the list $[1, 3, 0, 2]$. By using a list, after placing a (valid) queen in a row for a particular column (i), the algorithm can continue to try to place a queen in the next column ($i + 1$). In addition, it can keep track of the rows where there are queens, and avoid placing further queens in them.

12.1.2 Recursive structure

We can examine how a backtracking algorithm works by analyzing its recursion tree. Figure 12.3 illustrates it for a backtracking method that finds one solution to the four-queens puzzle. The root node represents the first call to the method, where the partial solution, drawn as a 4×4

board (but coded as a list), is empty. Subsequently, the order of the recursive calls is in accordance with a preorder traversal of the tree. The first recursive call, associated with the root's left child, places a queen (depicted as a shaded square) in the first row of the first column. The label "0" on the corresponding edge indicates the row where a queen is placed. Since this does not violate the constraints of the problem it can continue to place queens in the second column. In particular, it starts by avoiding placing a queen in the first row, since there is already a queen on it. Subsequently, the algorithm analyzes whether it can place a queen on the second row. Clearly it cannot, since the two queens would lie on the same diagonal. The associated partial solution would not be valid, and the algorithm does not proceed with expanding it. In other words, it will not carry out any further recursive calls with that (invalid) partial solution. This is illustrated in the diagram through dotted edges.

Subsequently, the method proceeds by placing a queen in the third row, which leads to a valid partial solution. Thus, the algorithm can continue to try to place a queen in the third column. The first free row is the second one, but this leads to an invalid partial solution. Similarly, placing a queen on the fourth row violates the constraints of the problem. Since there are no more options left, the algorithm "backtracks" to the node where there is only one queen on the first column and first row. Afterwards, the method explores all of the possible partial solutions that could be obtained when placing a queen on the fourth row of the second column. Since there are no valid solutions for this option either, the algorithm will end up backtracking to the initial node. Note that at this stage the method will have carried out an exhaustive search for solutions in which a queen appears on the first row of the first column. Subsequently, the algorithm continues exploring possibilities when starting by placing a queen on the second row of the first column. The same procedure is repeated until eventually the method is able to place four queens that do not threaten each other. In that case it can stop searching if the goal consists of finding only one solution, or it can continue to search for all of the solutions.

The recursion tree also illustrates several features of the algorithm. Firstly, the nodes correspond to recursive method calls where the partial solution (which will be an input parameter) is valid. The small dark nodes attached through dotted edges simply show how the tree would have been expanded had its corresponding partial solution been valid. Also, observe that an item is added to the partial solution at each recursive call. In other words, partial solutions that contain k items will be associated with nodes of depth k. Therefore, complete solutions are found when reaching leaf nodes, of depth n. Furthermore, a path from the root to a node specifies the partial solution, or a complete solution if the node is a leaf. For example, the solution found by the algorithm is the list $[1, 3, 0, 2]$, which is the sequence of labels associated with the edges of the path from the root to the leaf where the solution is found. In addition, since the values in the list represent different rows, the solution to the four-queens puzzle is actually a permutation of the rows, which are the first n nonnegative integers. A tree that explores all possible permutations would have $n!$ leaves, which can be very large, even for small values of n. However, the figure also illustrates that the recursive tree can be pruned considerably by not expanding invalid partial solutions, which is key to obtaining efficient algorithms.

Similar details will appear in other problems and examples in the chapter. In particular, Section 12.3 will cover the n-queens puzzle in depth, showing a specific coded algorithm that solves it. Lastly, the tree shown in Figure 8.18 could correspond to a recursion tree of an algorithm that searches for two-element variations with repetition of four items. The lists next to the nodes would be partial or complete solutions, while the labels next to the edges would indicate the element that is introduced in a partial solution. In practice, backtracking algorithms prune the recursion trees in order to gain efficiency by discarding invalid solutions.

12.2 GENERATING COMBINATORIAL ENTITIES

We have just seen that the solution to the n-queens problem is a permutation (of the first n nonnegative integers). This will also be the case in numerous problems. In addition, the goal in many other problems consists of finding subsets of n elements. Therefore, it is important to understand how to generate all of the permutations, and all of the subsets, of n distinct elements. This section presents specific algorithms in Python for generating these combinatorial entities. They are particularly beneficial since they can be used as starting points to build backtracking algorithms for many problems.

Lastly, these algorithms will not generate lists or other data structures in order to store all of the subsets or permutations. This would not be practical due to the large number of possibilities (there are $n!$ possible permutations, and 2^n different subsets, of n distinct elements). Instead, the methods will use a simple list (the partial solution) that rep-

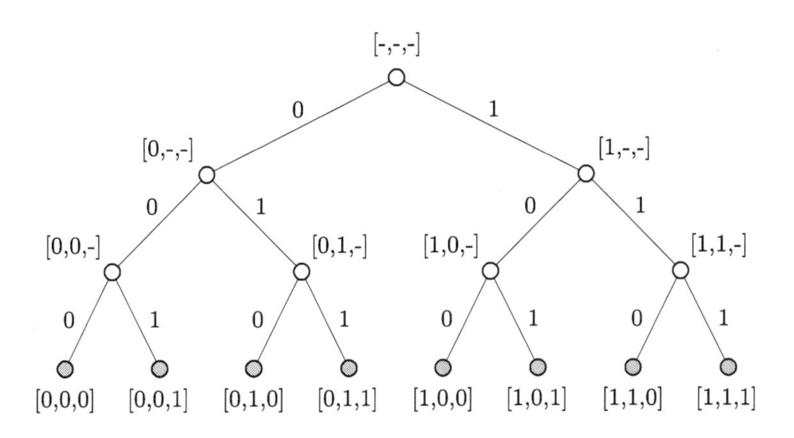

Figure 12.4 Binary recursion tree of an algorithm that generates all of the subsets of three items.

resents only one of the subsets/permutations, but which will be modified in order to contain all of them as the algorithms are executed. Thus, the methods will be able to process every subset/permutation (we will simply print them or count them), but will not return all of them in a single data structure.

12.2.1 Subsets

This section presents two strategies for generating all of the subsets of n (distinct) elements, which will be provided through an input list. In both methods the recursion tree will be binary, and the solutions (subsets) will be represented at its leaves. One method uses partial solutions of (fixed) length n, while in the other their length varies as the procedure carries out recursive calls.

12.2.1.1 Partial solutions of fixed length

Figure 12.4 shows the binary recursion tree of an algorithm that generates the eight subsets that it is possible to create with three distinct elements. Each subset is described through a binary list of n zeros or ones (naturally, Boolean values can be used as well). For example, given the list of elements $[a, b, c]$, the list [1,0,1] denotes the subset $\{a, c\}$. This binary list is the partial solution, which is complete when the method reaches a leaf. The binary digits are therefore the candidate elements that form it. As the method carries out recursive calls it simply appends

Listing 12.1 Code for printing all of the subsets of the elements in a list.

```
def generate_subsets(i, sol, elements):
    # Base case
    if i == len(elements):
        # Print complete solution
        print_subset_binary(sol, elements)
    else:
        # Generate candidate elements
        for k in range(0, 2):

            # Include candidate in partial solution
            sol[i] = k

            # Expand partial solution at position i+1
            generate_subsets(i + 1, sol, elements)

            # Remove candidate from partial solution
            sol[i] = None  # optional


def generate_subsets_wrapper(elements):
    sol = [None] * (len(elements))
    generate_subsets(0, sol, elements)


def print_subset_binary(sol, elements):
    no_elements = True
    print('{', end='')
    for i in range(0, len(sol)):
        if sol[i] == 1:
            if no_elements:
                print(elements[i], sep='', end='')
                no_elements = False
            else:
                print(', ', elements[i], sep='', end='')
    print('}')
```

a new candidate to the partial solution (the concrete binary value is shown on the edges of the tree), which is passed as a parameter to the method, and shown next to the nodes of the recursion tree. Initially that list has length n, but its entries are meaningless. When the algorithm reaches a base case it will have generated one of the 2^n possible subsets, at one of the leaves of the recursion tree. In those cases the lists contain n meaningful elements and will represent complete solutions.

Listing 12.1 shows an algorithm that complies with the recursion tree in Figure 12.4, and which prints each subset through the method `print_subset_binary`. Besides the initial list `elements` that represents the set of n items that it will group into subsets, the procedure also receives an input list `sol` that represents the partial solutions, and which has n components initialized to `None` in the wrapper method `generate_subsets_wrapper`. In addition, observe that the method includes a new candidate in the partial solution, at every level, as it descends through the recursion tree. Therefore, it uses a third parameter `i` that indicates the index in the partial solution where a new zero or one will appear. Furthermore, `i` also corresponds to the depth of the node associated with the recursive call responsible for introducing the new element in the partial solution. Since the recursive backtracking method `generate_subsets` begins introducing a zero or a one at index 0, `i` is initialized to 0 in `generate_subsets_wrapper`.

The method `generate_subsets` checks in line 3 if the partial solution in `sol` is really a complete solution, which occurs if `i` is equal to n. This would correspond to the base case, where the method simply prints the solution (line 5). In the recursive case the procedure has to perform two recursive calls in order to include more candidates in the partial solution. It accomplishes this through a simple loop where the variable `k` takes either the value 0 or 1 (line 8), and where the `i`-th component of the partial solution (`sol`) receives the value of `k` (line 11). The backtracking algorithms included in this book will also use a loop in order to consider every possible candidate that could (potentially) be included in the partial solution. The procedure then invokes itself (line 14) with the modified partial solution, and incrementing `i` in order to include more candidates (if the partial solution is not complete).

The code in line 17 is optional, since the method works correctly without it. Nevertheless, the procedure includes it for illustrative purposes, since similar instructions may be necessary in other backtracking algorithms. Observe that when a particular node in the recursion tree finishes executing, the control passes to its parent, where the value of the partial solution at the `i`-th position is meaningless. Therefore, we can express this explicitly through the code in line 17. However, the code works without it since this particular method only prints a subset when reaching a base case (where partial solutions are complete), and because the assignment in line 11 simply overwrites a zero when `k` = 1. The reader is encouraged to execute the program step by step with a

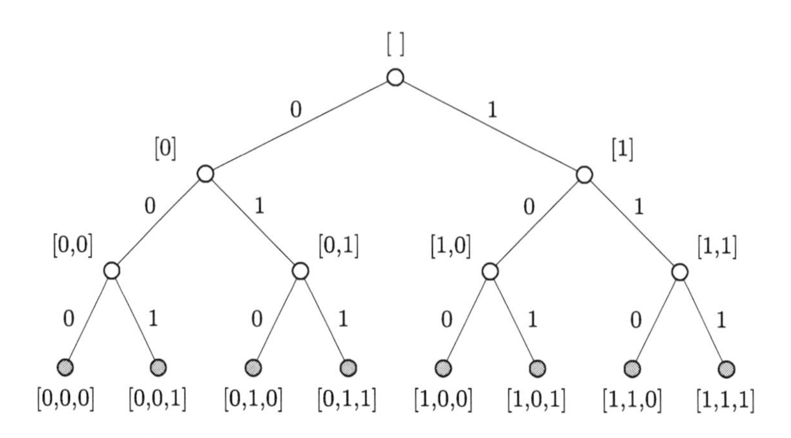

Figure 12.5 Alternative binary recursion tree of an algorithm that generates all of the subsets of three items.

debugger, with or without line 17, in order to examine the evolution of the partial solution.

Finally, a call to `generate_subsets_wrapper(['a','b','c'])` produces the following output:

```
{}
{c}
{b}
{b, c}
{a}
{a, c}
{a, b}
{a, b, c}
```

12.2.1.2 Partial solutions of variable length

In the previous algorithm the partial solutions always had the same length (n). Thus, the procedure can be used with lists or arrays that do not vary in size. In that case the parameter `i` was useful in order to specify the indices of the elements to be added to the partial solution. Instead, another possibility is to use partial solutions of variable size, where `i` is no longer needed. Figure 12.5 shows another recursive tree that has the same structure as the one in Figure 12.4, but the labels next to the nodes specify partial solutions at every method call where every element is meaningful.

Listing 12.2 Alternative code for printing all of the subsets of the elements in a list.

```
def generate_subsets_alt(sol, a):
    # Base case
    if len(sol) == len(a):
        # Print complete solution
        print_subset_binary(sol, a)
    else:
        # Generate candidate elements
        for k in range(0, 2):

            # Include candidate in partial solution
            sol = sol + [k]

            # Expand partial solution at position i+1
            generate_subsets_alt(sol, a)

            # Remove candidate from partial solution
            del sol[-1]


def generate_subsets_alt_wrapper(elements):
    sol = []
    generate_subsets_alt(sol, elements)
```

Listing 12.2 shows a method that is in accordance with this approach, which is very similar to the code in Listing 12.1. Firstly, it prints the solution if it contains n elements (base case). In the recursive case it uses the same loop, appends a zero or a one at the end of the partial solution, and subsequently invokes itself. In this case, the code in line 17 is necessary, where the method has to "undo" the change to the partial solution prior to invoking the method recursively. While the algorithm in Listing 12.1 overwrites a zero in the partial solution when incorporating a one, in this case the method needs to remove the zero from the list in order to include a one properly. Thus, calling the recursive method implies adding an element to the partial solution (line 14), and the algorithm deletes it (line 17) after returning from the recursive call. The code on line 17 of Listing 12.1 plays a similar role, but is not necessary in that method. Regarding efficiency, appending and removing elements from a data structure dynamically can be time-consuming. Thus, it is often more efficient to allocate a fixed amount of memory for the partial solution, and update it without altering its size. In this regard, the code

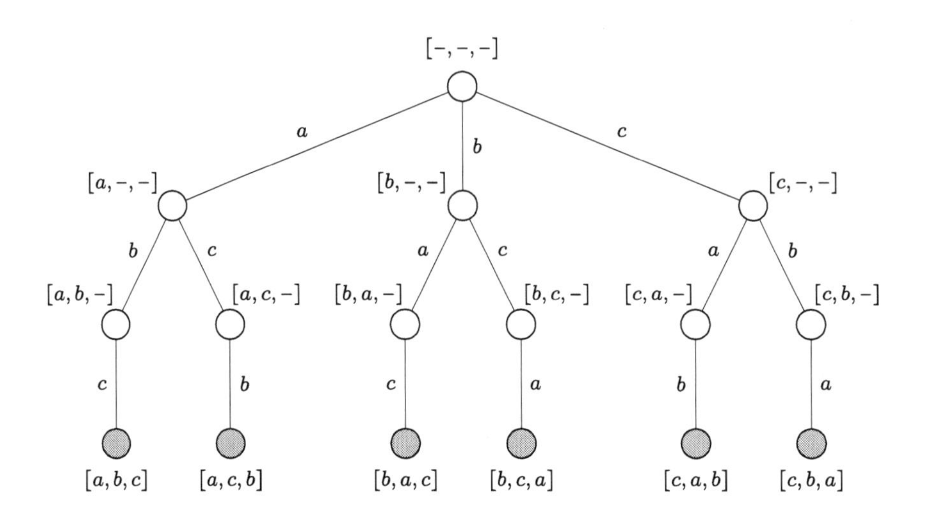

Figure 12.6 Recursion tree of an algorithm that generates all of the permutations of three items.

in Listing 12.1 runs faster than the method `generate_subsets_alt`. Finally, for this alternative algorithm the wrapper method must initialize the partial solution to an empty list.

12.2.2 Permutations

This section examines two similar algorithms that print all of the possible permutations of the n distinct elements in a given list. One way to represent a permutation is through a list of indices from 0 to $n-1$, which reference the locations of the elements in the input list. For example, given the list $[a, b, c]$, the partial solution $[1, 2, 0]$ would denote the permutation $[b, c, a]$. However, the following algorithms will simply use partial solutions formed by the items of the input list (and `None` values). Figure 12.6 shows the structure of their recursion trees, for the list $[a, b, c]$. Observe that the partial solutions have length n, but only the first items are meaningful. In the first call to the methods the partial solution is "empty," where all of its elements are set to `None`. The procedures also receive a parameter `i`, initially set to zero, which specifies how many candidate elements have been included in the partial solution. Thus, it indicates the position in the partial solution where the methods introduce a new candidate, and which is also equivalent to the depth of the node related to a method call in the recursion tree.

The methods begin by introducing one of the n candidate elements of the input list at the first position of the partial solution. Since there are n possibilities, the root node has n children. In the next level of the tree the methods invoke themselves $n-1$ times, since the element that appears in the first position cannot appear again in the permutation. Similarly, in the next level the partial solutions contain two elements and the methods invoke themselves $n-2$ times. This is repeated in general until they reach a base case, at a leaf of the recursion tree, where the partial solutions are complete permutations.

The next subsections describe both methods, where the main difference between them resides in how they test the validity of a partial solution.

12.2.2.1 Validity of partial solutions without additional data structures

Listing 12.3 shows a first method that complies with the recursion tree in Figure 12.6. Since `i` indicates the number of elements in the partial solution `sol`, the first `if` statement checks if the method has reached a base case. If the result is `True` the procedure simply prints the permutation stored in `sol`. If the algorithm has not reached a base case it uses a loop in order to try to include every candidate item of the initial input list `elements` in the partial solution. The loop's counter `k`, which serves as an index of `elements`, therefore receives values from 0 to $n-1$. Subsequently, the algorithm tests whether the `k`-th candidate has already been included in the partial solution (line 10). It is important to note that although the statement can be written in a single line, it requires examining `i` values in the worst case. If the `k`-th candidate is not in the partial solution the method includes it (line 13), and performs a recursive call incrementing the value of `i` in line 16. Lastly, the method does not require "undoing" the assignment in line 19 after carrying out the recursive call, since it simply overwrites the value `sol[i]`. In other words, it is not necessary to reset `sol[i]` to `None` after the recursive call. Nevertheless, the code includes the instruction so that its recursion tree conforms exactly to the one shown in Figure 12.6.

The method `generate_permutations_wrapper` is a wrapper procedure needed to initialize the partial solution with n `None` values, and call `generate_permutations` with `i` set to zero. Lastly, the code includes a basic procedure that prints a list, separating its elements by space bar characters.

Listing 12.3 Code for printing all of the permutations of the elements in a list.

```
def generate_permutations(i, sol, elements):
    # Base case
    if i == len(elements):
        print_permutation(sol)  # complete solution
    else:
        # Generate candidate elements
        for k in range(0, len(elements)):

            # Check candidate validity
            if not elements[k] in sol[0:i]:

                # Include candidate in partial solution
                sol[i] = elements[k]

                # Expand partial solution at position i+1
                generate_permutations(i + 1, sol, elements)

                # Remove candidate from partial solution
                sol[i] = None  # not necessary


def generate_permutations_wrapper(elements):
    sol = [None] * (len(elements))
    generate_permutations(0, sol, elements)


def print_permutation(sol):
    for i in range(0, len(sol)):
        print(sol[i], ' ', end='')
    print()
```

Finally, similarly to the rest of the algorithms included in the chapter, the method uses a loop in order to consider including every possible candidate in the partial solution. Without considering constraints in the internal nodes the resulting algorithm would lead to a full n-ary recursion tree (every internal node would have n children), which would contain n^n leaves. Naturally, we could test whether the partial solutions at those n^n leaves are valid. However, it is much more efficient to examine the validity of a candidate in an internal node, and discard unnecessary recursive calls as soon as the algorithm detects that a partial solution will never be valid. Thus, in line 6 the condition verifies that the candidate is

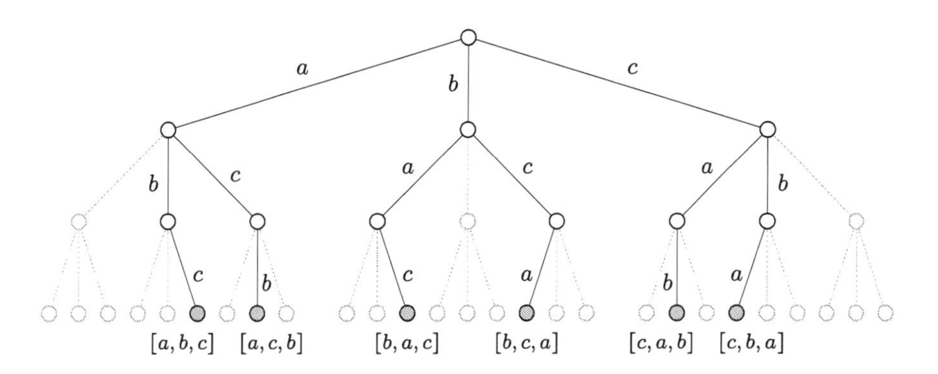

Figure 12.7 Pruning a recursion tree as soon as a partial solution is not valid.

valid, according to the constraints of the problem. This condition speeds up the algorithm dramatically by pruning the recursion tree, as shown in Figure 12.7. The lighter gray nodes and edges indicate method calls of the full ternary tree that are not executed when pruning the tree with the condition in line 6. Finally, the resulting recursion tree (depicted through dark nodes and edges) is identical to the one in Figure 12.6, and contains $n!$ leaves. Although it is a huge quantity even for relatively small values of n, it is much smaller than n^n.

12.2.2.2 Validity of partial solutions with additional data structures

Instead of examining the validity of a partial solution by analyzing its first `i` values (they have to be different than `elements[k]`), the code in Listing 12.4 uses the additional Boolean parameter list `available`, of size n. The idea consists of storing whether a particular item of the input list `elements` is not yet in the partial solution. The list is therefore initialized to n `True` values in a wrapper method. In this case, the condition in line 10 is much simpler, and can be evaluated in $\Theta(1)$ time. Subsequently, if the `k`-th element is indeed available the procedure first includes it in the partial solution (line 13), marks it as unavailable (line 16), and invokes itself incrementing the parameter `i` (lines 19 and 20). In addition, in this case it is necessary to undo the change to `available` in line 23. Observe that when the method returns from the recursive call it can carry out more iterations of the loop, inserting other elements at position `i` in the partial solution, but the list `available` has to be restored to its initial values. In particular, `available[k]` has to be set to

Listing 12.4 Alternative code for printing all of the permutations of the elements in a list.

```
def generate_permutations_alt(i, available, sol, elements):
    # Base case
    if i == len(elements):
        print_permutation(sol)  # complete solution
    else:
        # Generate candidate elements
        for k in range(0, len(elements)):

            # Check candidate validity
            if available[k]:

                # Include candidate in partial solution
                sol[i] = elements[k]

                # k-th candidate no longer available
                available[k] = False

                # Expand partial solution at position i+1
                generate_permutations_alt(i + 1, available,
                                          sol, elements)

                # k-th candidate available again
                available[k] = True


def generate_permutations_alt_wrapper(elements):
    available = [True] * (len(elements))
    sol = [None] * (len(elements))
    generate_permutations_alt(0, available, sol, elements)
```

`True`, since the algorithm has to include `elements[k]` at some point in a further position in the partial solution. Finally, although the method requires an extra input parameter, it is more efficient than the algorithm in Listing 12.3.

12.3 THE N-QUEENS PROBLEM

The n-queens puzzle was previously introduced in Section 12.1. Recall that it is a constraint satisfaction problem whose solutions correspond to permutations of the rows where the queens are placed. Therefore, we can use the code that generates permutations in Listing 12.4 as a start-

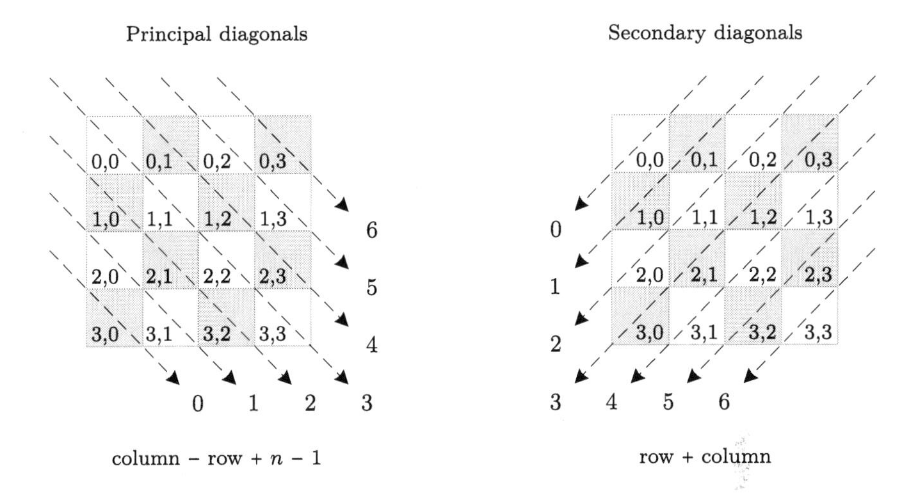

Figure 12.8 Indexing principal and secondary diagonals on a matrix or chessboard.

ing point for building the backtracking method that solves the problem. Although the final algorithm will introduce a few modifications, its structure will be very similar to the method that generates permutations.

Consider the method `generate_permutations_alt` and its input parameters. It permutes the items in the list `elements`, which can contain arbitrary numbers, characters, or other types of data. In this case the elements that we have to permute are the rows of the chessboard. Since they are simply the integers from 0 to $n-1$, we will be able to write the code omitting the list `elements`. However, we will use the list `sol` that represents a partial solution containing rows, the parameter `i` that indicates the column where a new queen will be placed, and the Boolean list `available` that specifies which rows (candidates) are free to be incorporated in the partial solutions. We will change the name of this list to `free_rows`, since even though there might not be a queen in a particular row, the row might not be *available* for inclusion in the partial solution if there is a conflict related to a diagonal.

In addition to those parameters, we can use two more Boolean lists in order to indicate whether there are queens on the diagonals of the chessboard. There are two types of diagonals: (a) principal diagonals, which are parallel to the main diagonal that runs from the top-left to the bottom-right corners of the board; and (b) secondary diagonals, which are perpendicular to the principal diagonals. There are exactly $2n-1$

of each kind, as illustrated in Figure 12.8 for $n = 4$. Thus, we can use the Boolean lists `free_pdiags` and `free_sdiags`, both of length $2n-1$, which will contain `True` values if there are no queens on the principal and secondary diagonals, respectively. The figure also shows how we can enumerate the diagonals, where the pair of numbers in a square indicate its row and column on the board. On the one hand, note that the sum of the column and row for squares along a specific secondary diagonal is constant. These sums range from 0 to $2n-2$, and can be used as indices to `free_sdiags`. On the other hand, we can subtract the row from the column in order to enumerate the principal diagonals. However, this provides values from $-(n-1)$ to $n-1$. Since we need to use nonnegative indices, we can simply add $n-1$ in order to obtain a valid index for a principal diagonal that will range from 0 to $2n-2$. These operations allow us to quickly determine the diagonals in which a square lies according to its row and column.

12.3.1 Finding every solution

Listing 12.5 shows a recursive backtracking algorithm that finds all of the solutions to the n-queens puzzle (for $n = 8$ there are 92 solutions, although only 12 are truly distinct in the sense that they cannot be obtained from others through rotations and reflections). In the base case the method processes a valid permutation if it is a complete solution (lines 6 and 7). For example, it could print it (see method `print_chessboard` in Listing 12.6), or draw queens on a chessboard image. The loop in line 12 is used to iterate through all of the possible candidates that could be appended to the partial solution. The variable `k` therefore represents rows of the chessboard. Subsequently, the `if` statement in lines 16 and 17 examines whether the `k`-th row is a valid candidate for the `i`-th column. In particular, it makes sure that the row and the corresponding diagonals are free (i.e., that they do not contain a queen). If the result is `True` the algorithm can proceed by incorporating the `k`-th row into the partial solution (line 20). Since this implies placing a queen on the chessboard, it is necessary to update the Boolean data structures in order to indicate that a row and two diagonals will no longer be free (lines 24–26). The method can then call itself (in lines 30 and 31) with the modified lists in order to continue placing queens on the next column (`i`+1). Lastly, after the recursive call the method needs to get ready to test whether it can place a queen in the next row, which would occur in the next iteration of the loop. Therefore, it is necessary

Listing 12.5 Code for finding all of the solutions to the n-queens puzzle.

```
def nqueens_all_sol(i, free_rows, free_pdiags,
                    free_sdiags, sol):
    n = len(sol)

    # Test if the partial solution is a complete solution
    if i == n:
        print_chessboard(sol)  # process the solution
    else:

        # Generate all possible candidates that could
        # be introduced in the partial solution
        for k in range(0, n):

            # Check if the partial solution with the
            # k-th candidate would be valid
            if (free_rows[k] and free_pdiags[i - k + n - 1]
                    and free_sdiags[i + k]):

                # Introduce candidate k in the partial solution
                sol[i] = k

                # Update data structures, indicating that
                # candidate k is in the partial solution
                free_rows[k] = False
                free_pdiags[i - k + n - 1] = False
                free_sdiags[i + k] = False

                # Perform a recursive call in order to include
                # more candidates in the partial solution
                nqueens_all_sol(i + 1, free_rows, free_pdiags,
                                free_sdiags, sol)

                # Eliminate candidate k from the partial
                # solution, and restore the data structures,
                # indicating that candidate k is no longer
                # in the partial solution
                free_rows[k] = True
                free_pdiags[i - k + n - 1] = True
                free_sdiags[i + k] = True


def nqueens_wrapper(n):
    free_rows = [True] * n
    free_pdiags = [True] * (2 * n - 1)
    free_sdiags = [True] * (2 * n - 1)
    sol = [None] * n
    nqueens_all_sol(0, free_rows, free_pdiags, free_sdiags, sol)
```

to restore the values of the Boolean lists, as if the `k`-th row had not been included in the partial solution. Finally, the algorithm modifies the lists specifying that the `k`-th row, and the diagonals related to the square at column `i` and row `k`, will be free of queens (lines 37–39).

12.3.2 Finding one solution

In many situations we are only interested in finding one solution to a problem. Moreover, some problems only have one solution (for example, the sudoku puzzle). Therefore, we can code backtracking methods that stop searching for solutions as soon as they find one. In Python it is possible to insert a `return` statement after finding a solution. For example, we can include it after the `print_chessboard(sol)` instruction in the `nqueens_all_sol` method of Listing 12.5. However, in many programming languages this is not possible.

There are many ways to implement a program that finds and processes one solution. Listing 12.6 shows one approach where the recursive method is a Boolean function, which returns `True` if it is able to find a solution. It is very similar to the code that finds all of the solutions. On the one hand, the method `nqueens_one_sol_wrapper` calls the recursive backtracking function with the `if` statement, and only prints the solution if the function is able to find it. Note that the list `sol` is mutable (it can be understood as a parameter passed by reference) and will end up storing the solution after the method call if it exists. Naturally, another option is to print the solution in the base case of the recursive function before returning. On the other hand, the function `nqueens_one_sol` first defines a Boolean variable `sol_found`, initialized to `False`, which indicates if the method has found a solution. The `for` loop is replaced by a `while` loop in order to stop iterating as soon as a solution is found. The body of the loop is identical to the one in the procedure `nqueens_all_sol`, but the result of the recursive call is assigned to the variable `sol_found`. Finally, the function can return the value in `sol_found` after executing the `while` loop.

12.4 SUBSET SUM PROBLEM

Given some set S of n positive integers, and a particular integer x, the goal of this problem is to determine whether the sum of some subset of elements of S is equal to x. Formally, if s_i denotes the i-th integer in S, the idea consists of checking whether there exists a subset $T \subseteq S$ such

Listing 12.6 Code for finding one solution to the n-queens puzzle.

```
def nqueens_one_sol(i, free_rows, free_pdiags,
                    free_sdiags, sol):
    n = len(sol)
    sol_found = False

    if i == n:
        return True
    else:
        k = 0
        while not sol_found and k < n:
            if (free_rows[k] and free_pdiags[i - k + n - 1]
                    and free_sdiags[i + k]):

                sol[i] = k

                free_rows[k] = False
                free_pdiags[i - k + n - 1] = False
                free_sdiags[i + k] = False

                sol_found = nqueens_one_sol(i + 1, free_rows,
                                            free_pdiags,
                                            free_sdiags, sol)

                free_rows[k] = True
                free_pdiags[i - k + n - 1] = True
                free_sdiags[i + k] = True

            k = k + 1

        return sol_found


def nqueens_one_sol_wrapper(n):
    free_rows = [True] * n
    free_pdiags = [True] * (2 * n - 1)
    free_sdiags = [True] * (2 * n - 1)
    sol = [None] * n

    if nqueens_one_sol(0, free_rows, free_pdiags,
                       free_sdiags, sol):
        print_chessboard(sol)


def print_chessboard(sol):
    for i in range(0, len(sol)):
        print(sol[i], ' ', end='')
    print()
```

that:

$$\sum_{s_i \in T} s_i = x. \tag{12.1}$$

In particular, we will design a backtracking algorithm that prints every subset T of S that satisfies (12.1). For example, if $S = \{1, 2, 3, 5, 6, 7, 9\}$ and $x = 13$, the method should print the following five subsets on the console: $\{6, 7\}$, $\{2, 5, 6\}$, $\{1, 5, 7\}$, $\{1, 3, 9\}$, and $\{1, 2, 3, 7\}$.

A naive brute-force solution to the problem consists of generating every possible subset T of S, and testing whether (12.1) is `True`. This exhaustive search can be carried out through a procedure similar to `generate_subsets` in Listing 12.1. The set S could be represented by the input list `elements`, and the method would contain an additional parameter that stores the value of x. Furthermore, the procedure would have to verify (12.1) in the base case before printing a subset. In other words, it would have to check (12.1) in each leaf of the recursion tree.

However, it is possible to speed up the search by using a backtracking approach. Listing 12.7 shows a particular implementation where a partial solution (`sol`) represents the subset T. In order to avoid calculating the sum of the elements in the partial solution in every recursive call, it uses an additional accumulator parameter `psum` that contains the sum, and which is obviously initialized to zero in the wrapper method. The key observation that leads to the algorithm is that we can prune the recursion tree as soon as the sum of elements related to the partial solution is equal to x (i.e., when (12.1) is satisfied); or when it is greater than x, where it will not be possible to satisfy (12.1) by expanding the partial solution, since the elements of S are positive.

Firstly, the method includes a base case (in lines 3 and 4) in order to print the subset associated with the partial solution if it satisfies (12.1). This allows us to prune the tree at an internal node, of depth `i`, of the recursion tree. In the base case T can only contain the first `i` elements of S. In other words, only the first `i` components of the partial solution are meaningful. If the n−`i` last entries can take arbitrary values, it would be necessary to pass the value of `i` to the printing method, and specify the "true" size of the partial solution. Instead, the algorithm calls the method `print_subset_binary` in Listing 12.1, since these last values will always be zero, indicating that the elements are not in the partial solution (T). This is achieved by initializing the partial solution with n zeros (in `print_subset_sum_wrapper`), denoting the empty set; and by updating the partial solution appropriately so that its n−`i` last items

Listing 12.7 Backtracking code for solving the subset sum problem.

```
1  def print_subset_sum(i, sol, psum, elements, x):
2      # Base case
3      if psum == x:
4          print_subset_binary(sol, elements)
5      elif i < len(elements):
6          # Generate candidates
7          for k in range(0, 2):
8  
9              # Check if recursion tree can be pruned
10             if psum + k * elements[i] <= x:
11 
12                 # Expand partial solution
13                 sol[i] = k
14 
15                 # Update sum related to partial solution
16                 psum = psum + k * elements[i]
17 
18                 # Try to expand partial solution
19                 print_subset_sum(i + 1, sol, psum, elements, x)
20 
21                 # not necessary:
22                 #psum = psum - k*elements[i]
23 
24         # Make sure a 0 indicates the absence of an element
25         sol[i] = 0
26 
27 
28 def print_subset_sum_wrapper(elements, x):
29     sol = [0] * (len(elements))
30     print_subset_sum(0, sol, 0, elements, x)
```

are always zeros when reaching a base case (this is implemented through the assignment in line 25).

If the partial solution does not satisfy (12.1) the method can simply terminate if the partial solution has n elements, without carrying out any operation. Thus, with the condition in line 5 the algorithm only continues expanding partial solutions if `i`$< n$. In that case it uses a loop in order to process the two possibilities of ignoring (`k`=0) or including (`k`=1) the `i`-th element of S in the partial solution. Since `k` is a number we can use it within the expression `psum + k*elements[i]` to indicate the new sum associated with the partial solution. Notice that when `k`=0 it does not change, while the method adds the `i`-th element of S when `k`=1.

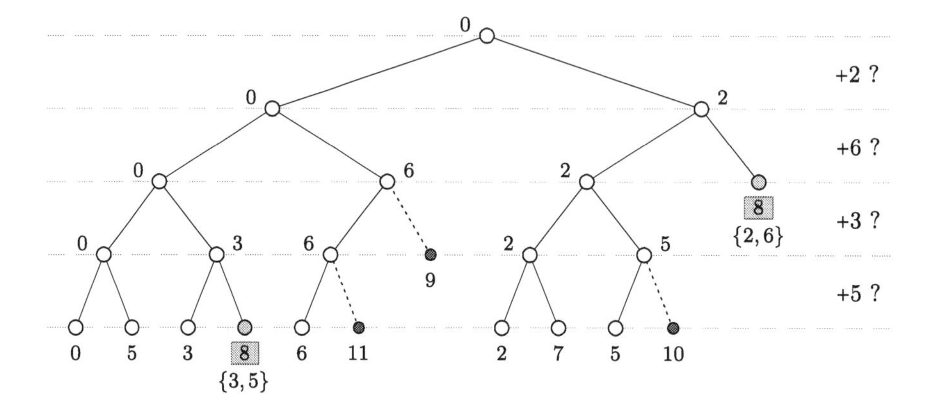

Figure 12.9 Recursion tree of the procedure that solves the subset sum problem for $S = \{2, 6, 3, 5\}$ and $x = 8$.

Therefore, the condition in line 10 makes sure that the new sum is less than or equal to x before including the element, and proceeding to make additional recursive calls. If the condition is `True` the method expands the partial solution with the corresponding candidate (line 13), updates `psum` (line 16), and carries out the recursive call (line 19). Afterwards, it is not necessary to restore the value of `psum` (in line 22) since in the first iteration of the loop `k`=0, which implies that the value of `psum` is not modified. Lastly, the algorithm sets the value of the partial solution to zero at position `i` when terminating, which is necessary in order to print the partial solutions correctly in the base case (its $n-$`i` last items must always be zero).

Finally, Figure 12.9 shows the recursion tree of the method `print_subset_sum`, for $S = \{2, 6, 3, 5\}$ and $x = 8$. The partial solutions are constructed at each level as in the methods in Section 12.2.1. When descending through a left branch the algorithm does not include the i-th element of S in T, but it incorporates it when advancing down a right branch. In this case the numbers next to the nodes indicate the sum of the items included in the partial solutions (i.e., `psum`) when invoking the method. Observe that the tree is pruned as soon as it finds a solution that satisfies (12.1), or if it detects that the sum of the elements in T is greater than x. Lastly, a call to `print_subset_sum_wrapper([2,6,3,5],8)` produces the correct result:

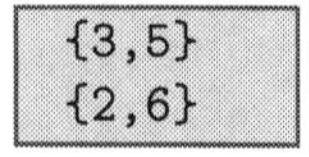
```
{3,5}
{2,6}
```

Finally, observe that without the assignment in line 25, the partial solution would hold a 1 in its third position when reaching the base case associated with the subset $\{2, 6\}$, and the method would print `{2,6,3}`.

12.5 PATH THROUGH A MAZE

This section describes a backtracking algorithm for finding a path through a maze, such as the one illustrated in Figure 12.10(a). A maze is defined as a rectangular array of cells that can either be empty or constitute a wall. We will code it by using a list of lists (of characters), denoted as `M`, where the first list will contain the cells of the top row of the maze (from left to right), the second list will represent the row below the first one, and so on. In order to define a maze a user can write each row as a line of characters, separated by a space bar character, as shown in (b). Specifically, an 'E' indicates an empty cell, which can be traversed by a path, and a 'W' indicates a wall. Since it is time consuming to enter large mazes manually, it is convenient to store the array of characters in a file that can be loaded automatically when running the program (this is also beneficial for debugging). The method will also receive parameters that indicate the cells where a path enters and exits the maze. In the figure these cells are the top-left and bottom-right ones, respectively. Thus, the initial cell in the example is `M[0][0]`.

The backtracking algorithm carries out an exhaustive search for paths that start at the initial cell and arrive at the final cell. Having reached some particular cell in an intermediate step, the algorithm needs to analyze every possible path when advancing to a neighboring cell going upwards, downwards, rightwards, or leftwards. Since the final cell is in the bottom-right corner of the maze the algorithm can begin by searching paths that start going downwards, then towards the right, subsequently upwards, and finally towards the left. For this particular order of search the method takes the steps shown in Figure 12.10(c). Observe that when arriving at an empty cell the method first tries to expand the path downwards. If it cannot advance it will continue trying to move towards the right, then upwards, and finally towards the left. If it cannot proceed in any of the four directions it must backtrack to a previous cell, and continue to search for other alternative paths. In the example the initial cell is at position $(0, 0)$, corresponding to row 0 and column 0. The method will mark that cell as being part of the current path, and update `M[0][0]` to 'P'. The first move advances to position $(1, 0)$, and also marks the corresponding cell with a 'P'. The method

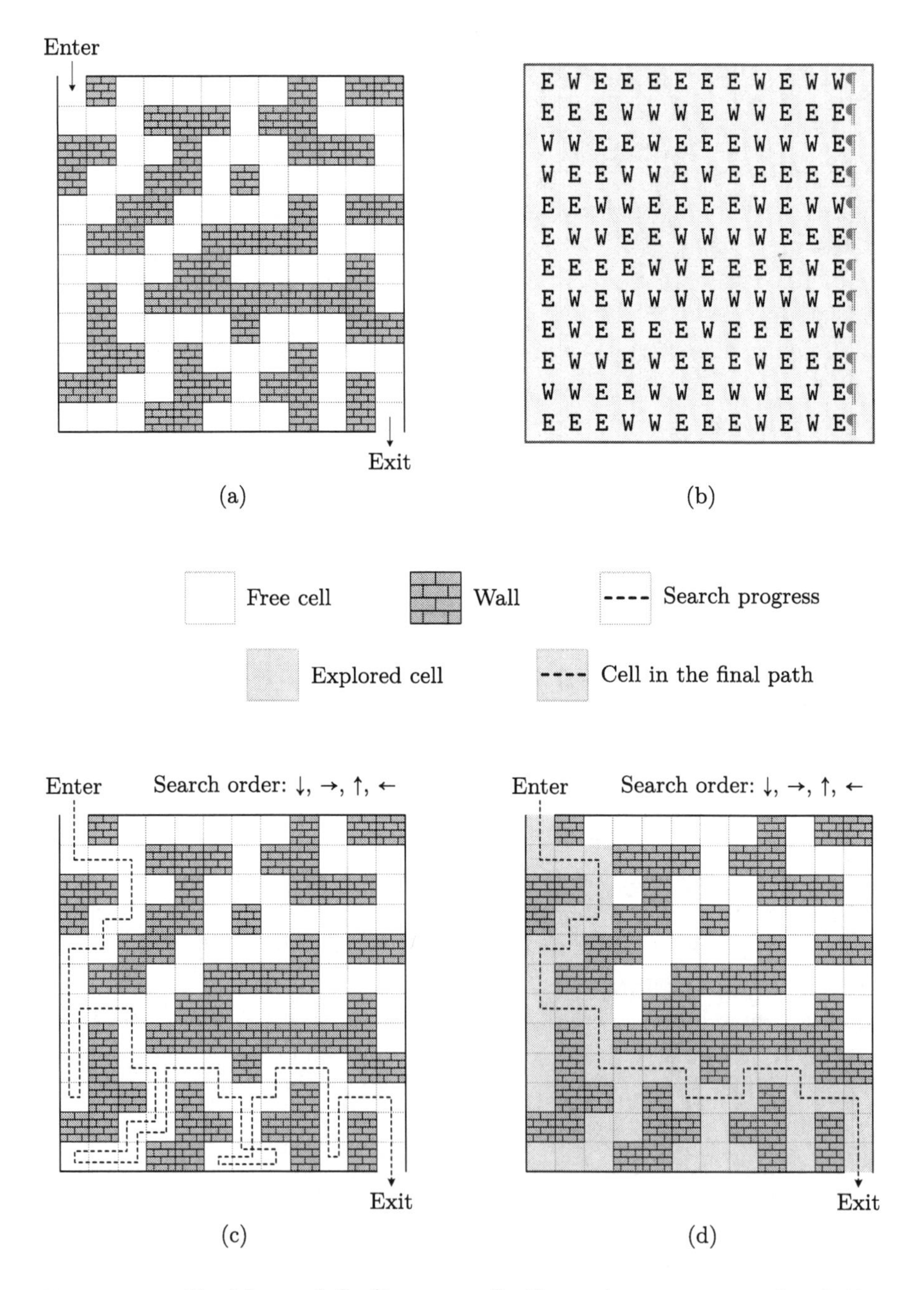

Figure 12.10 Problem of finding a path through a maze, and solution through backtracking when searching in a particular order.

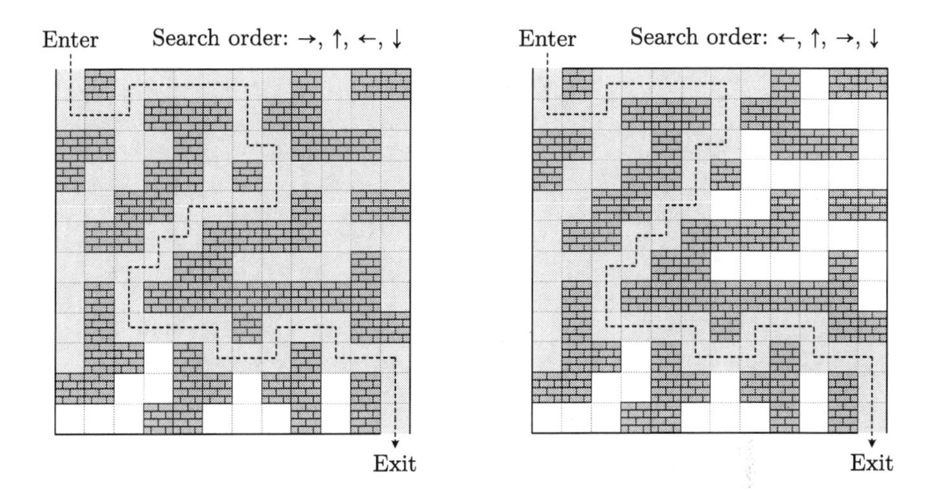

Figure 12.11 Different paths through a maze, depending on the search order.

then tries to advance downwards, but hits a wall located at $(2, 0)$. Since a path with that cell would not be valid, the algorithm continues by trying to advance towards the right, which it can do since the cell at $(1, 1)$ is empty. Eventually the path reaches the cell at (6,0), where the method will also begin by trying to advance downwards. The path ends up advancing to the cell at $(9, 0)$, where it hits a dead end. In that case, not only are there walls below and to the right, but going upwards is not allowed because the cell on top is already part of the path. Furthermore, going towards the left would mean exiting the limits of the maze. Since the algorithm cannot advance in any direction, it backtracks to the cell at $(8, 0)$, where it has not yet tried to advanced towards the right, upwards, or leftwards. However, these possibilities are also not allowed. This also occurs when returning to the cell at $(7, 0)$. After backtracking to the cell at $(6, 0)$, it explores new paths that go towards the right, after having exhaustively explored all of the possible paths going downwards. This process is repeated until the algorithm finds the exit cell. The solution is shown in (d), where the explored cells appear with a shaded background.

Backtracking algorithms can stop searching for solutions when they find the first one, or they can continue computing every solution. We will see an algorithm that halts as soon as it finds a path through the maze. In this regard, the order in which it examines the possible paths

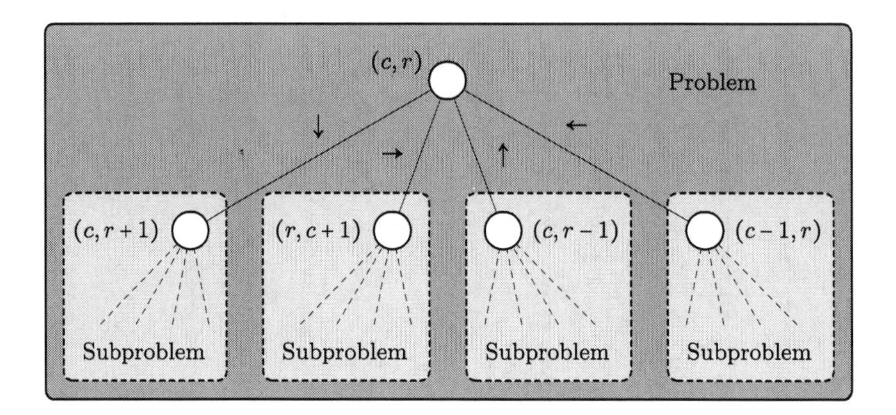

Figure 12.12 Decomposition of the problem of finding a path through a maze.

can lead to different solutions, as illustrated in Figure 12.11. Observe that the explored cells are different in each case.

The solution to this problem is clearly not a permutation of cells. This can be seen, for instance, since the length of a path (i.e., a solution) is not fixed. Furthermore, although a path consists of a subset of cells, these appear ordered in a solution. Therefore, it is not appropriate to model a solution as a subset. Instead, we will use a matrix of characters to code a solution, where the cells in a particular path will be marked with a 'P' character. Thus, a partial solution will be the input maze that also contains a path from the initial cell to some other free cell. Having arrived at this last cell, the method has to consider including four candidate cells in the maze. Therefore, the problem can be decomposed recursively as shown in Figure 12.12, where r and c denote a particular row and column, respectively. Every node of the recursion tree will have four possible children, which correspond to the four directions in which it is possible to advance. However, the algorithm will have to check that the new cell is free before expanding the path through a recursive call.

Listing 12.8 shows a possible implementation of a backtracking algorithm that solves the problem. Firstly, the parameters of the wrapper method are the initial maze, and four integers that indicate the coordinates of the entering (`enter_row` and `enter_col`) and exiting (`exit_row` and `exit_col`) cells. It also declares the list `incr`, which specifies the search directions. Each tuple (x, y) in the list indicates how to increment the column c and row r of a cell in order to advance to a neighboring

Listing 12.8 Backtracking code for finding a path through a maze.

```
def find_path_maze(M, row, col, incr, exit_row, exit_col):
    # Base case: check if path found
    if row == exit_row and col == exit_col:
        return True  # Solution found
    else:
        sol_found = False
        # Generate candidates
        k = 0
        while not sol_found and k < 4:

            # New candidate cell
            new_col = col + incr[k][0]
            new_row = row + incr[k][1]

            # Test candidate validity
            if (new_row >= 0 and new_row < len(M)
                    and new_col >= 0 and new_col < len(M[0])
                    and M[new_row][new_col] == 'E'):

                # Add to path (partial solution)
                M[new_row][new_col] = 'P'

                # Try to expand path starting at new cell
                sol_found = find_path_maze(
                    M, new_row, new_col, incr,
                    exit_row, exit_col)

                # Mark as empty if new cell not in solution
                if not sol_found:
                    M[new_row][new_col] = 'E'

            k = k + 1

        return sol_found


def find_path_maze_wrapper(M, enter_row, enter_col,
                           exit_row, exit_col):
    # search directions
    incr = [(0, 1), (1, 0), (0, -1), (-1, 0)]

    M[enter_row][enter_col] = 'P'
    return find_path_maze(M, enter_row, enter_col, incr,
                          exit_row, exit_col)
```

one. In other words, the new cell would be $(c + x, r + y)$. The order of search for this algorithm is therefore: down, right, up, and left. Observe that adding a unit to a row implies going down (since the first row is located at the top of the maze), while adding one to a column signifies moving towards the right. Lastly, the wrapper method indicates that the initial cell will form part of the solution path, and returns the (Boolean) result of calling the recursive backtracking method. The parameter `M` of `find_path_maze` is both the initial maze and the partial solution. The first two parameters of the recursive function indicate the coordinates of the last cell of the partial path `M`, and the algorithm will try to expand it advancing to one of its neighboring cells. Finally, the method receives `incr`, and the coordinates of the exiting cell.

The recursive function `find_path_maze` returns `True` if it finds a path through the maze, where the solution would be stored precisely in `M`. It declares the variable `sol_found`, initialized as `False`, which is set to `True` at the base case if the algorithm finds a complete solution. Otherwise, it uses a `while` loop to generate the four candidate cells, but one that terminates as soon as a solution is found. The variables `new_col` and `new_row` constitute the new candidate (lines 12 and 13). Subsequently, the algorithm examines whether it is valid. In particular, it has to be within the limits of the maze (lines 16 and 17), and it has to be empty (line 18). If the cell does not violate the constraints of the problem the method incorporates it to the path (line 21), and calls the recursive method in lines 24–26, where the output is stored in `sol_found`. If a solution is not found the cell will not belong to the path. Therefore, it sets its value back to 'E', indicating that it is empty (line 30). This is necessary since the path in the final maze is determined through the cells marked as 'P'. Without this condition all of the explored cells would contain a 'P' character. Lastly, the method can simply return the value of `sol_found` after exiting the loop (line 34).

Finally, Listing 12.9 shows additional code that can be used in order to execute the program and draw the maze. The method `read_maze_from_file` reads a text file defining a maze and returns the corresponding list of lists. The basic iterative procedure `draw_maze` uses the Matplotlib package to depict a maze. Finally, the last lines of the code read a maze from a file, and draw it only if there exists a path through it, where the initial and final cells are the top-left and bottom-right ones, respectively.

Listing 12.9 Auxiliary code related to the backtracking methods for finding a path through a maze.

```
import matplotlib.pyplot as plt
from matplotlib.patches import Rectangle

def read_maze_from_file(filename):
    file = open(filename, 'r')
    M = []
    for line in file.readlines():
        M.append([x[0] for x in line.split(' ')])
    file.close()
    return M

gray = (0.75, 0.75, 0.75)
black = (0, 0, 0)
red = (0.75, 0, 0)
green = (0, 0.75, 0)

def draw_maze(M, enter_row, enter_col, exit_row, exit_col):
    nrows = len(M)
    ncols = len(M[0])
    fig = plt.figure()
    fig.patch.set_facecolor('white')
    ax = plt.gca()

    if enter_row is not None and enter_col is not None:
        ax.add_patch(Rectangle((enter_col, nrows - enter_row),
                               1, -1, linewidth=0, facecolor=green,
                               fill=True))
    if exit_row is not None and exit_col is not None:
        ax.add_patch(Rectangle((exit_col, nrows - exit_row),
                               1, -1, linewidth=0, facecolor=red,
                               fill=True))

    for row in range(0, nrows):
        for col in range(0, ncols):
            if M[row][col] == 'W':
                ax.add_patch(Rectangle((col, nrows - row), 1, -1,
                                       linewidth=0, facecolor=gray))
            elif M[row][col] == 'P':
                circ = plt.Circle((col + 0.5, nrows - row - 0.5),
                                  radius=0.15, color=black, fill=True)
                ax.add_patch(circ)

    ax.add_patch(Rectangle((0, 0), ncols, nrows, edgecolor=black,
                           fill=False))
    plt.axis('equal')
    plt.axis('off')
    plt.show()

M = read_maze_from_file('maze_01.txt')  # some file
# Enter at top-left, exit at bottom-right
if find_path_maze_wrapper(M, 0, 0, len(M) - 1, len(M[0]) - 1):
    draw_maze(M, 0, 0, len(M) - 1, len(M[0]) - 1)
```

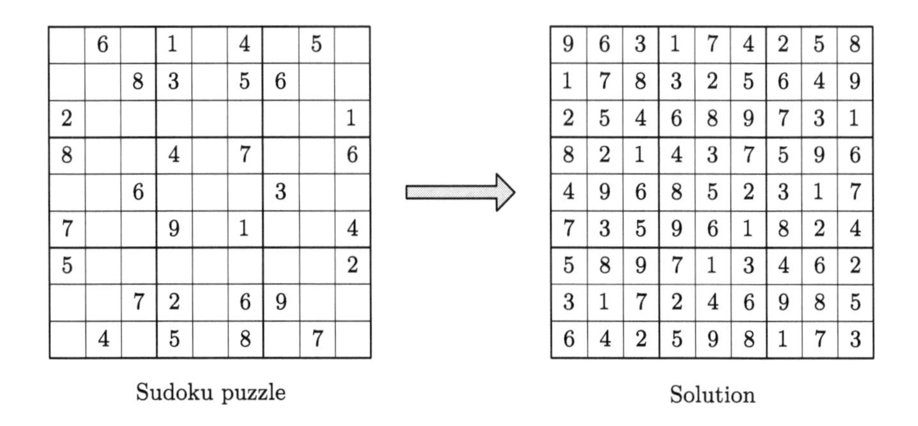

Figure 12.13 An instance of the sudoku puzzle and its solution.

12.6 THE SUDOKU PUZZLE

The sudoku puzzle is a constraint satisfaction problem where the goal is to fill a 9×9 grid with the digits from 1 to 9 (they can be any nine different symbols, since their numerical values are irrelevant). A particular digit can appear only once in each row, each column, and each of nine 3×3 non-overlapping subgrids, also called boxes, which cover the 9×9 grid. Figure 12.13 shows an instance of the problem, which consists of a partially filled grid with fixed initial digits. The problem is well-posed when the puzzle has a unique solution (the initial grid must contain at least 17 digits).

We will use a list of lists of digits to implement the grid, where a zero represents an empty cell. The partial solutions will be partially filled grids, and the algorithm must enumerate the cells somehow in order to expand the partial solutions with new digits. The algorithm described below considers expanding them in row-major order, where it will begin inserting valid digits (i.e., candidates) in the top row from left to right, then in the second row, and so on. Thus, the algorithm will have found a solution if it is able to place a valid digit in the bottom-right cell.

In a recursive case the method generates nine possible candidates to include in an empty cell. Thus, a node of the recursion tree could have up to nine potential children, as shown in Figure 12.14(a). However, the problem requires handling a scenario that has not appeared in previous problems. When expanding the partial solution some of the digits in the grid will be the initial fixed ones. When the algorithm processes a cell

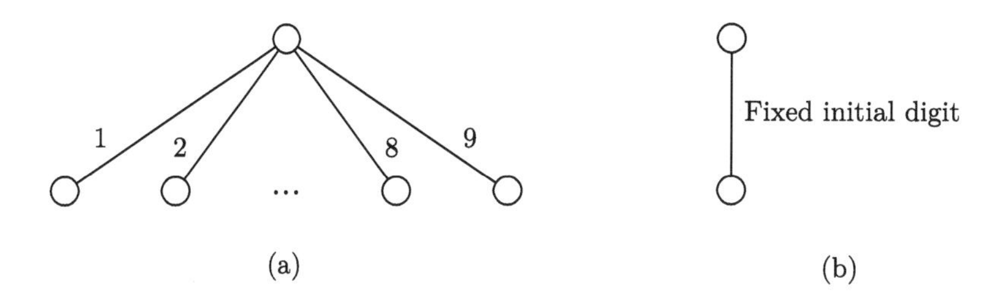

Figure 12.14 Recursive cases for the sudoku solver.

with one of these digits it must skip it (it cannot expand the partial solution for that cell), and carry out a recursive call that processes the following cell. This requires incorporating a second recursive case that is illustrated in (b).

Listings 12.10 and 12.11 show a backtracking procedure that solves the problem, together with several auxiliary methods. The recursive method `solve_sudoku_all_solutions` assumes that a sudoku might not be well-posed, and could potentially have several solutions (or even 0). Thus, it prints all of the valid solutions to a given sudoku grid. For example, if the top row of the sudoku in Figure 12.13 is replaced by an empty row, there are 10 different ways to fill the sudoku grid with digits.

The inputs to the recursive procedure are the coordinates (`row` and `col`) of a cell where it will try to include digits in order to expand the partial solution represented by the list of lists of digits `S`. Since the procedure expands the partial solution in row-major order, starting at cell $(0,0)$, it will have obtained a valid solution when `row` is equal to 9. In that base case it can simply print the sudoku grid (line 4). Otherwise, the method checks whether the current cell is not empty (i.e., if it contains one of the initial fixed digits). If the result is `True` it skips the cell, and continues by invoking the recursive method (line 14) with the coordinates of the next cell in the row-major order (computed in line 11).

In the second recursive case it uses a loop to generate the nine possible candidates to include in the empty cell. Afterwards, in line 20 it analyzes whether it is feasible to incorporate candidate `k` in the partial solution `S`, at cell (`row`,`col`). If it is, the method includes the candidate in the partial solution (line 23), and continues carrying out a recursive call with the next cell (line 29). When the loop finishes, the method needs to undo the changes made to the cell, leaving it empty (line 32). This is necessary for future explorations of solutions. Note that the cell has to

Listing 12.10 Code for solving a sudoku puzzle.

```
def solve_sudoku_all_sols(row, col, S):
    # Check if sudoku is complete
    if row == 9:
        print_sudoku(S)  # print the completed sudoku
        print()
    else:
        # Check if digit S[row][col] is an initial fixed symbol
        if S[row][col] != 0:

            # Advance to a new cell in row-major order
            (new_row, new_col) = advance(row, col)

            # Try to expand the partial solution
            solve_sudoku_all_sols(new_row, new_col, S)
        else:
            # Generate candidate digits
            for k in range(1, 10):

                # Check if digit k is a valid candidate
                if is_valid_candidate(row, col, k, S):

                    # Include digit in cell (row,col)
                    S[row][col] = k

                    # Advance to a new cell in row-major order
                    (new_row, new_col) = advance(row, col)

                    # Try to expand the partial solution
                    solve_sudoku_all_sols(new_row, new_col, S)

            # Empty cell (i,j)
            S[row][col] = 0


# Compute the next cell in row-major order
def advance(row, col):
    if col == 8:
        return (row + 1, 0)
    else:
        return (row, col + 1)
```

be free the next time the algorithm processes it after backtracking to an earlier cell.

Listing 12.11 Auxiliary code for solving a sudoku puzzle.

```
import math

# Check if the digit at cell (row,col) is valid
def is_valid_candidate(row, col, digit, S):
    # Check conflict in column
    for k in range(0, 9):
        if k != col and digit == S[row][k]:
            return False

    # Check conflict in row
    for k in range(0, 9):
        if k != row and digit == S[k][col]:
            return False

    # Check conflict in box
    box_row = math.floor(row / 3)
    box_col = math.floor(col / 3)
    for k in range(0, 3):
        for m in range(0, 3):
            if (row != 3 * box_row + k
                    and col != 3 * box_col + m):
                if digit == S[3 * box_row + k][3 * box_col + m]:
                    return False

    return True

# Read a sudoku grid from a text file
def read_sudoku(filename):
    file = open(filename, 'r')
    S = [[None] * 9] * 9
    i = 0
    for line in file.readlines():
        S[i] = [int(x) for x in line.split(' ')]
        i = i + 1
    file.close()
    return S

# Print a sudoku grid on the console
def print_sudoku(S):
    for s in S:
        print(*s)

S = read_sudoku('sudoku01_input.txt')  # Some file
solve_sudoku_all_sols(0, 0, S)
```

The code also includes: (a) a function that returns a tuple with the coordinates of the next cell in row-major order; (b) a function that determines whether the digit placed at some row and column violates the constraints of the problem, where the variables `box_row` and `box_col` are the top-left cells of the 3×3 boxes; (c) a function for reading a sudoku grid from a text file, where each row contains the nine initial digits of a sudoku row, separated by space bar characters; (d) a method for printing the grid; and (e) code for reading and solving a sudoku.

12.7 0-1 KNAPSACK PROBLEM

Backtracking algorithms are also used to solve optimization problems. In the 0-1 knapsack problem there are n objects with values v_i, and weights w_i, for $i = 0, \dots, n-1$. The goal of the problem is to introduce a subset of these objects in a knapsack with total weight capacity C, such that the sum of their values is maximized. Formally, the optimization problem can be written as:

$$\begin{aligned}
\underset{\mathbf{x}}{\text{maximize}} \quad & \sum_{i=0}^{n-1} x_i v_i \\
\text{subject to} \quad & \sum_{i=0}^{n-1} x_i w_i \leq C, \\
& x_i \in \{0, 1\}, \qquad i = 0, \dots, n-1.
\end{aligned}$$

The vector or list $\mathbf{x} = [x_1, \dots, x_n]$ is the variable of the problem, whose components can be either 0 or 1. In particular, the i-th object is introduced in the knapsack when $x_i = 1$. Therefore, $\mathbf{x}$ plays the same role as the binary list in the subset generation or subset sum problem. The sum $\sum_{i=0}^{n-1} x_i v_i$ is called the objective function, and simply adds the values of the objects that are in the knapsack. The constraint $\sum_{i=0}^{n-1} x_i w_i \leq C$ indicates that the sum of their weights cannot exceed the capacity C.

The following subsections describe two approaches that perform an exhaustive search for the **optimal solution**. The first is a standard backtracking algorithm that prunes the recursion tree when the partial solution violates the constraints of the problem (i.e., when the sum of the weights exceeds the capacity of the knapsack). The second approach uses an algorithm design paradigm known as **branch and bound**, which is closely related to backtracking. The method enhances the search by also pruning the recursion tree as soon as the algorithm detects that it will not be able to improve the best solution found in previous steps (by expanding a particular partial solution).

12.7.1 Standard backtracking algorithm

Since the solution is a subset of the n objects we can develop an algorithm whose recursion tree has the same binary structure as the one related to the algorithms for the subset generation or subset sum problem. Figure 12.15 illustrates the recursion tree for weights $\mathbf{w} = [3, 6, 9, 5]$, values $\mathbf{v} = [7, 2, 10, 4]$, and knapsack capacity $C = 15$, which will be input parameters to the recursive method. The nodes represent method calls with specific partial solutions as parameters, where the numbers depicted inside the nodes specify the remaining capacity in the knapsack, and the sum of the values of the objects included in it, as shown in (a). For a given index i in the partial solution, the algorithm can discard introducing the associated object in the knapsack (left child), or it can include it (right child). If it discards the object both of the numbers in the child node do not vary. However, if an element is introduced in the knapsack its remaining capacity decreases by the weight of the object w_i, while the total value of the objects in the knapsack increases by v_i.

The algorithm solves an optimization problem. Thus, it needs to keep track of the best solution found in previous steps, and must update it if it encounters a better one. The full recursion tree is shown in (b), where the shaded nodes indicate the method calls that update this best solution. Assuming that the initial optimal value is a negative number, the method would first update the optimal solution when reaching the first leaf. In that case the partial solution encodes the empty set, where the sum of values is obviously zero. Continuing the tree traversal, the method updates the best solution after introducing the last object in the knapsack ($w_3 = 5$, and $v_3 = 4$). Thus, the remaining capacity at the leaf node is 10, and the total value of the partial solution is 4. The process continues, and updates the best solution three more times (the corresponding best values are 10, 14, and 17). The last update naturally determines the optimal solution to the problem. In this case it consists of the subset formed by the first and third objects, where the optimal sum of values is 17. Lastly, it is also important to note how the method prunes the binary recursion tree. In particular, the remaining capacity associated with a valid partial solution (and node) cannot be negative.

Listing 12.12 shows a possible implementation of the backtracking algorithm. Firstly, similarly to other methods, its input parameters in-

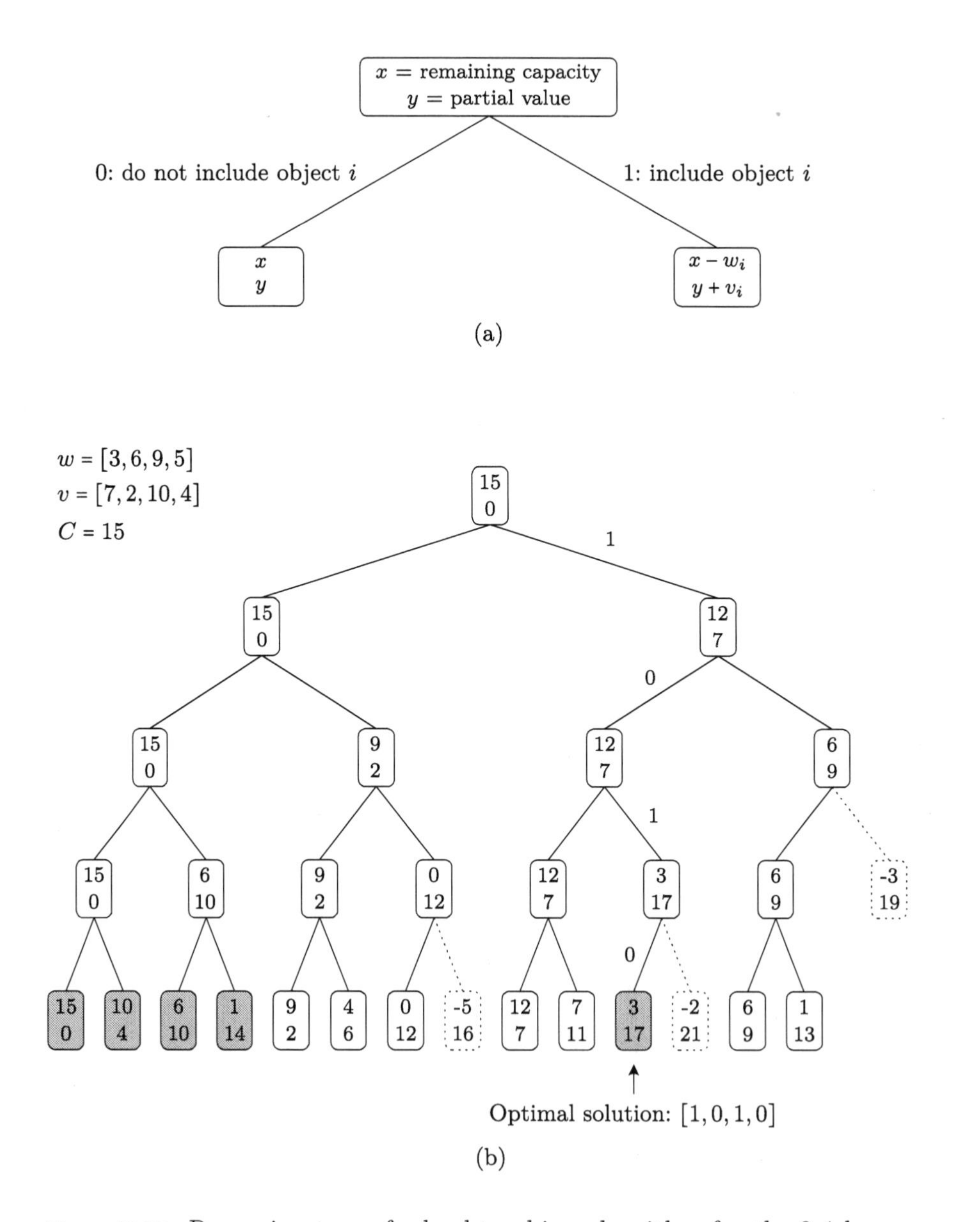

Figure 12.15 Recursion tree of a backtracking algorithm for the 0-1 knapsack problem.

Listing 12.12 Backtracking code for solving the 0-1 knapsack problem.

```
 1 def knapsack_0_1(i, w_left, current_v, sol,
 2                  opt_sol, opt_v, w, v, C):
 3     # Check base case
 4     if i == len(sol):
 5         # Check if better solution has been found
 6         if current_v > opt_v:
 7             # Update optimal value and solution
 8             opt_v = current_v
 9             for k in range(0, len(sol)):
10                 opt_sol[k] = sol[k]
11     else:
12         # Generate candidates
13         for k in range(0, 2):
14 
15             # Check if recursion tree can be pruned
16             if k * w[i] <= w_left:
17 
18                 # Expand partial solution
19                 sol[i] = k
20 
21                 # Update remaining capacity and partial value
22                 new_w_left = w_left - k * w[i]
23                 new_current_v = current_v + k * v[i]
24 
25                 # Try to expand partial solution
26                 opt_v = knapsack_0_1(i + 1, new_w_left,
27                                      new_current_v, sol,
28                                      opt_sol, opt_v, w, v, C)
29 
30     # return value of optimal solution found so far
31     return opt_v
32 
33 
34 def knapsack_0_1_wrapper(w, v, C):
35     sol = [0] * (len(w))
36     opt_sol = [0] * (len(w))
37     total_v = knapsack_0_1(0, C, 0, sol, opt_sol, -1, w, v, C)
38     print_knapsack_solution(opt_sol, w, v, C, total_v)
```

clude the partial solution sol and the index i related to the object that may be introduced in the knapsack. The method also receives the lists of weights **w**, values **v**, and the capacity C. With these parameters it is possible to compute the remaining capacity and the accumulated sum of values of the objects in the knapsack, in i+1 steps. However, it is more efficient to include two extra parameters that contain these quantities. In particular, `w_left` stores the remaining capacity, while `current_v` holds the sum of the values of the objects. The advantage is that they can be updated in a single step.

Furthermore, the method needs to store the best solution found as it is executed. The parameter `opt_sol` stores this information, which will be a copy of the partial solution when reaching a leaf of the recursion tree. Lastly, `opt_v` contains the optimum sum of values associated with `opt_sol`. This parameter is also redundant in the sense that it is possible to compute its value given `opt_sol`. However, it is more efficient to pass this information as a parameter.

As usual, the recursive function begins by analyzing the base case. If the method has completed a partial solution (line 4) it checks whether the current value of the partial solution is greater than the best value found at any previous step (line 6). If the result is `True` it updates `opt_v`, and overwrites the values of `opt_sol` with the ones in `sol`. The method will later return `opt_v` in line 31. The values of the list `opt_sol` will be available after executing the method since it is a mutable object.

In the recursive case the method generates the candidates (zero or one) to include in the binary partial solution list, in line 13. It then uses the condition `k*w[i]<=w_left` to check if it can prune the recursion tree. On the one hand, if `k` = 0 the object is not introduced in the knapsack and the condition is always `True` since `w_left` is nonnegative. However, if `k` = 1 the algorithm can only include the object in the knapsack if it fits. In other words, w_i must be less than or equal to the remaining capacity of the knapsack. The method would continue by updating the partial solution, the remaining knapsack capacity, and the sum of the values of the objects in the knapsack. The function carries out a recursive call with these new parameters, storing the result in `opt_v`, which ends up returning in line 31.

Regarding the wrapper function, `i` is initially 0, `w_left` stores the total capacity C, while `current_v` is 0 since the knapsack is empty. The optimal value `opt_v` can be either 0 or some negative number. In Figure 12.15 we have assumed that it is negative. This simply implies that the method will update the optimal solution even when it consists of an empty set (at the leftmost leaf).

Finally, Listing 12.13 shows a simple iterative code that can be used to print a solution to the problem. Its optimal value can be computed from the partial solution and the list of values. However, the method includes such value as a parameter since the function `knapsack_0_1`

Listing 12.13 Auxiliary code related to the 0-1 knapsack problem.

```
def print_knapsack_solution(sol, w, v, C, opt_value):
    n = len(sol)
    k = 0
    while k < n and sol[k] == 0:
        k = k + 1

    total_weight = 0
    if k < n:
        print('(', w[k], ',', v[k], ')', sep='', end='')
        total_weight = total_weight + w[k]

        for i in range(k + 1, n):
            if sol[i] == 1:
                total_weight = total_weight + w[i]
                print(' + ', sep='', end='')
                print('(', w[i], ',', v[i], ')', sep='', end='')

    print(' => ', '(', total_weight ,
          ',', opt_value , ')', sep='')


w = [3, 6, 9, 5]  # List of object weights
v = [7, 2, 10, 4]  # List of object values
C = 15  # Weight capacity of the knapsack
knapsack_0_1_wrapper(w, v, C)
```

returns it, and is therefore readily available in `knapsack_0_1_wrapper`. The last lines simply define an instance of the problem, and compute its solution.

12.7.2 Branch and bound algorithm

Branch and bound can be viewed as a variant of backtracking that performs a more efficient exhaustive search of solutions to discrete and combinatorial optimization problems. The idea consists of using bounds in order to prune the recursion tree, not only when a partial solution does not satisfy the constraints of the problem, but also when it is certain that expanding the partial solution will not lead to a better solution than the best one found in previous steps.

The key to the branch and bound algorithm that we will examine for the 0-1 knapsack problem relies on using an extra parameter to store the maximum sum of values that could be obtained by expanding a partial solution. If at some method call this value is smaller than the best value found by the algorithm in previous steps, it will not continue to expand the partial solution, since it will not be able to obtain a better solution. This allows us to prune the recursion tree at more nodes, which can lead to a considerably more efficient search.

Figure 12.16 illustrates the recursion tree for the same weights, values, and knapsack capacity as in Figure 12.15. In this case, the numbers inside the nodes indicate the partial value, and a bound on the maximum possible value that we can obtain by expanding the associated partial solution, as shown in (a). Initially, the partial value is 0, and the bound is the sum of all of the values of the objects (i.e., $23 = 7 + 2 + 10 + 4$), as illustrated in (b), since at first we have to contemplate the case where every object fits in the knapsack.

Each internal node of depth i can have two children. Descending through the left branch implies not introducing object i in the knapsack. Therefore, the partial value does not change, but the bound is decreased by v_i, since the object will not contribute its value to the total sum of a complete solution. Instead, the object is introduced in the knapsack when descending through the right branch. This implies adding v_i to the partial value, but the bound remains unaltered.

Similarly to the example in Figure 12.15, the method updates the best solution found so far in the shaded leaves. In addition, the nodes drawn with a dotted contour indicate method calls that are not carried out because the sum of the weights of the objects exceeds the knapsack's capacity. Furthermore, the figure shows a new type of node with a lighter dashed contour. The algorithm also discards the method calls associated with these nodes since the value of the bound is smaller than the best sum of values encountered previously. For instance, the best value after reaching the fourth leaf is 14. Afterwards, consider reaching the node with partial value 2 and bound 16. Discarding the third object implies reducing the bound by 10. This means that the maximum sum of values that it is possible to obtain by expanding the partial solution is $16 - 10 = 6$. Since $6 < 14$ (depicted underneath the node), the algorithm avoids calling the method, pruning the recursion tree. Observe that it has the same structure as the recursion tree in Figure 12.15, but it contains less nodes since it prunes the tree on more occasions. In practice, this enhancement can have a dramatic effect regarding efficiency.

Listing 12.14 shows a branch and bound code for solving the problem that is very similar to the one in Listing 12.12. The main difference between them is the new (fourth) parameter `max_v` that stores the bound.

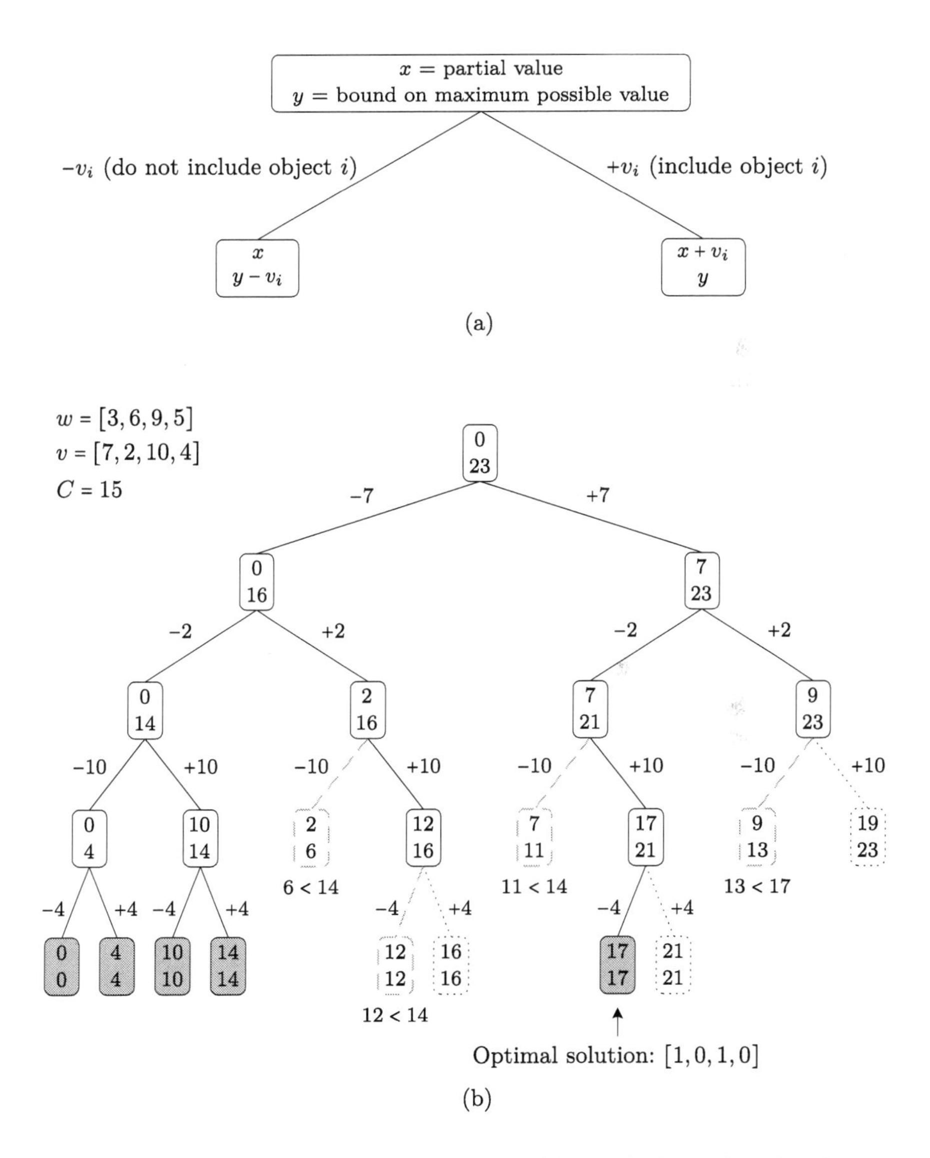

Figure 12.16 Recursion tree of a branch and bound algorithm for the 0-1 knapsack problem.

Listing 12.14 Branch and bound code for solving the 0-1 knapsack problem.

```
def knapsack_0_1_bnb(i, w_left, current_v, max_v, sol,
                     opt_sol, opt_v, w, v, C):
    # Check base case
    if i == len(sol):
        # Check if better solution has been found
        if current_v > opt_v:
            # Update optimal value and solution
            opt_v = current_v
            for k in range(0, len(sol)):
                opt_sol[k] = sol[k]
    else:
        # Generate candidates
        for k in range(0, 2):

            # Check if recursion tree can be pruned
            # according to (capacity) constraint
            if k * w[i] <= w_left:

                # Update maximum possible value
                new_max_v = max_v - (1 - k) * v[i]

                # Check if recursion tree can be pruned
                # according to optimal value
                if new_max_v > opt_v:

                    # Expand partial solution
                    sol[i] = k

                    # Update remaining capacity
                    # and partial value
                    new_w_left = w_left - k * w[i]
                    new_current_v = current_v + k * v[i]

                    # Try to expand partial solution
                    opt_v = knapsack_0_1_bnb(i + 1, new_w_left,
                                             new_current_v,
                                             new_max_v, sol,
                                             opt_sol, opt_v,
                                             w, v, C)

    # return value of optimal solution found so far
    return opt_v
```

Listing 12.15 Auxiliary code for the branch and bound algorithm related to the 0-1 knapsack problem.

```
def knapsack_0_1_branch_and_bound_wrapper(w, v, C):
    sol = [0] * (len(w))
    opt_sol = [0] * (len(w))
    total_v = knapsack_0_1_bnb(0, C, 0, sum(v), sol,
                               opt_sol, -1, w, v, C)
    print_knapsack_solution(opt_sol, w, v, C, total_v)


w = [3, 6, 9, 5]  # List of object weights
v = [7, 2, 10, 4]  # List of object values
C = 15  # Weight capacity of the knapsack
knapsack_0_1_branch_and_bound_wrapper(w, v, C)
```

Note that it is initialized to the sum of all of the values of the objects in the wrapper method. In the recursive case, after making sure that a partial solution with a new candidate is valid (in line 17), the method computes the new value of the bound `new_max_v` in line 20. Observe that when `k` = 0 the new bound is decreased by v_i, while when `k` = 1 it remains unaltered. Subsequently, the method uses another `if` statement to check if it can prune the tree according to the new bound and the optimal value of a solution computed in earlier calls (line 24). The rest of the code is analogous to the backtracking algorithm. Finally, Listing 12.15 contains an associated wrapper method, defines an instance of the problem (through $\mathbf{w}$, $\mathbf{v}$, and C), and solves it.

12.8 EXERCISES

Exercise 12.1 — There are numerous ways to build algorithms that generate subsets of elements. The methods described in Section 12.2.1 were based on binary recursion trees. The goal of this exercise is to implement alternative procedures that also print all of the subsets of n items provided in a list. However, they must generate a subset at each node of the recursion tree illustrated in Figure 12.17. Observe that there are exactly 2^n nodes, and the labels 0, 1, and 2 represent indices of the elements of an initial input list $[a, b, c]$. Furthermore, instead of using binary lists to indicate the presence of items in a subset, the partial solutions will contain precisely these indices. For example, the list $[0, 2]$

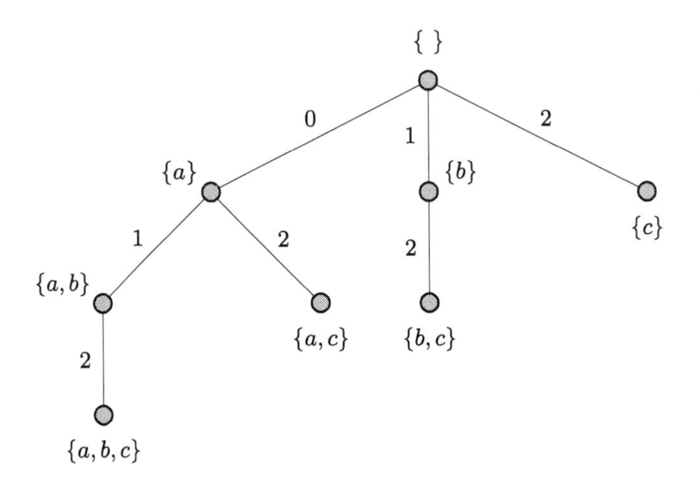

Figure 12.17 Alternative recursion tree of an algorithm that generates all of the subsets of three items.

Figure 12.18 One solution to the four-rooks puzzle.

will represent $\{a, c\}$. Therefore, partial solutions will also correspond to complete solutions.

Exercise 12.2 — Implement a backtracking algorithm that prints all of the solutions to the n-rooks puzzle. It is analogous to the n-queens puzzle, but uses rooks instead of queens. Since rooks can only move vertically and horizontally, two (or more) rooks cannot appear in the same row or column. However, several can appear on a same diagonal. Figure 12.18 shows a solution to the puzzle.

Exercise 12.3 — Implement a backtracking function that counts the number of valid solutions to a sudoku puzzle that may not be well-posed (i.e., it may have 0, 1, or more solutions).

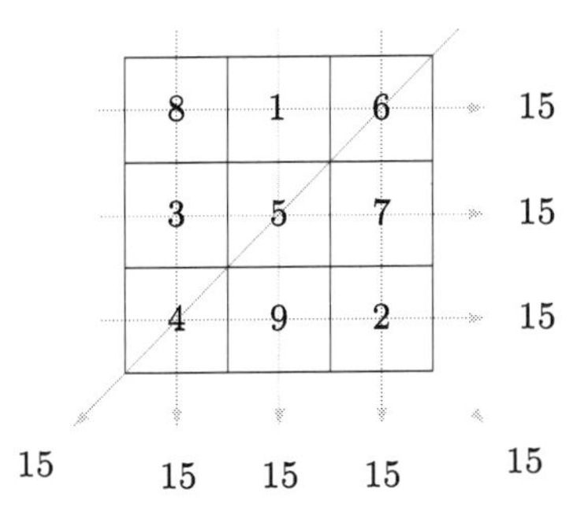

Figure 12.19 A 3×3 magic square.

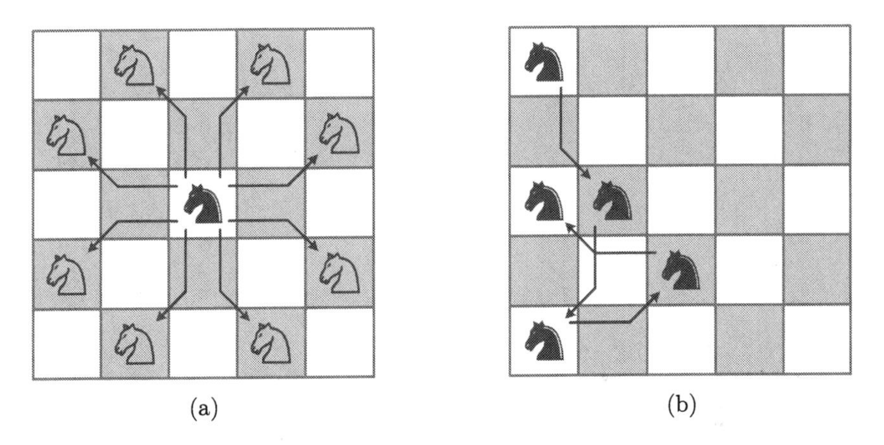

Figure 12.20 Chess knight moves.

Exercise 12.4 — A magic square is an $n \times n$ grid of the first n^2 positive integers such that the sum of the elements in each row, column, and diagonal is equal. The particular sum is called the "magic constant," and is $n(n^2+1)/2$ for a general n. Figure 12.19 shows an example of a 3×3 magic square, where the magic constant is 15. Design a backtracking algorithm that prints every possible 3×3 magic square. Note: it is not necessary to implement a method that finds magic squares for a general n. What problem could we run into when computing magic squares of larger values of n?

Exercise 12.5 — A knight can jump from a square to another one in L-shaped patterns, as shown in Figure 12.20(a), where (b) shows a sequence of four moves. The "knight's tour" problem consists of determining a sequence of moves of a knight on an $n \times n$ chessboard, when staring from a specific square, such that it visits every square only once. Implement

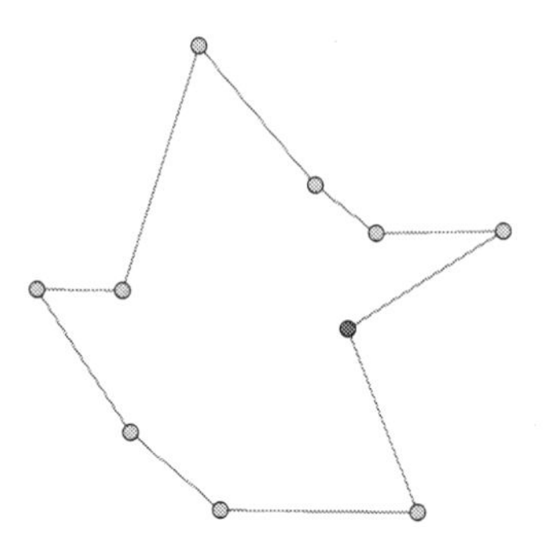

Figure 12.21 An instance and solution to the traveling salesman problem.

a backtracking algorithm that provides one knight's tour, and test it for $n = 5$ and $n = 6$. Besides n, the method will receive the coordinates of an initial square on the chessboard. Finally, it is not necessary to search for a "closed" tour that ends at a square that is one move away from the initial square.

Exercise 12.6 — The "traveling salesman problem" is a classical discrete optimization problem. Given n cities, the goal is to find the shortest path that visits each one exactly once and returns to the starting city. Figure 12.21 shows the shortest path for 10 cities. Implement a backtracking algorithm that solves the problem. Assume that the locations of the cities are specified in a text file, where each line contains the x and y coordinates of the cities on a two-dimensional map. The file can be read with the `loadtxt` method from the NumPy package. The starting city will be the one specified in the first line of the file. In addition, consider Euclidean distances between cities (the salesman travels from one city to another by following a straight line). These can be computed, for example, with the method `pdist` in the SciPy package. In particular, the recursive backtracking method should receive a matrix of distances between the cities (their locations are not necessary when using this distance matrix). Finally, test the code for $n \leq 10$.

Exercise 12.7 — The following problem is known as "tug of war." Given a nonempty set of n numbers, where n is even, the goal consists of dividing it into two subsets of $n/2$ elements in order to minimize the absolute difference between the sum of the numbers in each subset. For example, given the set $\{3, 5, 9, 14, 20, 24\}$, the optimal way to partition

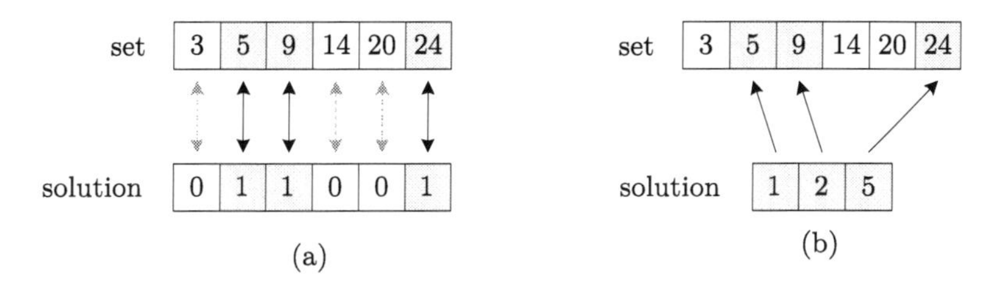

Figure 12.22 Two ways to represent a solution for the tug of war problem.

it leads to the two subsets: $\{5, 9, 24\}$, and $\{3, 14, 20\}$. The sums of their elements are 38 and 37, and the absolute difference between these sums is 1.

Implement a backtracking algorithm based on generating subsets that solves this problem. Assuming that the input set is coded as a list, the solution would be a binary list of length n with $n/2$ zeros, and $n/2$ ones. The positions of the zeros (or ones) would indicate the locations of the elements of a particular subset. In the example, one solution could be the list $[0, 1, 1, 0, 0, 1]$ (the list $[1, 0, 0, 1, 1, 0]$ would be equivalent), representing the subset $\{5, 9, 24\}$.

In addition, implement a more efficient strategy where the solution is a list **s** of length $n/2$ whose elements appear in increasing order, and correspond to the locations of the elements of a particular subset. For example, the subset $\{5, 9, 24\}$ would be represented by the list $[1, 2, 5]$. Figure 12.22 shows these two ways to represent a solution.

Exercise 12.8 — In the problem described in Section 12.4 the goal consisted of printing every subset T of a set of n positive integers S, for which the sum of the elements of T was equal to some integer value x. Design an alternative recursive function that instead computes the (valid) subset T with the least cardinality (i.e, number of elements). In particular, it will store the optimal subset in a list, and will return the optimal cardinality. For example, for $S = \{1, 2, 3, 5, 6, 7, 9\}$ and $x = 13$, the method would store the subset $\{6, 7\}$, and return the value 2. In addition, note that in this optimization problem the optimal subset may not be unique, but the optimal cardinality is. For instance, for $S = \{2, 6, 3, 5\}$ and $x = 8$, both $\{2, 6\}$ and $\{3, 5\}$ would be optimal subsets. Code the partial (and optimal) solution as a list of binary digits, similarly to the method in Listing 12.7. Also, prune the tree considering the validity of a solution and the best cardinality found. Finally, code a wrapper method that computes the optimal subset, and prints it if it exists (it may not be possible to find a subset T of S whose elements add up to x).

Further reading

MONOGRAPHS IN RECURSION

This book has provided a broad coverage of recursion, containing essential topics for designing recursive algorithms. However, the reader can find additional aspects, examples, and implementation details in other references. In particular, the following book:

- Jeffrey Soden Rohl. *Recursion Via Pascal.* Cambridge Computer Science Texts. Cambridge University Press, 1st edition, August 1984

contains numerous examples in Pascal of recursive algorithms on data structures (linked lists, trees, or graphs) implemented through pointers. In contrast, the current book avoids pointers since they are not used explicitly in Python. The suggested reference includes backtracking algorithms for generating additional combinatorial entities such as combinations, compositions, and partitions. Lastly, it contains a chapter on recursion elimination, which offers low-level explanations on how to transform recursive programs into equivalent iterative versions.

Another book that focuses entirely on recursion is:

- Eric S. Roberts. *Thinking Recursively.* Wiley, 1st edition, January 1986,

which contains examples in Pascal, and a more recent edition:

- Eric S. Roberts. *Thinking Recursively with Java.* Wiley, 1st edition, February 2006,

where the code is in Java. The book contains a chapter on recursive data types, and another on the implementation of recursion from a low-level point of view.

Java programmers can also benefit from:

- Irena Pevac. *Practicing Recursion in Java.* CreateSpace Independent Publishing Platform, 1st edition, April 2016,

which contains examples related to linked lists, linked trees, and graphical problems.

DESIGN AND ANALYSIS OF ALGORITHMS

The current book contains problems that can be solved by applying algorithm design techniques such as divide and conquer, or backtracking. There are numerous excellent texts that analyze more advanced problems, describe other design techniques like greedy algorithms or dynamic programming, or rely on advanced data structures in order to construct efficient algorithms. The following list of books is only a small subset of the broad literature on algorithm design (and analysis) techniques:

- Thomas H. Cormen, Charles E. Leiserson, Ronald L. Rivest, and Clifford Stein. *Introduction to Algorithms.* The MIT Press, 3rd edition, 2009.
- Robert Sedgewick and Kevin Wayne. *Algorithms.* Addison-Wesley Professional, 4th edition, 2011.
- Anany V. Levitin. *Introduction to the Design and Analysis of Algorithms.* Addison-Wesley Longman Publishing Co., Inc., Boston, MA, USA, 3rd edition, 2012.
- Michael T. Goodrich, Roberto Tamassia, and Michael H. Goldwasser. *Data Structures and Algorithms in Python.* Wiley Publishing, 1st edition, 2013.

More ambitious readers can explore texts aimed at training for programming competitions (e.g., the ACM International Collegiate Programming Contest, or the International Olympiad in Informatics). Well-known references include:

- Steven S. Skiena and Miguel Revilla. *Programming Challenges: The Programming Contest Training Manual.* Springer-Verlag New York, Inc., Secaucus, NJ, USA, 2003.
- Steven S. Skiena. *The Algorithm Design Manual.* Springer Publishing Company, Incorporated, 2nd edition, 2008.

Other textbooks focus exclusively on the analysis of algorithms, such as:

- Jeffrey J. McConnell. *Analysis of Algorithms: An Active Learning Approach.* Jones and Bartlett Publishers, Inc., USA, 1st edition, 2001

Among these,

- Robert Sedgewick and Philippe Flajolet. *An Introduction to the Analysis of Algorithms.* Addison-Wesley Professional, 2nd edition, 2013

is considerably more advanced, and is used by the author in a massive open online course offered by *coursera.org*. Similarly, this and other online learning platforms offer excellent courses on algorithms.

FUNCTIONAL PROGRAMMING

Recursion is omnipresent in functional programming. Thus, programmers should have mastered the contents of this book in order to be competent in this programming paradigm. Popular references include:

- Harold Abelson and Gerald J. Sussman. *Structure and Interpretation of Computer Programs.* MIT Press, Cambridge, MA, USA, 2nd edition, 1996.
- Richard Bird. *Introduction to Functional Programming Using Haskell.* Prentice Hall Europe, April 1998.
- Martin Odersky, Lex Spoon, and Bill Venners. *Programming in Scala: A Comprehensive Step-by-Step Guide.* Artima Incorporation, USA, 1st edition, 2008.

The book by Odersky et al. is used in a highly recommended course also offered at *coursera.org*.

PYTHON

It is assumed that the reader of this book has some programming experience. The following popular texts can be useful to readers interested in learning more Python features, looking up implementation details, or developing alternative and/or more efficient recursive variants of the examples covered throughout the book:

- Mark Pilgrim. *Dive Into Python 3.* Apress, Berkely, CA, USA, 2009.
- Mark Summerfield. *Programming in Python 3: A Complete Introduction to the Python Language.* Addison-Wesley Professional, 2nd edition, 2009.
- Mark Lutz. *Learning Python.* O'Reilly, 5th edition, 2013.

RESEARCH IN TEACHING AND LEARNING RECURSION

The current book incorporates several ideas that stem from research in teaching and learning recursion. In particular, it focuses on the declarative abstract level of problem decomposition and induction, instead of focusing on a computational model (e.g., a trace mental model) or iterative/imperative thinking. This is suggested, for example, in:

- D. Ginat and E. Shifroni. Teaching recursion in a procedural environment – how much should we emphasize the computing model? *SIGCSE Bull.*, 31(1):127–131, 1999.
- R. Sooriamurthi. Problems in comprehending recursion and suggested solutions. *SIGCSE Bull.*, 33(3):25–28, 2001.

The diagrams and methodology (the template is similar to the one in Sooriamurthi's paper) used in Chapter 2 have been introduced precisely in order to focus on declarative thinking. In addition, the structure of the book also reflects this goal. Elements such as tracing, recursion trees, the program stack, or the relationship between iteration and tail recursion are not introduced until Chapters 10 and 11.

The book also includes examples from articles where I have addressed teaching mutual recursion, and tail (and nested) recursion through function generalization:

- M. Rubio-Sánchez, J. Urquiza-Fuentes, and C. Pareja-Flores. A gentle introduction to mutual recursion. *SIGCSE Bull.*, 40(3):235–239, 2008.
- Manuel Rubio-Sánchez. Tail recursive programming by applying generalization. In *Proceedings of the Fifteenth Annual Conference on Innovation and Technology in Computer Science Education*, ITiCSE '10, pages 98–102. ACM, 2010.

Finally, there is rich literature on research in teaching and learning recursion. The following recent surveys contain hundreds of references related to the topic:

- Christian Rinderknecht. A survey on teaching and learning recursive programming. *Informatics in Education*, 13(1):87–119, 2014.
- Renée McCauley, Scott Grissom, Sue Fitzgerald, and Laurie Murphy. Teaching and learning recursive programming: a review of the research literature. *Computer Science Education*, 25(1):37–66, 2015.

Index